D1287252

The Realization of Star Trek Technologies

Mark E. Lasbury

The Realization of Star Trek Technologies

The Science, Not Fiction, Behind Brain Implants, Plasma Shields, Quantum Computing, and More

Franklin Lakes Public Library
470 DeKorte Drive
Franklin Lakes, New Jersey

Mark E. Lasbury
Indianapolis, IN
USA

ISBN 978-3-319-40912-2 ISBN 978-3-319-40914-6 (eBook)
DOI 10.1007/978-3-319-40914-6

Library of Congress Control Number: 2016943409

Printed on acid-free paper

This Springer imprint is published by Springer Nature
The registered company is Springer International Publishing AG Switzerland

To accomplish great things we must dream as well as act.

—Anatole France

But the truth is, it's not the idea, it's never the idea, it's always what you do with it.

—Neil Gaiman

The human race is a remarkable creature, one with great potential, and I hope that Star Trek *has helped to show us what we can be if we believe in ourselves and our abilities.*

—Gene Roddenberry

To my wife, who doesn't know the difference
between Star Trek and Star Wars.
To our children, whose sense of curiosity
ushers in the next generation
of advances.
Hi Ruth.

Preface

As man reaches into space, he will acquire new drugs, food, minerals, and knowledge which he cannot conceive of now.

—Gene Roddenberry, *The Making of Star Trek*, p. 200

Two popular genres in American television of the 1960s were science fiction and westerns. The most popular science fiction show was probably *The Twilight Zone*, which ran from 1959 to 1964, but live action and animated programs were both popular throughout the decade. While *The Outer Limits*, *The Jetsons*, *Lost in Space*, *Jonny Quest*, *Thunderbirds*, and *Voyage to the Bottom of the Sea* all had short runs in the 1960s, one should not forget the most successful of all the science fiction/fantasy of that time, *Doctor Who*, which originally ran from 1963 to 1989 on the BBC. Into this landscape of space- and cattle ranching-based story visual telling came *Star Trek* in 1966.

Series creator Gene Roddenberry intended *Star Trek* to be a "Wagon Train in the cosmos," a wild-west show with starships. Gene was also a fan of technology, so science soon came to the forefront in his new fiction series. Flight in deep space and between planets was a staple of *Star Trek*, with the starship *USS Enterprise* as a critical character in the show. This was a direct result of Roddenberry's experiences in the air. Gene originally studied law enforcement at Los Angeles City College, switching to aeronautical engineering just before the USA entered World War II.

Roddenberry earned his pilot's license in 1941 and shortly thereafter volunteered for the US Air Corps. He flew 89 combat missions, earned the Distinguished Flying Cross the Air Force Medal during the war, and was shot down twice. While stationed in the South Pacific, he submitted stories and poetry for publication with considerable success. After the war ended, Gene flew as a commercial pilot for Pan-Am World Airways, running the longest routes they had. He retired soon after a 1947 crash in Africa that claimed the lives of several passengers. It was then that he decided to pursue writing for the new medium of television, returning to Los Angeles with his new found love for engineering and flight.

Six television series (five live action and one animated), thirteen movies, and 50 years later, we are now ready to celebrate the golden anniversary of the original Star Trek challenge to, "Go where no man has gone before." The stories, characters and technology are as popular as ever, and because of this, the franchise continues to grow. The latest J.J. Abrams produced film debuted this summer (2016), and a new series that links to the original television show is in production, scheduled to debut in 2017. While both Doctor Who and Star Trek have thrived over the years, the two series differ in their approaches to technology. Doctor Who is closer to the fantasy end of the spectrum, focusing more on interpersonal stories rather than technology, although there will certainly be those fans who would argue the point. Star Trek is known for its social commentary, yet it is the technology and the science behind the high tech toys (implied, casual or plausible) that separates Star Trek from Doctor Who and many other science fiction shows.

One can point to any number of toys and tools that *Star Trek* either introduced or more thoroughly explained. In many cases, the explanations were cursory or developed only over time, and several were convoluted combinations of jargon and science concepts. Regardless, it has been shown over the decades that *Star Trek* made sense, both as storytelling and science. Roddenberry hired Harvey P. Lynn of the Rand Corporation as a part-time consultant to vet the science before the original series debuted. He made suggestions as to terminology to use on the show and let Roddenberry know when the writers' vision did not match the known science of the day. The subsequent production teams continued relying on consultants to ensure that the science and technology were at least plausible, and the writers had technical guides to make sure that there was consistency in the gadgets and the explanations of their function.

The universal translator allowed species that had never met before to converse without subtitles, although the communication itself sometimes became the story. Other technologies were meant to save the crew from mundane tasks of day-to-day survival. Being able to replicate food eliminated the need for growing crops and raising animals, or for interrupting a five-year mission to restock bullion cubes and hot pockets every few months. The tricorder began as a mechanism to provide a certain female yeoman an excuse for being on screen more often, according to SE Whitfields's and Roddenberry's book, *The Making of Star Trek* (p. 169). But the majority of the technological equipment served to highlight the humanity of the characters (Scotty's babying of the dilithium crystals to keep the ship going, Geordi's visor, Data, and the Borg) or help the stories to be exciting and cutting edge (transporters, phasers, deflectors, cloaking devices, and replicator). In the production sense, some technologies provide information to the viewer or save time in order to move the plot along (tricorder, universal translator, communicator, and transporter).

Since *Star Trek* aired for only three seasons in its first run, most people fell in love with the USS Enterprise via syndicated reruns. New episodes were not produced again until 1987, so a whole generation of viewers (1969–1987) grew up with only the 79 episodes. Despite this limitation, fifty years of science training has been influenced by the hard technology of this series. In speaking to dozens of researchers for this book, almost all were familiar with *Star Trek* and most indicated that the series had an effect on either their desire to study science or on their research track. Dr. Asier Mazo, of Pamplano, Spain, told me that his group definitely had *Star Trek* in mind when they built and named their holographic acoustic tractor beam, and Dr. Tracy Canfield does not just study computational linguistics, but she is also an expert in Klingon and a science fiction author in her own right. *Star Trek* has transcended the small and big screens; research laboratories around the world are turning Trek science fiction technologies into science fact.

Many of the franchise's technologies existed in literature before Roddenberry starting contemplating scattering McCoy's atoms across the universe. Tractor beams were first described as attractors or repulsors in several Buck Rogers and other short stories of the late 1920s and early 1930s, but *Star Trek* gave us the graviton as an explanation of how they work. George Gamow described faster than light speed in a series of scientific fables in the 1880s, thirty years before Einstein told us it was not possible. Fortunately, creative writers ignored Einstein's speed limit and have invented many mechanisms for moving astronauts beyond the speed of light. *Star Trek* introduced warp drive and plasma subspace bubbles, once again providing a more scientific framework for a previously described fictional technology. Most famously, Isaac Asmiov introduced R. Daneel Olivaw and his positronic brain in the Robot series of 1940. The explanation for the workings of Daneel's brain was sorely lacking; Asimov said only that it was constructed of a platinum/iridium alloy. Rodenberry and his successors provided a bit more of the mechanism behind Mr. Data's positronic neural net, and seeing the lights and connections during his maintenance gave it a computer-like appearance, mimicking human synaptic activity.

Other technologies from *Star Trek* were explained in much more detail and have been shown to have parallels in our world. This book describes several fictional devices that have a distinct possibility of being realized in the next decades as well as the research that is bringing those functions to life. In some cases, the tools of *Star Trek* are mimicked outright, trying to achieve the same exact function in an engineered fashion. In other cases, the technologies being developed now are a stylized or scaled version of those seen in the series and movies. Some of the most amazing stories play out when the fictional mechanisms of action for *Star Trek* devices become science fact, in either direct fashion, as with many of the wireless sensor technologies of the tricorder, or in the spirit of the *Star Trek* tool. For example, Chap. 1 discusses optical phased arrays, which are a way to control laser beams; this is connected to the phaser because *Star Trek* described it as a "phased array emitter" decades ago. Amazingly, new research suggests that optical phased arrays are going to be important in developing directed energy weapons like the phaser.

In 2002, science and science fiction author Gregory Benford wrote in *The Magazine of Fantasy and Science Fiction* about the use of "wuantum mechanics" in some science fiction stories (103(4–5); 187, 2002). Used as a writing and plot crutch, a wuantum mechanics device is any piece of technology never spoken of before, but which some character pulls out of nowhere in order to survive the impending catastrophe and save the day. Or perhaps a known piece of technology is quickly modified using previously unknown terminology and mechanisms that have no basis in reality. True, Geordi and Picard sometimes made things all better when they "modified the main deflector to emit an inverse tachyon pulse" so they would, "be able to scan beyond the subspace barrier," (*TNG: All Good Things*) but examples of quantum mechanics are less common in *Star Trek* than in other series or stories. The believability factor was a big part of Roddenberry's vision of *Star Trek* and was the reason he hired Harvey Lynn, the reason the writers had technical manuals, and the reason that many franchise technologies are coming true.

Star Trek was/is prophetic in so many areas, from communicators turning into mobile phones, to talking computers that we now call Siri. In many cases, the transition from science fiction to science fact has been so direct that devices and products hit the market with *Star Trek* names. There is a 3D printer called a replicator, a nonlethal laser weapon called a PHaSR, a ten million dollar design prize for making a tricorder, and the NASA design for a warp drive ship we currently cannot build is called the INX Enterprise. Commercials, websites, and the news media constantly compare emerging technologies to those things we have seen on *Star Trek*. Some comparisons are apt, while others are exaggerations intended only to increase sales or viewership. The descriptions and uses of technology by Kirk and others are so vivid that they have stuck in the public consciousness, so that modern research may quickly evoke memories from *Star Trek*. But perhaps more importantly, current science advances show how forward thinking and accurate many *Star Trek* technologies were. Roddenberry was not just a writer and producer; he was a futurist of the first order.

So if *Star Trek* predicted so many advances and was so accurate in its technology, why aren't we telling our replicator what we want for dinner, bouncing bullets off deflector shields, and hiding our dirty dinner dishes behind cloaking devices? In some cases, the reason is that the research has not yet reached a crucial point. Scientists have not isolated the graviton particle yet, so we cannot make a true *Star Trek* deflector or tractor beam. If and/or when the graviton is discovered, it might be decades before man could harness it to do work, if at all. For other technologies, the problem is a matter of scale. While today's lasers are powerful enough to imitate handheld phasers, laboratories are nowhere close to producing lasers that can disintegrate a building or planet. Some devices need to get bigger; current tractor beams can only move miniscule objects. Other things need to get smaller—the visual prostheses that are available today require outside peripherals, cameras, and lots of wires, while many particle accelerators cover 20 or more square miles—too big for a handheld weapon.

This book is divided into nine chapters, one for each selected technology. Each chapter describes how *Star Trek* used a device or technique and the science that the writers used to explain the workings of each one. With this information as a base, the remainder of the chapter surveys the state of the research that mimics or parallels the *Star Trek* version and the research that may allow for a true approximation or realization of the technology. In some cases, the research is a new twist on old engineering methods, such as how the magnetic field around Earth can be simulated to protect astronauts as a deflector shield or how gravity itself can be used to tractor asteroids away from a collision course with Earth. In other cases, completely new engineering and physics has afforded science the opportunity to pursue *Star Trek* gadgets. Metamaterials, quantum dots, plasma toroids, retinal mapping, statistical machine learning—these have all contributed to the progress science has made toward *Star Trek* technological magic.

When episodes or movies are referenced to make a point about how a technology is used or how it functions, an effort has been made to identify the series and episode in which that canonical information can be found. The series will be abbreviated as follows: *Star Trek: The Original Series* will be abbreviated *TOS*; *Star Trek: The Next Generation* as *TNG*; *Star Trek: Deep Space Nine* as *DS9*; *Star Trek: Voyager* as *VOY*; and *Star Trek: Enterprise* as *ENT*. If the references are from a film, the entire title will be used. Other reference materials are used as well, and the technical manuals from several of the series help explain the technologies within the *Star Trek* universe.

The latest scientific research is used to explain how people are pursuing the technologies exploited in *Star Trek*. The pertinent papers, books, or websites and press releases are listed at the end of each chapter. The research is cutting edge, so some additional information explaining background or interesting related points is pointed out in text boxes in each chapter. Additionally, emerging technologies may appear in more than one chapter. These connections are highlighted, and the sections of other chapters are stated. For example, metamaterials research is applicable for the realization of cloaks, deflectors, and tractor beams, while lasers are important for phasers, replicators, tractor beams, and tricorders. Finally, the words of the researchers are used where ever possible to explain the concepts behind their work and in relating the influence of *Star Trek* on their work.

The book is not comprehensive by any means. Each chapter focuses on a single theme, while other topics are not described for reasons of space or due to the need of significant additional background. While each topic is explained so that a non-professional can follow the research, the text does include significant numbers of technical terms to help the reader seek additional information. Likewise, the list of *Star Trek* technologies discussed is not exhaustive. Most of the included technologies were chosen for their position in science right now. Communicators were omitted because society has gone beyond them by now with smartphones and bluetooth. PADDs are likewise commonplace now; millions of people use similar tablet computers every day.

Some technologies are not in the book because they have not gotten to the experimentation stage yet. Yes, warp drive has moved from NASA's conjecture stage to their reasoned speculation level, but humans only have experience with conventional rockets and a few new technologies that might allow us to go marginally faster. No current line of research can produce anything like warp drive or even a warp drive precursor. Twenty years ago, Mexican physicist Miguel Alcubierre did some mathematics that suggested that warp drive could be possible if someone can develop an energy/density field that is lower than that of a vacuum. What does that mean? Science needs to find a way to engineer a negative mass and something called negative energy. Since these new entities are nowhere in sight, the technology will have to mature before a discussion of warp drive is appropriate in this type of forum.

Artificial intelligence (AI) is not given a chapter because the science is also too immature. Though Chap. 6 discusses statistical machine learning and neural nets in the context of artificial intelligence, this is not true AI by a long shot. Futurist Neil Gershenfeld is of the opinion that mankind will soon be able to merge brains and computers. Perhaps this will help us learn enough to mimic intelligence in an artificial network, but there is no data to suggest such a merging is yet possible. No matter how determined the media is to report that AI is either here or just around the corner, true machine learning without significant human input and sentience in a machine is probably half a century to a century away at best.

Holodecks are not included because they involve several future technologies that are all beyond present science. We do have some types of virtual reality that let you see additional objects or surround you with different scenery; there are even some that transmit sensations or movements in those realities and allow you to manipulate virtual objects. However, none of these technologies allow you to pick objects up like in a *Star Trek* holodeck or have it respond creatively if you go off program. The holodeck uses replicators, holograms, transporters, and AI, none of which we have yet. Therefore, it would be hard to intelligently discuss holodecks that use them all. Transporters and replicators do have their own chapters, so one can look at the progress science is making with them. Then, throw in a bit of optical phased arrays from the phaser weapon chapter to include holograms, and together, these will give one an idea of where holodeck research might be headed.

The legacy of *Star Trek* and the role it has played in driving research and researchers must be honored. *Star Trek* may be entertainment, but it is entertainment with a sociological and technological edge. The public needs to be aware of the amazing science that is behind the sci-fi technologies that entertain and amuse them. Even if some of this research would have been pursued had *Star Trek* never made it to television, there is no doubt the franchise has spurred research that has expanded knowledge and inspired people to enter the world of science. Even if much of the work being done in laboratories all over the world produces only personal tools that we recognize from the series, *Star Trek* is also important because it stimulates work that helps mankind to better explain his universe and his place in it. Science is truly going where no person has gone before, and *Star Trek* continues to play a role in that journey.

Indianapolis, USA Mark E. Lasbury

Acknowledgements

Thank you to all the scientists and engineers who endured endless follow-up phone calls, e-mails, and questions. Thank you to Mary James at Springer Scientific Publishing for helping me to understand the process of publishing a book. Thank you to Pamela J. Durant for the journal articles searches, the editing, and the words of encouragement.

Contents

Chapter 1
Phasers Weapons of Less Destruction

She wants to know if it hurts? Of course it hurts. It's supposed to hurt. It's a phaser.

—Quark
DS9: Profit and Loss

1.1 Introduction

Space is a violent place. The temperature can range from almost 300 °C (572 °F) in the sunlight to −100 °C (−148 °F) in the shade of the International Space Station. Space suits have both heaters and air conditioners to keep astronauts from freezing solid or boiling. In deep space it gets even colder, just 2.7 °K above absolute zero (−270.45 °C/−454.81 °F). Exposure to the vacuum of space will cause bubbles to form in all the tissues of the body; the worst imaginable case of the bends. Lungs will collapse and any air in a body space will expand and explode. If this doesn't do the poor astronaut in immediately, the radiation of space will blind them, burn their skin and mutate their DNA in a thousand different ways. Space is not to be messed with.

It could be worse. Consider the possibility of roasting in a solar flare or being fenestrated by a fleck of cosmic dust traveling 27,576 kph (17,135 mph). Even worse, your home planet or space station might be obliterated by a passing asteroid or meteoroid. Outer space can be cruel to living things, and humans have only made the situation worse. Debris from past missions and satellites have come crashing to Earth. There is an entire program that tracks manmade space debris so that NASA or the ESA won't fly a spacecraft into a discarded wrench. Most disappointing, we have slowly begun to weaponize space. China used an antisatellite missile against one of their orbiting weather observers in 2007 and the US destroyed spy satellites with missiles in the 1980s and again in 2008. Russian cosmonauts still carry handguns on all missions, although discharging them is probably the worst thing they could do in a confined spaceship.

M.E. Lasbury, *The Realization of Star Trek Technologies*,
DOI 10.1007/978-3-319-40914-6_1

Have faith—*Star Trek* shows us how space can become less violent over time. Even though weapons are prevalent in *Star Trek*, their usage goes down as the different series progress. When phasers first come to *Star Trek* in 2233 (*Star Trek* 2009), they fire them every chance they get. Kirk and company of the original series speak of or use phaser weapons an average of 5.05 times per episode, but 24th century space explorers use them less often. *Star Trek: The Next Generation* series has 2.40 phaser mentions per episode, *Deep Space Nine* only 1.52, and *Voyager* has 2.12 phaser references in each hour long show.

Star Trek: *Enterprise* isn't included on the list because they don't have phasers, and they only make 1.12 references/show to phase pistols or phase rifles, the precursors to the phasers of the 23rd century (*TNG: A Matter of Time*). On the other hand, *Enterprise* uses their torpedoes more often than any other series (1.26 mentions/episode). Captain Archer doesn't have photon torpedoes at his disposal, only the Klingons have them before Kirk and Spock battle Nero in *Star Trek* (2009). Instead, the NX-01 Enterprise has spatial torpedoes (*ENT: Fight or Flight*) that are considerably less powerful than even the antimatter photonic torpedoes that immediately precede true photon torpedoes (*ENT: The Expanse*).

Photon torpedoes, and later quantum torpedoes, are used at a similar rate in the rest of the series (0.75/episode for *TOS*, 0.90 for *TNG*, 0.59 for *DS9* and 1.03 for *VOY*). One could argue that with less phaser usage and steady photon torpedo usage, the late 24th century universe is a safer place than the middle 2200s. Captain Janeway even makes mention of the wild days of Kirk's time, "Space must have seemed a whole lot bigger back then…they were a little slower to invoke the Prime Directive, and a little quicker to pull their phasers." (*VOY: Flashback*) It would be wonderful if humans could avoid the Wild West days of space altogether, but at least we have a model for reducing violence in space over time. Everything we need to know about living together in space we can learn from *Star Trek*.

Box 1.1. A Note on Photon Torpedoes

The photon torpedo is a conventional kinetic weapon, it just has a more exotic delivery method and payload than traditional torpedoes. Kinetic weapons, either with or without high explosives, do damage via projectiles and shock waves. Even nuclear weapons are fairly conventional, the nuclear fission reaction is just a larger, more efficient way of producing shock waves and projectiles.

The payload—made from mixing small amounts of matter and antimatter. The matter and antimatter are kept separate until the torpedoes arm themselves in mid-flight, and the impact of the torpedo against a target mix the deuterium and anti-deuterium. When they annihilate one another they produce an explosion and a lot of radiation (*VOY; Good Shepherd*).

24th century photon torpedoes have many small capsules of matter and antimatter that intermix at arming but remain segregated by magnetic separators. The separators degrade at impact and the reactant mix. They annihilate with a significantly higher yield because the matter and antimatter are more interspersed (Next Generation Technical Guide, p. 128). Notice that nowhere does *light* enter the reaction; the "photon" in the name is misleading.

Why they're called torpedoes—Aircraft fire missiles in air; ships fire torpedoes in water. There's no water in space, so why are photon torpedoes so named—because spacecraft are more often referred to as ships (spaceships, starships) than planes. This might be because the term "spaceship" in literature predates the invention of the airplane. J.J. Astor's 1894 novel, *A Journey in Other Worlds*, makes reference to a spaceship, while the Pall Mall Gazette (1880) uses the hyphenated word space-ship when referring to the "projectile" Jules Verne wrote of in *From The Earth To The Moon* (1865). These both predate the first airplane demonstration in 1906.

Space terminology borrows heavily from naval jargon; crew, bridge, course, admiral. The term torpedo once referred to what we would call mines, explosives set in place to be hit by unsuspecting vessels. This occurs in *Star Trek* as well, photon torpedoes are teletransported to designated locations in space and rigged to explode when struck (*VOY: Year of Hell, Dark Frontier*). One could conclude that torpedoes are fired into the ocean of space.

One indication that *Star Trek* crews were not eager to kill or do violence is that they could "set phasers to stun," to do the least amount of damage necessary. Over the centuries, phaser weapons gain more precision to give the user more chances to avoid killing. The earliest type 1 (handheld) phasers have eight settings, ranging from stun to disrupt or vaporize (Star Trek: The Next Generation Technical Guide, p. 135). The 24th century type 2 and type 3 phasers have 16 settings; they aren't damaging until level four, and one must use setting seven for them to be deadly (TNG Technical Guide, p. 136). Even photon torpedoes have power settings, from fireworks (level 1) to a level 10 that is so powerful that it violates strategic arms treaties (Sternbach and Okuda 1994, p. 22).

Sublethal or nonlethal versions of weapons give the user a chance to contemplate his/her actions—just how much force do they need to exercise? This already makes space in the *Star Trek* universe a safer place than Earth in the 21st century. No handgun or rifle built today can be dialed down from lethal to nonlethal. The challenge is clear—can humans follow *Star Trek's* lead and choose to develop directed energy weapons for use on Earth and in space that can be nonlethal? Happily, the answer seems to be yes, though we have to keep in mind that nonlethal energy weapons can often become lethal at the flick of a switch. The amazing thing is just *how* researchers are harnessing energy in weapons; there are even ways to bend a lightning bolt inside a laser beam around a corner to get at an enemy—even *Star Trek* didn't think of that one.

1.2 Particle Beams as Directed Energy Weapons

Directed energy weapons do not rely on explosives or large projectiles to do damage; their destructive power comes from highly focused energy in the form of waves (electromagnetic wave weapons) or particles (particle beam weapons). Though you might not recognize them, particle beams are a part of everyday life, your older television set and several new cancer treatments employ particle beams. Ramp up the speed and number of those protons, electrons, or neutral atoms and they could be considered weapons. In the other grouping, light beams can refer to any type of electromagnetic wave focused in one direction. Lasers, X-rays, infrared radiation, microwaves, and gamma rays can all be targeted to do damage. These are all massless beams and are therefore different than those weapons that shoot small mass particles. The fact that they are different presents a problem for *Star Trek* fans because at different points in time phasers are described as both electromagnetic energy weapons (*Star Trek Generations*) and as particle weapons (*Star Trek: First Contact*).

1.2.1 Were Star Trek *Phasers Particle Beam Weapons?*

In those episodes and films where phasers are defined as particle beam weapons, they are said to fire a pulse or stream of subatomic particles (*TNG: The Mind's Eye*). The *Memory Alpha Star Trek* website states that plasma (ionized gas atoms) is passed through a "phase emitter" that stimulates a discharge of subatomic "rapid nadion" particles—not just nadions, *rapid* nadions. Inside that description are references to plasma and phase emitters that we will talk about below, although those subatomic rapid nadions are completely science fiction—for now.

There are most certainly subatomic particles that science has not discovered yet. Perhaps someday a particle lurking only in plasma streams will be identified, and perhaps they will name it the nadion in honor of *Star Trek*. Speculating even further, perhaps that nadion will carry energy that can be used to do work, like in an energy beam tool or weapon. There's no evidence for it now, but *Star Trek* has been right so many times before (communicators, tricorders, tractor beams, teleportation, etc.) that it can't be ruled out. If we confine ourselves to the *Star Trek* universe, then nadions exist, have energy, and are massed particles unlike light photons or gravitons. After firing a phaser, residual nadion particles are left on surfaces to be picked up by tricorder sensors (*VOY: Phage*) like gunpowder found on surfaces by crime scene investigators, so the phaser can't be just an energy beam.

Phasers come in several types, three of which can be held in the hand, the others being ship mounted. Type 1 phasers are small, about the size of a garage door opener. Type 2 phasers look like handguns, with grips and a trigger, although they become more stylized and look more like a sleek universal remote in later designs (*TNG; DS9*, and *VOY*). Type 3 phasers are rifles with a stock, a long barrel, and some type of sighting device. Kirk and cohorts carry all three at times, and the pistol and rifle forms are found in both *Star Trek: The Motion Picture* (1979) and *Star Trek* (2009). Particle beam phasers are quasi-conventional in that they fire projectiles, although they don't rely on explosives to give the projectiles energy. Most of the recent projectile weapons in Earth's history have relied on explosives, but the work being done now is moving from bullets to billions of very fast moving atom-sized projectiles.

If one accelerates protons, neutral atoms, or electrons to high energies, say 20 % the speed of light, they become very destructive. Atoms with an electrical charge (ions) are easy to accelerate using magnetic repulsion, though they are hard to keep moving in a straight line because they tend repel each other. To keep ion beams from naturally diverging or being deflected by a magnetic field, the charges are balanced by a transfer of electrons with gas in a chamber, producing neutral particles. In one type of neutral particle beam generator, light hydrogen (one proton, one electron, and no neutrons) or deuterium (one proton, one electron, and one neutron) gases are subjected to a significant electrical field. This turns the gas atoms into ions that can be accelerated through a magnetic field or an electrical field in a vacuum. Circular particle accelerators are preferred because more than one circuit can be used to get the particles up to speed without increasing the size of the accelerator. Once they reach the desired energy, the atoms are neutralized in the gas chamber and fired at the target. Most particle accelerators on Earth are extremely large in order to generate more power; however, a particle beam was launched into space and fired in the 1980s.

1.2.1.1 Neutral Particle Beams

In July of 1989, the Los Alamos National Laboratory fired a rocket into space that carried a neutral particle beam accelerator (NPB) as part of the Strategic Defense Initiative (Star Wars) project. The BEAR (**B**eam **E**xperiment **A**board **R**ocket), mission achieved low Earth orbit and a 1 meV hydrogen beam was produced using a hydrogen injector, a linear accelerator, and a xenon gas neutralizer. The entire experiment was over six minutes after liftoff, with the NPB firing 50 microsecond pulses of 1 meV particles at 5 Hz. The rocket was recovered and the NBP was fired again back in the laboratory. The paper describing the mission concluded with the following prophetic statement, "This first operation of an NPB accelerator in space uncovered no unexpected physics." (O'Shea et al. 1990) The ability to use conventional particle beam technology outside Earth's atmosphere meant that the race for space weapons was on.

Not much has been heard of neutral particle beams since the BEAR test, suggesting that they were deemed either ineffective or too expensive… or perhaps they worked very well and are being kept secret. Neutral beams would not be deflected by the Earth's magnetic field, so it has been predicted that they would make better space-based particle weapons than proton (+ charged) or electron (− charged) beams. Then again, it's likely that if they were sufficient for weapons they would be sufficient for other applications. Unfortunately, the scientific literature of the 1990s and 2000s does not mention NBPs much at all. This may be changing, as a recent theoretical use was described wherein NPBs might be used as a propulsion method (Mole 2013). Perhaps enough progress has been made that NBPs are again being considered. Other types of particle beams haven't suffered from the same lack of attention. Just as when Commander Chakotay cuts through rock with a phaser drill to free a shuttlecraft (*VOY: Once Upon a Time*), ion beams are finding nonmilitary, destructive uses.

1.2.1.2 Proton Beams

Cancerous tissues are the current targets of proton beams. Proton beam therapy is similar to commonly used X-ray radiation treatments, yet some significant differences make protons more appealing. Focused beams of X-rays are much wider and therefore less precise than proton beams. This spares normal tissues that might be damaged by X-rays and allows doctors to use higher doses of protons and still avoid side effects. Also, the proton beam can be tailored to individual tumors based on shape, location, tissue density and depth, so that the protons have just enough energy to reach the tumor and induce cell death via ionizing radiation and then go no further.

A proton beam treatment starts with the generation of plasma from hydrogen gas (Box 4.1). The positively charged protons are collected using a negative charge and are funneled into a vacuum tube pre-accelerator. At this point, the protons have about 2 meV of energy. The circular radio frequency accelerator delivers more energy to the protons with each transit so that by the time they leave the accelerator they have been boosted to between 50 and 300 meV. The final energy level can be controlled precisely so that the protons will travel the right distance into the tissue before giving up their energy. The nozzle from which the beam is discharged before it enters the body works to collimate and widen the beam so it can irradiate the target in a timely fashion.

The above description points out several issues that have thus far prevented ion beams from being used as weapons, especially for hand-held weapons like type 1–3 phasers. One, the equipment is so large that the beam is generated in a room separate from the patient. Two, the beam is only several protons wide, perhaps a millimeter diameter at most—could a pinhole take down a plane or stop a tank? Three, the energy is so low that it spares most tissue, much too low to be an efficient weapon. And finally, the beam travels a very short distance before reaching its target, much closer than a soldier would come to an enemy tank or that two starships would come to each other. To make weapons, or to even make devices that are more efficient, cheaper, and more manageable will require both miniaturization and a major boost in power.

1.2.1.3 Electron Beams

Electron particle beams have more uses than proton beams. Cathode ray tubes of old TVs and computer monitors used streams of electrons to stimulate phosphors in 2D arrays. Electron microscopes use beams of electrons to image very small structures based on their transmission or scattering of the particles. The energy from a beam of electrons can be used to fuse build materials in a specific type of 3D printing (Sect. 3.3.1.3), and electron beam lithography can be used to etch patterns onto polymer surfaces that are both small (sub 10 nm or 3.94×10^{-7} in.) and precise patterns for nanoelectrode gates on chips or microcantilever sensors for point of care testing devices (Sect. 9.6.1). These are all non-military uses; if they are to be used for weapons they will have to be accelerated to even higher energies to overcome their small mass (1/1836th the mass of a proton). In general, particle beam-based phasers will need very powerful accelerators, regardless of the type of particle used.

This is a problem because the most powerful commercial and research accelerators are huge. The CERN Large Hadron Collider (LHC) particle accelerator is a circle with a circumference of 27.35 km (17 miles)! All that distance has been put to good use, the LHC recently accelerated particles to energies of 6.5 TeV—where 1 TeV equals one trillion electron volts (Fig. 1.1). In May, 2015, CERN researchers accelerated two protons in opposite directions to 6.5 TeV each and then

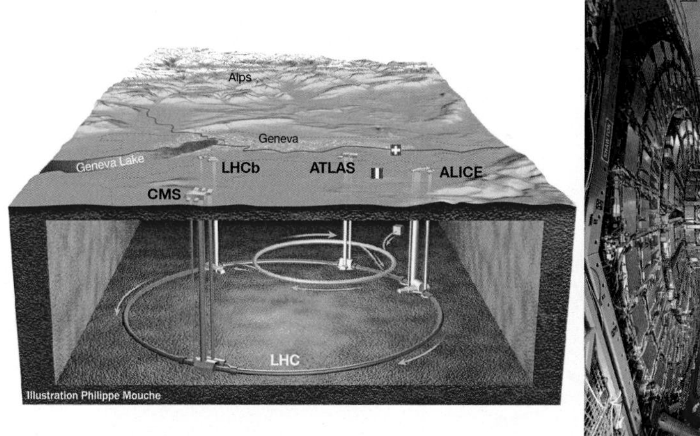

Fig. 1.1 The *left image* shows the layout of the Large Hadron Collider at CERN on the Switzerland-France border. The CMS, LHCb, ATLAS, and ALICE experiment stations are marked around the 27.4 km (17 mile) circumference of the tunnels that are located between (150–500 ft) underground. Despite the gigantic size of the loop, particles speeding through the supermagnetic powered tunnels will make 40,000 circuits per second. The CMS detector supermagnet ring is show on the *right*. The diameter of the detector apparatus is more than 15 m (49.2 ft). (*Image credits* © 2015–2016 CERN, two images merged)

put them on a collision course. The resulting impact had the energy of 13 TeV, enough that a portion of the public worried that a black hole might form (Ali et al. 2015). Each particle might be very light, yet their high velocity (high energy) makes it like crashing huge trains traveling hundreds of miles per hour straight into one another. Impressive, but to make significant progress in powerful particle beam miniaturization the accelerators are going to have to be greatly reduced in size. Devices called laser accelerators might just be the answer.

1.2.2 Using Lasers to Accelerate Particle Beams

A laser accelerator uses the energy of light to speed up the particles of a beam. Speed can be gained in a couple of different ways, depending on the design of the accelerator. One version is the size of a microchip, though the laser still helps to accelerate particles to levels sufficient for many devices. It uses silicon waveguides doped with titanium (Peralta et al. 2013; Breuer and Hommelhoff 2013) to accelerate electrons in two stages. The first stage is an electron gun that raises their speed to perhaps 2 % of light speed, while the second increases their energy while keeping their speed constant. It is in this second stage that the electrons are focused through a 0.5 μm (2×10^{-5} in.) channel that contains the silicon guide with ridges on each side. The shape of the apparatus is like looking down on two Lego blocks on their sides with their studs almost touching one another. When an infrared laser is shone on the metal-doped silicon, it sets up an electrical field along each stud of

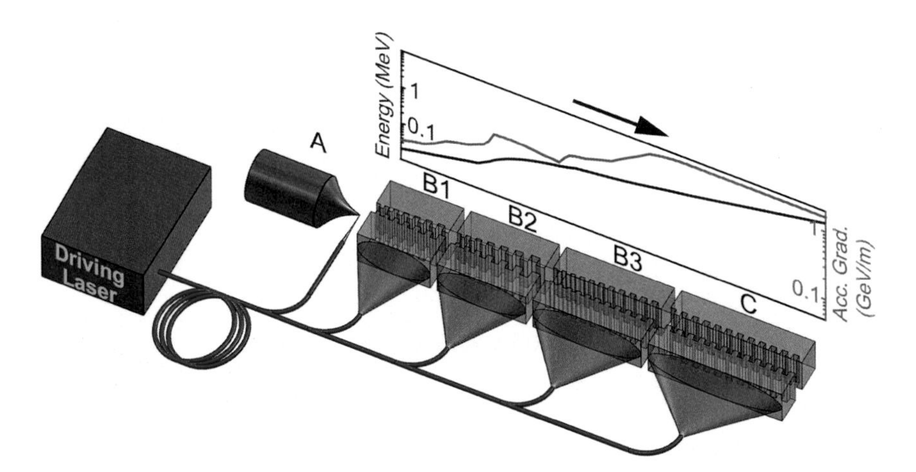

Fig. 1.2 This schematic demonstrates one type of laser accelerator. The gun (*A*) fires a beam of electrons between the silicon waveguides (*B1–3*, *C*). The infrared laser induces an electrical field in the titanium-doped guides, which speeds up the electrons as they pass. The increase in electron energy (speed) is shown by the *blue line*, while the *right line* traces the amount of acceleration in each guide. [*Image credit* Breuer and Hommelhoff (2013)]

this dielectric laser accelerator (DLA). The electrons that fly through the middle gap gain energy when they interact with the electrical fields (Fig. 1.2).

The initial tests of this fruitfly sized chip accelerator achieved accelerations to 0.3 GeV per meter of accelerator. If the chips were lined up to be 0.3 m (1 ft) long, they could match energies with some 3.2 km (2 mile) long linear accelerators, and the chip version can deliver billions more of these high speed electrons per second. An additional advantage of the DLA is very short pulse duration for the bunched electrons, on the order of a few attoseconds (one attosecond is 1×10^{-18} s). Short pulses make very fast electronic switches possible, while packaging the pulses together can make for powerful beams. This makes the on-chip accelerators scalable; the power can be turned down to operate microcircuitry or turned up to crash particles together for research. While this is good enough for many applications, they will have to be made even more powerful if they are to be used in phasers that can stun or kill.

1.2.3 Plasma Laser Particle Accelerators

The Department of Energy's Lawrence Berkeley National Laboratory (Berkeley, CA) has been at the forefront of developing laser accelerators to build more powerful particle beams. They recently accelerated electrons to greater energy than any previous small accelerator, approximately 4.25 GeV, using one of most powerful lasers in the world. In this instance, they included plasma in the mix (Leemans et al. 2014). In the plasma laser accelerator method, the laser pulses induce waves in a cloud of plasma like the wake of a speedboat in water. The propagating waves have peaks and troughs that create strong electric fields. Electrons passed through this field basically skip off the peaks of the waves, gaining energy with each interaction. Powerful accelerations are generated in a matter of centimeters with laser wakefield accelerators, the accelerator in the Berkeley demonstration was just nine centimeters long but generated more energy than conventional radio frequency accelerators that are kilometers in length.

Inventive engineering has reduced the size of particle accelerators from miles long to inches long with almost no loss in power. Laser plasma accelerators will continue to get more powerful and might eventually be able to function in a handheld phaser, *if* the size of the laser can be reduced to the same degree. Most of the powerful lasers that function in particle accelerators are as big as cars. While this is the major hindrance to handheld particle beam devices at the present time, laser research is increasing power and reducing size at an astonishing pace. While most laser research focuses on the using lasers on their own, the advances will be applicable to laser accelerators as well.

1.3 Light Beams as Directed Energy Weapons

Lasers are the truest examples of directed energy weapons; they hit a target with pure energy. No projectiles are involved, and the challenge to using lasers as weapons lies in this lack of mass. Heavy projectiles don't have to travel very fast to carry lots of energy, while massless photons from lasers must be extremely powerful to make up for their lack of mass. Powerful lasers require vast amounts of energy, and this requires more equipment, more space and an adequate supply of electricity. These issues are being addressed in the public and private sectors, and science must keep an open mind as to which directions to follow. *Star Trek* gives us the obvious path, from pistols, to electrical shock phase pistol (*ENT: Broken Bow*) to laser- or particle-based phasers. However, that doesn't mean that this is the only path. As we shall see, 21st century science may move toward laser weapons, but don't count out a side step to focused lightning weapons.

1.3.1 The Phaser Goes from Maser to Laser

The original scripts called for *Star Trek* characters to carry both phasers and lasers (*TOS: The Cage*), but that didn't last long. Series creator Gene Roddenberry became concerned early in filming that the public would not accept lasers as weapons; the lasers of 1965–1966 just weren't very strong. A commercial laser pointer of today would have been considered cutting edge technology when Mr. Spock raised his eyebrow at the camera for the first time. As the laser became public knowledge, Roddenberry worried that the audience would reject the show based on unbelievable technology. At first the writers balanced the lasers with phasers (*The Menagerie, part II*); later they decided to go just with phaser technology. In fact, lasers were spoken of as ancient technology by the second season (*TOS: A Private Little War, Patterns of Force*).

Roddenberry knew that a device called the maser had preceded the laser by a decade and that they were more powerful than the lasers of that day. The name is an acronym for **M**icrowave **A**mplification by **S**timulated **E**mission of **R**adiation, in the first papers describing them in 1954 and 1955. Microwave emitters themselves are older than masers, though masers are much more than just emitters just as lasers are more than just light bulbs. Einstein posited in 1917 that if one stimulated molecules with energy, they could emit radiation on their own (Einstein 1917). The type of electromagnetic radiation that molecules emit is based on their structure; different molecules can be stimulated to produce different types of waves.

Just after World War II, the molecules in which scientists were most interested emitted sub-centimeter microwaves, waves a bit shorter and more powerful than the radio waves of radar. Charles Townes of Columbia University hypothesized that if he held a microwave-emitting compound inside a reflective (resonant) cavity, the microwaves emitted would bounce around and hit more of the compound's atoms and stimulate even more microwaves—a self-sustaining and amplified emission. Townes used ammonia molecules as the emitter and produced a 10 nW maser (Gordon et al. 1955), extremely weak by today's standards but a proof of the concept.

It wasn't long after the maser became an important scientific tool for radio telescopes that Townes and others starting thinking about duplicating the amplification with radiation at even shorter wavelengths (ie. higher frequencies and higher energies). Infrared light, visible light, and even X-ray amplification followed in the 1960s. Townes wanted these light-based amplifiers to be called "optical masers," but a graduate student at Columbia named Gordon Gould suggested *laser*, where **L**ight, replaced **M**icrowave in the acronym. In fact, Gould wanted each frequency group to have their own acronym; gamma ray lasers would be called grasers, and X-ray lasers would be xrasers, etc.—they never caught on. It was casually agreed that microwave or longer wave emitters would be called masers, and infrared or short wavelength emitters would be called lasers.

Box 1.2 How Lasers Work

Electron energy levels—Because an electron travels in a spherical cloud, it must be careful not to cancel its own energy by having its wave peaks meet troughs on a second, third, etc. orbit. This means that only very specific energies are allowed for each electron in each orbit around a nucleus. The size and constituents of the nucleus and its number of atom's electrons determine its allowable energy levels. An electron can only absorb the precise amount that lifts it to the next energy level, or perhaps two or three levels. Here we will consider the idealized situation of one ground state and one excited state; multiple level jumps have messy fast and slow emissions and metastable states.

Photon emission—After absorbing energy from light or electricity (pumping energy), the electron will seek to return to the lowest (most stable) energy level. When it drops a level, the electron emits a photon of light. The light will be of a specific wavelength, and is called *photon emission*. The frequency of the emitted photon depends on the structure of the molecule accepting the energy.

Amplification—The emitted photon has exactly the energy needed to raise another electron of the target substance to the higher level to stimulate emission of another photon. The emitted photons will disperse in any direction, so are unlikely to hit another atom, but if you bounce the photons off mirrors at each end of the target compound, then many more photons will strike atoms and will *amplify* the emission. One of the mirrors is only partially silvered, so some of the photons will escape as a beam, the type depends on the frequency of light emitted. Higher frequency emitters go by the acronym *laser* (**L**ight **A**mplification by **S**timulated **E**mission of **R**adiation); lower frequency emitters are masers (substitute **mi**crowave for **l**ight). Higher frequency and more photons make a more powerful beam.

Characteristics of lasers—(1) every emitted photon has the same frequency, so the light is *monochromatic*; (2) the mirrors of the resonant cavity are parallel, so the photons line up and don't diverge much when the beam is emitted; (3) several electrons may be raised an energy level (called population inversion) and drop to ground state in short order. This keeps the photons in phase in time (*temporal coherence*) and space (*spatial coherence*).

Lasers can be continuous or pulsed depending on the input pumping energy and timing of the population inversion dump. Typically, pulsed lasers have higher peak energy outputs. The most powerful peak output pulse laser in 2015 is 2 PW (2×10^{15} W) while your laser pointer is 1–5 mW (1×10^{-3} W).

1.3.2 Star Trek *Phasers Are Really Just Lasers*

The name "phaser" was a play on words by the show's production team, but one with several possible meanings. A photon is the basic unit of light, so the phaser is basically a "photon maser," another name for a laser. The portmanteau of the two words is plain enough, yet it sounded like a completely alien technology in 1966. However, as soon as the writers began to favor phasers over lasers, they started writing in different explanations of what they were and how they worked. A 1968 book about the making of the series admitted that phasers were really just lasers. It said that they, "Could be set to pulsing frequencies that would interfere with the wave patterns of any molecular structure." They called the pulsing "phasing," so the weapons were given the name phaser (Whitfield and Roddenberry 1968, p. 193). This might refer to the change in the phase velocity of light (Sect. 2.4.1) in the center of a laser beam, although Roddenberry never says it specifically.

The evolution of the name doesn't stop there. Phaser was also said to be a mash up of the words "phased array emitter" (*Star Trek*: The Next Generation Writers Technical Guide, Fourth Season Edition, 1990, p. 14), although by the time the writer's technical guides were expanded into the published *Star Trek: The Next Generation Technical Manual*, the expansion of the acronym had been changed to **PHAS**ed **E**nergy **R**ectification (p. 123). The manual submits that rectification refers to transducing the stored energy into a beam without going through an intermediate stage. Just how that would work in a weapon is unclear. Rectifying a wave usually refers to turning alternate current electricity to direct current. However, one could make a good case that the "phased array emitter" definition of a phaser was a bit of sci-fi genius. There are indeed phased array emitters in the most advanced lasers these days, and they do have implications for using lasers as weapons.

1.3.2.1 Optical Phased Arrays

A phased array emitter usually refers to an antenna that has several transmitters mounted in a specific pattern, each one broadcasting waves of a slightly different phase. If waves are in phase, their peaks and troughs line up; if they don't line up the waves are said to be out of phase. Interacting waves of different phases create interference patterns (Sect. 5.3.3.2). If two waves of the array have positive interference, the resulting wave will have twice the amplitude and four times the energy. The geometry of the array and the interference pattern determines the direction of the positive interference patterns and therefore the direction of the broadcast. Radar (**RA**dio **D**etecting **A**nd **R**anging) sweeps can be managed with phased array emitters without moving a transmitter back and forth, just by manipulating the phases of the waves from the different emitters. Optical versions of the phased arrays are similarly used for the light equivalent of radar, called LIDAR or LADAR (with **LI**ght or **LA**ser replacing radio).

By controlling intensities and phases of arranged light emitters, a beam in a single direction is produced. The beam is stronger than the original light beam and can be varied in intensity and direction based on the strength and geometry of the interferences. LIDAR systems measure the time it takes a light beam to return to a source after reflecting off an object in the beam's path. The speed of light is almost constant, so the timing of the signal's return is directly related to distance the light traveled and the distance to the object is half the total distance traveled. Bounce many beams off of the object in a short time and one can track its movement. Unfortunately, a set of mechanical gears and motors to sweep a LIDAR emitter is too slow and sloppy to take advantage of the light speed signals. Controlling the direction of the LIDAR beam using an optical phased array is much faster and more accurate.

Researchers at MIT recently produced a nanoscale optical phased array for lasers on a 576 μm^2 (8.9×10^{-7} in.^{2}) microchip, about twice the size of the period at the end of this sentence. The array had 4096 laser emitters arranged in a 64×64 pattern. Each emitter was the equivalent of a pixel, but the interference patterns from the laser beams allowed for much more intricate and detailed images than could be produced with that number of individual pixels. In this array, the phase of each emitter's beam was static, so demonstration array could produce only one image, an MIT logo (Sun et al. 2013). The goal is to be able to project any image at any point in a large field, so the team also produced a smaller array to demonstrate this capability (Fig. 1.3).

The second chip was made up of a 4×4 array of laser emitters, with a heat source that modulated the phase of the wave from each emitter (Sun et al. 2013). Tuning the phase of each laser beam allowed for complex control of the interference patterns. Images or beams could be changed and moved in real time. This demonstrator could change direction as for a LIDAR sweep or a targeting system on a laser weapon. The large demonstrator array could be made tunable as well, although to do so would have required thousands more electrical connections than the graduate student in charge of the work was prepared to solder. Nevertheless, the

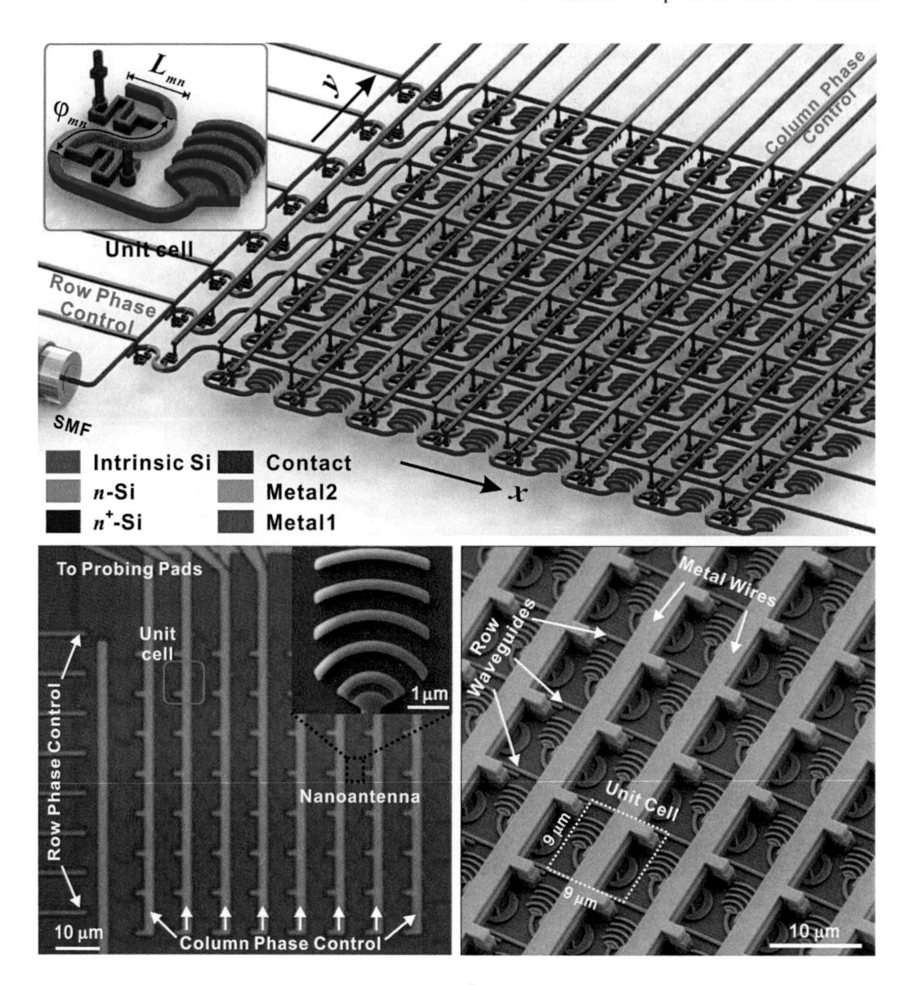

Fig. 1.3 The *top diagram* shows the 8 × 8 optical phased array layout. The *inset* shows one unit cell where the length L of the directional coupler determines the direction that the individual beam will be sent, and the configuration of the optical delay line (*purple*) determines the phase of the light from that unit cell. The *red line* with semicircles is the nanoantenna for that unit cell. The *lower left* photomicrograph shows the production array with the contacts and input laser in orange and an individual nanoantenna in the *inset*. The *lower right* photomicrograph is a higher magnification of the same array with the dielectric silicon dioxide etched away. [*Image credit* Sun et al. (2014)]

small array proves that nano-optical phased arrays work, while the large array is the biggest ever demonstration of the concept.

Using these chips as a starting point in their research, another group demonstrated a LADAR phased array in May of 2015 that can sweep 51° of space 100,000 times per second. That's more than 10,000× faster than current mechanical sweeping systems (DARPA press release, May 21, 2015a). The SWEEPER (**S**hort-range **W**ide-field-of-view **E**xtremely agile **E**lectronically steered **P**hotonic

Emitter**R**) system is a very effective tracker despite its horribly convoluted acronym. Most optical arrays use individual emitters and individual controllers, each of which is susceptible to failure. The SWEEPER overcomes this problem by using a single solid-state mechanism that is stable and can compensate for changes across its area. The array is housed on a single microchip, can be deployed in drones or handheld LIDAR units, and even has implications for production of a handheld laser weapon since it is small enough to be housed in a type 1 phaser.

Optical phased array technology affords the ability to control the direction and power of a laser without moving the source in any way. This is reminiscent of phasers in two aspects. One, phasers are scalable in power, as witnessed by the many times that Kirk, or Picard or even security chief Tasha Yar calls out to set phasers on stun (*TNG: Angel One*). Likewise, an optically phased laser of sufficient power could be anything from annoying or lethal just by controlling the degree of positive interference of the array. Second, the direction and shape of a phased array beam is controllable, even while the laser is held steady. This is similar to when Deep Space Nine's phasers perform sweeping maneuvers, changing direction and shape to cover large swathes in a single pass (*DS9: The Way of the Warrior*).

It is safe to assume that handheld and ship-mounted phasers use optical phased arrays for both scaling and directional control, and that **PHAS**ed array emitt**ER** is an apt explanation for the name of the weapon. It certainly fits better than a phased energy rectifier, whatever that is, although it less technically correct than the original "photon maser" explanation. It is good enough that any future handheld, scalable laser weapon that employs optical phased arrays could easily find itself called a phaser. Interestingly, optical phased arrays could be used in other *Star Trek* technology as well. The holodeck uses holograms, along with other technologies we don't have yet, to produce alternative reality environments that can be touched and maneuvered through. The strong positive and negative interferences induced through optical phased arrays are excellent ways to produce holograms, so it is really just a matter of time before we are being entertained in a solid 3D hologram that moves and changes with our actions.

1.4 Masers Weapons

The above discussion highlights the fact that *Star Trek* has described several different mechanisms of action of phasers, from the untestable nadion particle beam generation to the completely plausible optical phased array. The original photon maser explanation is the easiest to accept when one surveys the directed energy weapons being developed today. The *Star Trek* canon even includes a decidedly overdone description of how the maser beam would be produced. Supposedly, plasma is pumped into a pre-fire chamber of superconducting lithium-copper alloy where it undergoes a rapid nadion effect so that the strong nuclear force is stripped from the plasma. The explanation goes on—the protonic force of the now liberated subatomic particles interacts with the lithium-copper to generate an EM beam (TNG

Technical Manual, pp. 123–125). If one ignores all the jibber jabber in the middle, the crux of the matter is that energy is being transduced to a coherent electromagnetic wave pattern starting from plasma. We saw above that plasmas can be tied to lasers in particle accelerators. Now we shall see how masers and lasers are being made into scalable weapons—some even use plasma.

1.4.1 The Adaptive Denial System Is a Microwave Emitter Weapon

Even though masers are used as timers in atomic clocks and for astronomical research, they are poor cousins to the laser when it comes to high-energy directed energy weapons. Masers are handcuffed by the conditions and equipment needed to produce them. Most require significant vacuum and cryogenic cooling; however, your microwave oven doesn't feel cold because it isn't a maser, it's just a microwave emitter. The difficulties in producing masers that are small, light, and cheap limits their potential as directed energy weapons, but that doesn't mean that a microwave emitter itself can't be used as a deterrent. Microwaves have less energy than light waves so they don't penetrate the body very deeply; the most they can do is heat up your skin and hurt like heck, which is often good enough. Current microwave-based weapons can't be set to stun or kill, just pain (Fig. 1.4).

Imagine a local insurgent with bad intentions mixing in with a crowd of refugees approaching a parked Humvee column. The individual acts as though he/she doesn't understand the verbal warnings to stay back and keeps walking forward. One Humvee points what appears to be a large radio transceiver at the person and in very short order his skin starts to feel uncomfortably hot—stingingly hot. He backs off immediately, and the threat is alleviated. If the person wasn't a threat, there are no worries because no permanent damage was done, but if he tries to avoid the beam and keeps coming at the column, the soldiers know to take more aggressive action. This is the Army's nonlethal Active Denial System (ADS) at work.

Microwaves (100 MHz–300 GHz) cause polar molecules to rotate quickly. Water (H_20) is a polar molecule, with a partial positive charge on the hydrogens and a partial negative charge on the oxygen. The water molecules try to align themselves with the EM field, so they rotate. An EM wave field oscillates, so the water molecules keep rotating as they constantly try to align themselves with the changing field. The rotating water molecules bump into other molecules and transfer some of their energy in the form of heat. In a microwave oven, the food heats up but so does a glass or ceramic bowl the food is in because the microwaves aren't tuned specifically to water. They rotate any polar molecules, including the ones in ceramics and glass.

Fig. 1.4 The active denial system is a nonlethal personnel deterrent. It radiates a microwave beam that heats the skin without doing damage. There are two versions of the system; the larger system is town by a large truck, while troop carriers are smaller with a similar set up. (*Image credit* US Department of Defense)

Adult humans are 50–65 % water, so microwaves have the potential to heat them up. Is there a danger of cooking an innocent person who approaches a convoy looking for help? Not really. The ADS emits microwaves at 95 GHz, strong enough to penetrate only 1/64th of an inch (0.4 mm) into the skin (Human Effects Advisory Panel Final Report 2008). In more than 13,000 volunteer doses delivered by the ADS, injury occurred in less than 1/10th of 1 percent of individuals. The ADS1 from Raytheon Corp. (Waltham, MA) and the US Department of Defense was deployed in Afghanistan in 2010, although there are no reports of it having been used prior to withdrawal in 2011. A smaller ADS2 system is in development, and Raytheon produced a much smaller system under the name Silent Guardian for civilian use. The Los Angeles County Jail was set to install a Silent Guardian System for inmate crowd control in 2014, but the project was postponed over concerns that its use could be seen as a torture technique or cruel and unusual punishment. The fact that many opponents refer to it as the "pain ray" might have something to do with its less than overwhelming support.

1.5 High Energy Lasers as Directed Energy Weapons

Lasers are the best hope for making workable directed energy weapons. Some *Star Trek* phasers are huge ship-mounted systems, able to render an entire planet's populace unconscious but otherwise unharmed (*TOS: The Trouble With Tribbles*), or destroy entire starships (*Star Trek II: The Wrath of Khan*). Other phasers are so small that they can fit in the palm of a hand, yet can reduce a humanoid like Jo'Bril to ash (*TNG: Suspicions*) or bore a tunnel in solid rock (*TNG: Chain of Command, part I*). Phaser energy weapons are powerful, yet tunable, and they can be aimed accurately on a single spot, or beamed wide and move in a pattern (*DS9: Way of the Warrior*). The question is whether current science can match those qualities in a laser weapon.

Universities, commercial labs, and the military have produced lasers that are powerful and lasers that are handheld, but no lasers that are both powerful *and* handheld. The lasers of LIDAR ranging and targeting systems are not powerful enough in and of themselves to do damage. Lasers in medicine and biology can be finely tuned as to the amount of power they deliver, yet have limited wattage based on their purpose. It should be a small matter to combine these attributes and produce a laser weapon, so why don't we have police with laser pistols, a military with laser rifles, and ships or aircraft with laser beam weapons to destroy enemy material?

The most basic reason we don't have laser weapons yet is that we haven't needed them—Ronald Reagan's Strategic Defense Initiative excepted. Other than that, it turns out that there are myriad issues to deal with when trying to create a laser weapon. Current technology can make strong lasers; however, they get bigger as they get more powerful. How does one fit something the size of an RV on a tank? In addition, strong lasers are power drains; if one wants to fire a laser from a fighter jet, just where is all that power supposed to come from? While lasers may travel at the speed of light, it remains difficult to keep a laser locked precisely on one small spot for the time needed to destroy a relatively fast maneuvering missile or drone. Finally, environmental conditions drain laser power and defocus their aim, so lasers in air have even more problems that space-based lasers would have. Yes, there are indeed issues to be dealt with in trying to build laser weapons. However, to paraphrase the title of William Shatner's personal survey on *Star Trek* technologies —they're working on that (Shatner and Walter 2004, title page).

Box 1.3 Phasers in Space
Star Trek stories are visual—rich in color, motion, and meaning. What we see helps tell the story, including battle scenes where photon torpedo detonations and phaser beams light up space. Could we really see and hear phaser beam shots in space and would photon torpedoes produce huge explosions and fireballs that roll out into space?

Seeing phasers—The phasers in the original series are red (*TOS: The Changeling*), blue (*TOS: The Galileo Seven*) or sometimes green (*TOS: Wink of an Eye*). At other times they are invisible (*TOS: Whom Gods Destroy*). In truth, only a few episodes got it right—only those that had invisible phasers in space were scientifically accurate.

We see light for only two reasons. Either the source beams travel directly to our eye or they are reflected from a surface to our retina. A regular light beam diverges greatly from its source because the generated waves disperse in all directions and rays reflected of objects scatter as well. Some waves may travel directly to your eyes and you can see the color. Therefore, rays that travel from source to target without striking something are invisible unless something gets in their way and reflects rays to your eyes.

Dust, water vapor and other impurities in air will scatter the beam, and make it visible in many environments—smoky nightclubs with cheesy laser shows for instance. In other cases, Rayleigh scattering can make a laser visible if the wavelength is correct. Our sky is blue because blue light is scattered by the air molecules (air has roughly 10^{19} atoms/cm^3). Rayleigh scattering is dependent on the inverse of the wavelength (λ^{-4}), so blue light is scattered much more than red or green light. If a phaser is blue, then it will scatter to a greater degree in air and be more visible than green or red lasers.

Dust particles are rare in space, perhaps one part dust per million, so scattering of a beam by impurities is unlikely. Rayleigh scattering is also impossible, since there is only 0.1–1.0 atoms/cm^3. Therefore, only those episodes in the original series that have invisible phaser beams are scientifically plausible. As for the noises phasers make—noises in free space? Sound waves are mechanical waves; they can propagate in air but not the vacuum of space. There would be no pew! pew! when attacking the Klingons. The truth is—lights and sound make good TV, so ignore the problems and just enjoy.

The strongest handheld laser that is legal to purchase is about 3500 mW. These laser pointers on steroids look a bit more like light sabers than phasers, but they can pop balloons, ignite paper, or do serious and permanent damage to a retina. While thirty-five hundred milliwatts may sound like a lot, current efforts by the military are realizing 300 kW to 1 MW lasers. That's a laser 85,700-285,700× more powerful than the one you can buy. Of course, these are meant to be anti-tank and anti-ship weapons. Lasers to be used against enemy troops can be somewhat less powerful, and are scalable so that they can be used as a deterrent rather than a lethal weapon, just like the diplomatic Captain Picard informs Q while in orbit around Farpoint Station (*TNG: Encounter at Farpoint*).

1.5.1 The Protocol on Blinding Laser Weapons and Dazzler Lasers

It is important that we consider matters of scale when designing laser weapons; they have the potential to do so much damage. Even some laser pointers can burn photoreceptors and retinal neurons (Sect. 7.2.1) if shone in the eye for too long. As it became apparent in the 1980s that countries would try to weaponize lasers, diplomats called for a ban on lasers that could blind combatants. The Protocol on Blinding Laser Weapons, drafted as a 1995 protocol to the 1980 Convention on Certain Conventional Weapons by the International Committee of the Red Cross (ICRC), called for prohibition of laser weapons whose sole purpose is to cause blindness. It also recommended that all feasible precautions be taken to spare visual systems when designing and using lasers (ICRC website). The Protocol took effect in July of 1998, and 105 countries are a party to the treaty. Finland was the first country to sign in 1996, while Algeria agreed to the protocol in late 2015.

To stay within the protocol, the US military is developing some laser weapons that are nonlethal and specifically designed not to injure retinal tissues. The GLARE laser dazzler (BE Meyers Advanced Photonics, Redmond, WA) has been used by all branches of the US military since 2009. This is a warning weapon, called a visual disruptor. When a suspicious individual approaches a prohibited area, the rifle mounted GLARE is shone in the person's eyes. This type of laser is designed to diverge more than high power lasers, so that a larger area is covered and the power is reduced to non-injurious levels. Not injuring the person doesn't negate blinding them temporarily. The intense beam of the laser depletes the photoreceptors of the retinal pigments that stimulate neural impulses, so the person will have trouble seeing until the pigments can recover. This is just like what happens when you look at the Sun or someone takes your photograph using a flash. Also, the glare of the green laser provides a distinct and unmistakable message that does not rely on language and can be used before the person is close enough to require more lethal force. The message is clear—this area is off limits, yet no one is hurt.

To ensure that the dazzlers stay with the Protocol on Blinding Lasers standards, the dazzlers have a nominal ocular hazard distance (NOHD). For the GLARE laser weapon, if a person is more than 30 m (98 ft) away from the source, then the power transmitted to the target will be meet FDA and World Health Organization standards (10 fold below minimal levels that cause damage). The British military has developed a 100 mW dazzler called GLOW, that has a NOHD of only 10 m (33 ft). This means the GLOW laser is of lower power as compared to the GLARE system; the target can be much closer to the source and still not sustain damage. However, the issue then becomes whether the soldiers are comfortable with allowing someone with 10 m before taking the measures have a stronger affect. The concept of NOHD is an indicator that these dazzlers have passive scalability, the closer the target to the source, the greater the power delivered. Other systems being developed are actively scalable, like the very Star Trek-sounding PHaSR system from the US Department of Defense.

The PHaSR (**P**ersonnel **H**alting **a**nd **S**timulation **R**esponse) is a self contained laser gun, not a barrel attachment like the dazzlers. Intended for the military and law enforcement, the rifle shape is more like *Star Trek* phasers in that it has two settings that the user can switch between, with the shorter wavelength setting being more powerful. Another of the technological advances included in this weapon is an infrared ranger finder that will mark the distance of the target and modulate the power of the laser to keep it above the NOHD. The prototypes were tested in 2005 to great success, although a search of the internet doesn't show new information on these weapons after 2006.

The most recent advances integrated into nonlethal lasers are found in power modulation and in power supply (stronger and longer lived batteries), but this only goes so far. No amount of research using the current paradigm will result in lasers powerful enough to disable a tank or an incoming missile. In many ways, high energy lasers (HEL) are their own worst enemy. The increased power has a tendency to destroy the components used to produce the laser light itself; the lasers destroy themselves as they are used. The mirrors of the laser are especially vulnerable to damage or decreased function as power is increased.

1.5.2 Mechanisms to Reduce Laser Thermal Blooming

As power goes up, so does heat. The heat and energy conveyed to the mirrors of the resonant cavity (Box 1.2) can deform them. This leads to losses in focus, steadiness, and power, and an increase in divergence. Until the 2000s, deformed mirrors were a major impediment to development of higher intensity lasers. The solution came when scientists figured out how to make the deformations work for them instead of against them. Deformable mirrors—reflective surfaces that can change shape—can now compensate for changes in mirrors due to thermal expansion or contraction. As a bonus, computers control mirror deformation can adjust for pockets of differing air density between the source and the target. In standard lasers, the changes in air density cause the beam to diverge or travel over the surface of the target.

The field of adaptive optics encompasses the many ways to improve optical systems by reducing wave front distortions, including ways to correct mirror aberrations with deformable mirrors. One of the most commonly used adaptive mirror mechanisms is the piezoelectric control of ceramic electrodes glued to the mirror. When a voltage is applied to a specific ceramic segment, it changes length and creates tension in the mirror, causing a local curvature. There are also segmented mirrors, wherein a computer controls each segment's tip and tilt independently. One of the more interesting deformable mirror designs uses 10 nm (3.94×10^{-7} in.) diameter ferromagnetic particles in a reflective liquid. The magnetic particles change shape when exposed to magnetic fields of different geometries and this will in turn alter the mirror's shape. A ferromagnetic liquid mirror array can compensate for thermal expansion or other affects that can sap a laser's power.

A powerful laser's thermal effect on its mirrors is often surpassed by the havoc that the atmosphere can have on beam focusing. Atmospheric turbulence caused by environmental conditions such as wind, humidity, dust levels or air pressure creates different volumes of air that have different densities. Light is refracted as it passes through media of different densities, so the laser will lose focus if the atmosphere isn't consistent from source to target. Unfortunately, the situation gets worse. Just as the environment can affect the atmosphere and degrade the laser beam, the laser itself can affect the atmosphere. A high power laser beam will heat the air through which it travels (called blooming) and can even strip electrons from the air molecules (forming a plasma). The heating changes the air density and defocuses the beam, while producing a plasma drains the power of the laser.

Adaptive optics and optical phased arrays can both help overcome environmental effects and thermal blooming in HELs, but they aren't the only solutions. Two additional mechanisms are being pursued to reduce blooming and increase beam focus. Ultrashort pulse lasers (USPL) were pursued first by industry as a way to cut precise shapes on very small scales, followed closely by space administrations that believe the short pulses used will protect electronics from radiation damage. Recent US Army research results (Pomerleau 2015) indicate that laser pulses of femtosecond (10^{-15}) to 10 ps (10^{-11}) duration require less power and undergo less blooming, so their designs for laser weapons are reflecting shorter pulse times and higher pulse rates. In combination with USPL, arrays of small lasers are being investigated as more efficient and more focused than large single beams. Arrays of multiple small beams tend to produce less heat, bloom less, and can be optically combined for more powerful resultant beams with less energy input. A defense technology website recently stated that laser arrays are a major focus of new US Air Force and Navy research. Appropriately, several of the HEL weapons described below employ arrays of short pulse lasers.

1.6 Examples of Weaponized Lasers

For many years, the military pointed to the ABL (airborne laser system) as proof that mobile laser systems could work. The ABL was a 1 MW behemoth mounted on a 747 and designed to destroy cruise and intercontinental ballistic missiles. While it was successful in its testing period, there were drawbacks. The system was large and a huge power drain, it could never be mounted on a fighter or bomber and had to be located very close to the launch site to destroy ballistic missiles—unlikely at best. The ABL was a victim of budgetary cuts in 2012 as other systems became available.

Current projects are in the various states of readiness. The US Navy's LaWS (**La**ser **W**eapon **S**ystem), a ship-mounted system has already been deployed in the Persian Gulf, while the Solid State Laser Technology Maturation system from the Office of Naval Research is still in the design stage. In between there is the ATHENA 30 kW laser that burned a hole through a truck engine block in a 2015 video, and the 60 kW truck-mounted fiber optic laser called RELI (**R**obust **E**lectric **L**aser **I**nitiative) from the US Army. Other countries are just starting research for laser weapons. The British Ministry of Defense didn't initiate a laser weapon program until March of 2015, so the systems discussed below are all of US design.

1.6.1 The US Navy LaWS

As the first ship-mounted system, LaWS represents several advances in laser weaponry. As installed on the amphibious transport ship USS Ponce in mid-2014, the LaWS was the first HEL system to be tied directly into the ship's power grid. This was an important step forward because the LaWS could be fired without degrading the ships other capabilities; the power it drew did not blackout the rest of the ship. It uses an array of several small lasers to reduce the needed power, yet it is a powerful weapon due to beam combining (current version is 30 kW) (Fig. 1.5). The Navy estimates that the price per firing is a mere 59 cents (Davenport 2014), and it won't run out of ammunition as long as ship has power. Compare that to $50–70 for just one second's firing of the conventional USS Ponce weapon, the Phalanx CIWS 6-barrel gun.

Another advantage is that the LaWS is a solid-state laser. The first "lased" materials (the compounds whose electrons are stimulated to give off light) of the 1960s were cylindrical rubies, while gases became more popular in 1970s and 80s.

Fig. 1.5 The LaWS laser system is deployed on a US Navy vessel and has been test fired successfully. The 30 kW laser is hooked into the ship's power grid and can fire as long as the ship has power. (*Image credit* US Department of Defense)

Now there is move back to solid state lasers because they have a higher density and therefore more electrons to stimulate. This makes them more efficient and powerful lasers, as well as cheaper and simpler to build. The lased material for LaWS is an array of optical fibers doped with rare Earth metals like ytterium and neodymium. The fiber laser array is powerful and fast enough to attack unmanned aircraft or swarms of small boats, but is not yet able to down missiles or planes.

The LaWS test involved damaging a moving powerboat with a wooden hull. The main purpose of the exercise was to assess several of the system's capabilities, including the tracking and targeting systems, the tip/tilt adaptive optics system compensation for atmospheric turbulence, and scalability of the laser from dazzler to 30 kW. It was so successful that LaWS was deployed on the Ponce to the Persian Gulf in December 2014 as a defense only weapon. Even though on active duty, the second generation LaWS is already being built. The project calls for increasing the power of the system to 150 kW by 2020 and this power increase is hoped to make it functionally lethal to drones, missiles, and enemy battleships.

1.6.2 The US Air Force HELLADS Laser

The HELLADS laser (**H**igh **E**nergy **L**aser **A**rea **D**efense **S**ystem) is a minimum 75 kW airborne laser project initiated by DARPA and the Air Force Research Laboratory. It can be deployed on planes because its footprint, weight and power draw have been reduced. These three components—size, weight and power consumption (known collectively as SWaP) are the most important factors in realizing mobile laser packages. The HELLADS system uses an electrically pumped laser that is only one tenth the weight of typical models. The system hit crucial SWaP goals mid-2015 that triggered the initiation of airborne tests at White Sands Missile Range starting in 2016 (DARPA Press release 2015b).

HELLADS can fire a continuous beam or long pulses without overheating or destroying its mirrors because it uses a liquid cooled system that reduces heat production. Longer firing times reduce the time needed to destroy a target that subsequently reduces the demands on the tracking system, although its Lockheed Martin targeting and tracking system represents a new generation that uses phased arrays to maximize energy absorption at a single point and is impressive on its own. Another major advance is the modular laser array for scalability; the system can work at 75 kW, 150 kW or 300 kW depending on the number of modules activated. Taken together, these advances mean the HELLADS will be 5–10× more powerful than the LaWS weapon while only 1/10th the size and weight, and will be powered by rechargeable lithium-ion batteries rather than drawing power from the plane, tank, or ship on which it is mounted. Initial tests are being run on large planes, but the plan is to install the Generation 3 HELLADS on the Predator C Avenger drone by 2017.

Deploying the HELLADS is much harder on fighter jets or fast moving drones. Their speed and banking create more vibrations that must be compensated for by the tracking and targeting systems. The Generation 3 HELLADS will meet all criteria for deployment on drones. It will maintain steady tracking via the phased array tracking system and the smaller lithium batteries will deliver five to seven shots from a single charge (GA-ASI press release, April 8, 2015). Yet the entire 300 kW system will have a 0.26 m^3 (9 ft^3) footprint, about the size of a mini-fridge.

1.6.3 The DARPA Excalibur Project

The HEL portion of Excalibur's mission is to develop an optical phased array laser that can compensate for atmospheric turbulence and still target fast moving objects while the laser itself is mounted on a supersonic jet fighter. The phased array system also uses beam combining of small, electrically powered, fiber lasers whose positive interference pattern will have the power to do damage to armored vehicles, ships, and missiles. In early 2014, DARPA fired a 21-laser array in which each module of seven lasers was only 10 cm wide. It destroyed a target 4.3 miles (7 km) away from 164 ft (50 m) in the air, where the air turbulence is greatest. A newly designed algorithm senses the quality of the beam and then communicates with the optical phased array and adaptive mirrors to maintain control of beam combining, resulting in high beam focus and maximum power at the center of the beam. The algorithm assesses the turbulence in the air within the beam's path and adjusts the adaptive optics several thousand times each second (Fig. 1.6).

The short-term goal is to develop a 100 kW solid state modular laser that is twenty times lighter than conventional chemical lasers and can destroy air and land

Fig. 1.6 DARPA's Excalibur laser array is made up of multiple individual electrical lasers. The seven fiber laser arrays measure only inches in diameter and can modify direction and strength in real time to accommodate for changes in the air between the laser and the target. Multiple units of the arrays can also combine for a more powerful laser. Test versions have used 1–3 of these arrays in combination. (*Image credit* DARPA)

targets. A longer term goal, as part of the Endurance project, will allow Excalibur to destroy incoming missiles from manned or unmanned aircraft using the phased array targeting and beam quality algorithms. Interestingly, the phased array optics will permit the Excalibur to have non-weapon functions. The same tracking and laser components will work together as a functioning LADAR, as a laser-based communication system, and as airborne defense to track incoming missiles and aircraft.

1.7 Emerging High Power Laser Technologies

Excalibur, HELLADS, ATHENA, RELI, GLARE, PHaSR, and LaWS represent weapons that fire coherent beams of energy, have additional functions, are scalable in range and power, and can be ship-mounted to destroy enemy vessels and material. The parallels between these developing real world technologies and *Star Trek* phasers are striking, including the really cool names. If DARPA can miniaturize these features to be handheld (a daunting task in the least), one could argue that the phaser would be truly realized. On the other hand, maybe we don't need phasers; 21st century engineers are building some energy weapons that *Star Trek* never even imagined.

1.7.1 Free Electron Lasers and X-ray Free Electron Lasers

What if it wasn't necessary to lase a solid, liquid or gas to produce a laser beam? True, the photons must come from somewhere, but in a new type of laser, a electron beam of free electrons is the lased material. The free electron laser (FEL) is a hybrid particle beam and laser, with some unique attributes as well. Electron beams can be accelerated to near the speed of light by traditional radio frequency or laser accelerators (Sects.1.2.2 and 1.2.3), yet speeding electrons don't emit photons just because they're moving fast. The electron beam moves through something appropriately called an undulator, made from a periodic series of supermagnets with alternating poles that force the electrons into a zig-zag path. This generates electrical fields that elevate the electrons to higher energy levels. As they return to lower states the electrons will emit photons. This is very different from lasing atoms that have electrons in particular orbits and will accept and emit energy of specific wavelengths. These are free electrons, not associated with any nuclei, so they can be tuned to emit beams of various wavelengths, from infrared to soft X-ray (Fig. 1.7).

Since the electrons pass through the undulator only once, it doesn't seem that coherent light waves of the same phase would be possible and therefore it would not produce a true laser. However, the photons undergo a process called self-amplified spontaneous emission (SASE) in the undulator. At the front end of

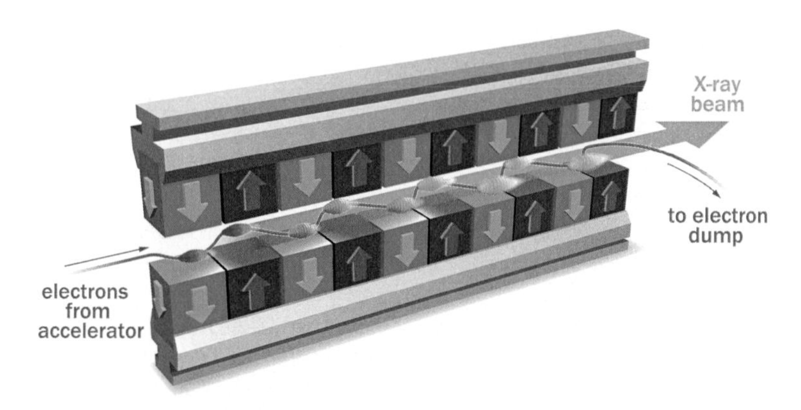

Fig. 1.7 In a free electron laser, an electron particle beam is guided through an undulator of supermagnets. The alternating poles of the magnets (*blue* and *green*) force a zig-zag pattern in the electron beam. The electrons emit photons of X-rays when stimulated by the magnets. The bunching of the electrons (self-amplified spontaneous emission) when they emit X-rays induces coherence in the waves, making it a laser, not just an X-ray beam. (*Image credit* European XFEL)

the undulator the electrons are equidistant from one another and emit photons in all directions (incoherent light). However, further along the undulator path the electric fields and the electrons' oscillations begin to align the photons into microbunches. The waves of each bunch are in phase and all the microbunches are separated by exactly one wavelength (the distance from crest to crest), so all the emitted photons of all bunches end up in phase with one another. The SASE process results in the beams with the highest measured coherence ever (Vartanyants et al. 2011).

An FEL can produce infrared, visible, ultraviolet, or even X-ray laser beams, but it is the X-ray lasers that represent a significant advance. Two issues had heretofore made them exceedingly difficult to generate. One, the energy required to excite a threshold number of electrons into a population inversion increases with the cube of the frequency; it takes eight times more energy to produce a beam of twice the frequency. Second, traditional mirrors don't reflect X-rays, they pass right through them. To build a resonant cavity for X-rays takes very expensive metal mirrors still don't work very well because they deform and can only reflect X-rays at low angles. Therefore, FELs, since they use undulators instead of mirrors, are the best way to produce X-ray free electron lasers (XFEL, be sure to read that as a FEL that produces X-rays, not as an electron laser that is free of X-rays). Unfortunately, the single pass issue does require that an undulator be significantly longer than a resonant cavity, so the laser will be bigger.

The idea of developing X-ray lasers for military purposes began with Ronald Reagan's Star Wars in the 1980s, though the Navy was the only branch of the military that actively pursued a FEL for any length of time. By 2010 they had a working prototype that was to be deployed by 2018, but government accountants felt the size and power requirements for FELs would never be reduced sufficiently. They concluded that conventional lasers provided a greater potential for deployment and defunded the FEL in 2012. However, the advent of a compact XFEL with peak power output of 10 GW (10,000× more powerful than the strongest current military laser) and a 10 femtosecond (about one millionth of one billionth of a second) duration (Ishikawa et al. 2012) may draw renewed interest from the military. The compactness and femtosecond duration of this new XFEL reduce the SWaP to the point of making a weapon feasible.

SWaP may be further enhanced by the development of a terahertz electron accelerator to speed the electrons into the undulator (Nanni et al. 2015). Accelerator size scales with wavelength; terahertz accelerators will reduce the footprint of XFELs from over three meters to less than one because terahertz wavelengths are 500–900× shorter than radio waves. This is complemented by a new XFEL that has a pulse time in the attosecond range (one billionth of one billionth of a second), further reducing the power requirement (Sadler et al. 2015). Attosecond lasers are faster than chemical reactions, so they will help image chemical intermediates and individual electron movements within atoms. It is a nice juxtaposition that a technology originally sought for war in space will finally unlock many of the secrets of inner space.

1.7.2 Laser Induced Plasma Channels

Even more amazing (and a bit scarier) than FELs and XFELs is a laser that ionizes the air and permits a lightning bolt to be aimed at a target. **L**aser **I**nduced **P**lasma **C**hannels (LIPC) were first developed in 2008 as a way to channel lightning in a thunderstorm (Kasparian et al. 2008), but lately they have received more attention as potential weapons. The US Army fired a prototype in 2012 that electrified an entire car (Kaneshiro 2012). LIPC weapons work by exploiting the very blooming that optical phased array lasers and adaptive optics seek to overcome. The powerful laser ionizes the air through which it travels, ripping electrons from air molecules and creating a channel of electrically conductive plasma. An electrical discharge within the beam will stay within conductive channel and zap whatever lies in its path. A LIPC weapon is basically a high tech taser with a laser lead instead of wires, and two laser LIPCs can create a closed circuit so that much higher voltages can be transmitted to the target. A recent demonstration of LIPC used two lasers for another reason; a short pulsed laser created the plasma channel while a longer pulse laser held it open longer to allow the electrical charge to travel farther. This demonstration shot a man made lightning bolt over 33 ft (10 m) and hit a small target, showing that a lightning bolt produced in this way can be aimed precisely (Scheller et al. 2014) (Fig. 1.8).

Several enhancements to laser and plasma technologies may give us a LIPC that is even more bizarre. In one rather shocking display, a research team at the University of Missouri fired a toroid plasma beam in free air—no laser required. The plasma "smoke ring" traveled about two feet in air and was a foot or more in diameter (Curry et al. 2013). A similar result was produced by Clint Seward at Electron Power Systems, Inc., as an explanation of the origins of ball lightning in

Fig. 1.8 The plasma channel of an LIPC device is not visualized well. The laser that ionizes the air does not use visible light and the massive electrical charge is very bright. Since a plasma channel is made of positive and negative ions, it acts as a good conduit for electricity in free space; until something large gets in the way, like this car. (*Image credit* US Army via Picatinny Arsenal)

the atmosphere (Seward 2014). If the team can reduce the diameter and project the toroid further, one can envision a plasma channel for a lightening bolt that can be fired without the need for a blooming laser. It may be sheer speculation at this point, but some amazing things have started out as sheer speculation (Fig. 1.9).

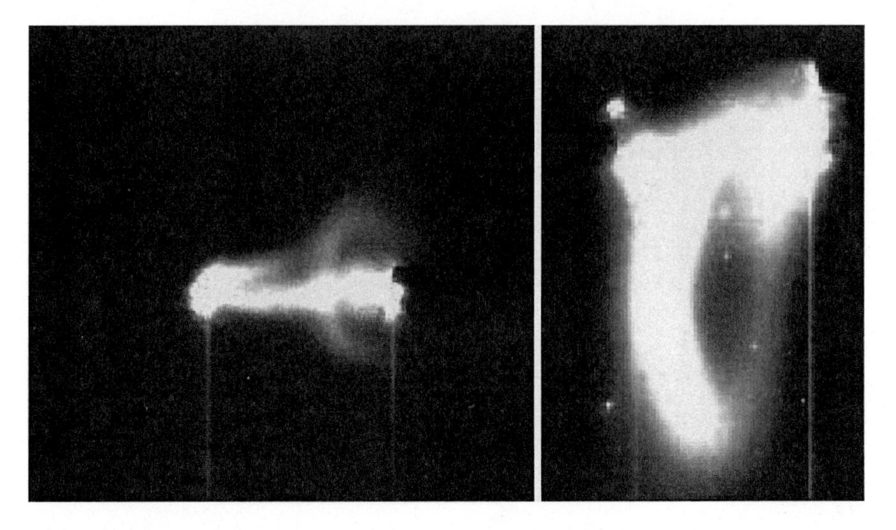

Fig. 1.9 The *left image* is a forming plasma toroid (spheromak) around an electric arc between two electrodes. This is basically a free air version of the tokamak (see Sect. 8.6.3). The *right image* is the same spheromak as it progresses through air. These are mimics of the ball lightning in nature, and are neutral plasmas as they contain equal amounts of electrons and ions. With added energy from magnetic fields, it is hoped that these spheromaks could aid in development of 1000 MW colliding plasma toroid fusion reactors. (*Image credit* C. Seward/Electron Power Systems, Inc./G. Vesperman. Ball Lightning Fusion Reactors, 2014. http://www.padrak.com/vesperman/Ball_Lightning_Fusion_Reactors)

If that isn't sci-fi enough, how about a laser and plasma channel that can bend around objects or corners? The astronomer Sir George Airy predicted in the 1830s that light beams could be bent and yet still not diffract. An Airy beam can reform after encountering an object in its path and diffracts less than other beams, just like the Bessel beams we will discuss in regards to tractor beams (Sect. 5.3.3.2). The key to forming an Airy beam is sending it through a spatial light modulator, basically thousands small pieces of glass of varying thicknesses (Siviloglou et al. 2007) or crystals of varying refractive indices (Ellenboegen et al. 2009). The result is a beam that interferes with itself; the positive interactions can produce a curved portion of maximum intensity.

Even more amazing is that the beams can take the form of arcs or S-curves, and if the lasers are strong enough, they could bloom the air and create a curved plasma channel (Polynkin et al. 2009). An electric charge sent through a curved plasma channel will follow the path of least resistance and bend too; one could in essence fire a

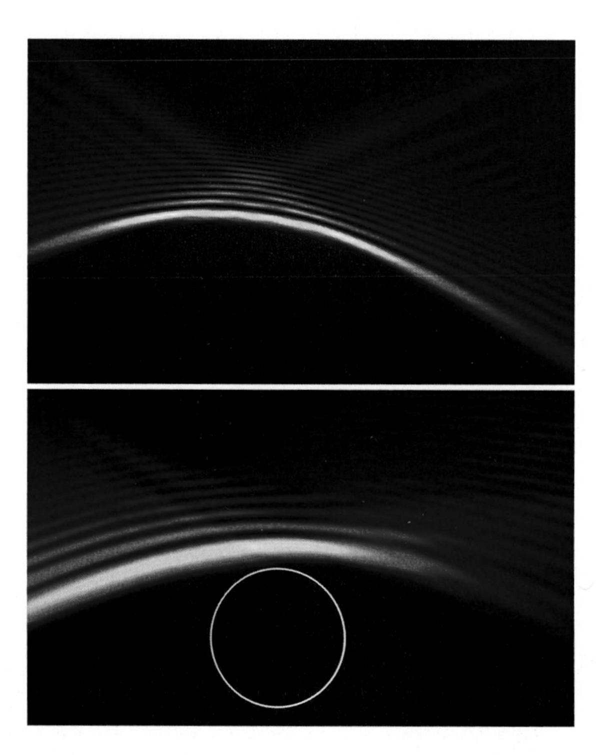

Fig. 1.10 Airy beams are a way to curve light without diffraction. They were first achieved in 2007 by a group at Univ. Central Florida and used to move small particles in microfluidics. The images here are numerical simulations of an Airy beam generated across a metal surface as a surface plasmon polariton (SPP) (Sect. 2.5.1) using a light modulator on a gold film. These two SPP light beams were generated from a single source; they can be controlled in direction and curvature real time. SPP Airy beams may be important in controlling optical nanochip circuits in the future. In theory, these non-diffracting light beams could be increased in energy to become curving laser weapons. [*Image credit* Zhang (2011)]

lightning bolt around a corner. The electrical charge can even jump over a disrupting object (Fig. 1.10) and return to the reformed plasma channel behind it (Clerici et al. 2015). Imagine a LIPC weapon that can zap a tank hiding behind a hill or a person crouching behind a bush. The LIPC could be nonlethal or lethal depending on the voltage used, just like a phaser. And in the spirit of *Star Trek*, it could be used for good as well. A LIPC fired into the sky could channel atmospheric lightning away from buildings and individuals, saving lives and property.

1.8 Conclusion

Star Trek phasers represent the best and the worst of science fiction. Weapons for space use means that humans have not learned to live together in peace despite acquiring all the technology and knowledge needed to travel beyond Earth' atmosphere, but their ability to go from nonlethal to lethal means that the Federation of Planets does have regard for life. In another sense, the multiple explanations for phaser function make it harder to buy into the technology, yet several of the explanations are becoming more plausible through research being carried out today. Connections between 1960s explanations and 2000s advances indicate that the original *Star Trek* was science fiction of the highest order.

Despite sometimes referring to phasers as particle beams, it appears that lasers will come closest to matching phaser capabilities and forms. Phaser banks on a ship like the *USS Enterprise* would be possible based on today's technology. Gigawatt XFELs could approximate the power shown in several of the episodes, and the size of the units would not represent a problem. It is only when considering their use on smaller craft that the SWaP issues come to the fore.

Particle accelerators are shrinking, yet remaining powerful. Lasers are achieving smaller footprints while increasing in power. The problems that high power beams bring have been dealt with through phased arrays and adaptive optics. Even the issues of power consumption and supply are coming under control for use on tanks, trucks, boats, and planes, and drones. The US military is deploying or is ready to deploy several energy weapons on mobile platforms, yet SWaP issues for handheld, lethal weapons have not been solved. Non-lethal handheld weapons fall within the Protocol on Blinding Laser Weapons standards, but anti-material lasers that can be carried by soldiers will require power sources that are still too large and lasers that are too heavy and run too hot.

The available research shows that directed energy weapons may have non-military uses, and their dual use status makes getting information on them and people to talk about their work much harder. The Department of Defense has a clearing office to vet all inquiries on emerging technologies. The purpose of a project and the subjects and details involved must be approved before any researchers can be approached for comments or questions. Therefore, most the information for this chapter was acquired from defense department sources or from scientific reports generated before subjects were brought under the military umbrella. Unfortunately, there may be many technologies under development or issues relevant for energy weapons that we don't know about because they are held close to the vest by the government. The reasons for this are evident, although one hopes that defense issues is not hampering peaceful research into energy and particle beams.

It is a good sign that the US military is actively seeking scalable energy weapons that can have nonlethal levels of function. Keith Jadus, acting director of the lethality portfolio for the Office of the Deputy Assistant Secretary of the Army for Research and Technology, said in a 2014 interview that, "If somebody is not pointing a gun at us, we can use a non-lethal device to disperse them. If they respond aggressively we can escalate to a lethal interaction, but it gives us the option not to have to." (McNally 2014) Perhaps we have already learned the lesson provided us by *Star Trek*. Let's hope we can remember the lesson when we move into space.

References

AF Ali, M Faiza, and MM Khalil. Absence of black holes at LHC due to gravity's rainbow. *Physics Letters B* 743; 295-300, 2015. doi: 10.1016/j.phyletb.2015.02.065. http://www.sciencedirect.com/science/article/pii/S0370269315001562

J Breuer, and P Hommelhoff. Laser-induced acceleration of nonrelativistic electrons at a dielectric structure. *Physical Review Letters* 111 (13); 134803, 2013. doi: 10.1103/PhysRevLett.111.134803. http://journals.aps.org/prl/abstract/10.1103/PhysRevLett.111.134803

M Clerici, Y Hu, P Lassonde, C Milian, A Couairon, DN Christodoulides, Z Chen, L Razzari, F Vidal, F Legare, D Faccio, and R Moandotti. Laser-asssited guiding of electric discharges around objects. *Science Advances* 1(5); e1400111, 2015. doi: 10.1126/sciadv.1400111. http://advances.sciencemag.org/content/1/5/e1400111.full

RD Curry, A Lodes, W Brown, and M Schmidt. Investigation of a toroidal air plasma under atmospheric conditions. *Plasma Science (ICOPS), 2012 Abstracts IEEE International Conference on* July 8-13, 2012, Edinburgh, Scotland, pg. 2P-177, 2012. doi: 10.1109/PLASMA.2012.6383564

C Davenport. The Pentagon's newest weapons look like something out of Star Wars. *Washington Post* December, 19, 2014. Accessed December 25, 2015. https://www.washingtonpost.com/news/the-switch/wp/2014/12/19/the-pentagons-newest-weapons-look-like-something-out-of-star-wars/

A Einstein. The quantum theory of radiation. *Physikalische Zeitschrift* 18; 121-128, 1917. http://web.ihep.su/dbserv/compas/src/einstein17/eng.pdf

T Ellenboegen, N Voloch-Bloch, A Ganany-Padowicz, and A Arie. Nonlinear generation and manipulation of Airy beams. *Nature Photonics* 3(7); 395, 2009. doi: 10.1038/nphoton.2009.95. http://www.nature.com/nphoton/journal/v3/n7/full/nphoton.2009.95.html

JP Gordon, HJ Zeiger, and CH Townes. The maser – new type of microwave amplifier, frequency standard, and spectrometer. *Physics Review* 99; 1264, 1955. doi: 10.1103/PhysRev.99.124. http://journals.aps.org/pr/pdf/10.1103/PhysRev.99.1264

Human Effects Advisory Panel. A Narrative Summary and Independent Assessment of the Active Denial System. Penn State Applied Research Laboratory, Feburary 11, 2008. http://jnlwp.defense.gov/Portals/50/Documents/Future_Non-Lethal_Weapons/HEAP.pdf

International Committee of the Red Cross. Protocol on Blinding Laser Weapons (Protocol IV to the 1980 Convention), 13 October, 1995. Accessed December 24, 2015. https://www.icrc.org/applic/ihl/ihl.nsf/Treaty.xsp?documentId=70D9427BB965B7CEC12563FB0061CFB2&action=openDocument

T Ishikawa, H Aoyagi, T Asaka, Y Asano, N Azumi, T Bizen. A compact X-ray free-electron laser emitting in the sub-angstrom region. *Nature Photonics*, 6; 540, 2012. doi: 10.1038/nphoton.2012.141. http://www.nature.com/nphoton/journal/v6/n8/full/nphoton.2012.141.html

J Kaneshiro. Picatinny engineers set phasers to "fry." June 21, 2012. www.army.mil The official homepage of the United States Army. Accessed December 27, 2015. http://www.army.mil/article/82262/Picatinny_engineers_set_phasers_to__fry_/

J Kasparian, R Ackermann, Y-B Andre, G Mechain, G Mejean, B Prade, Ph Rohwetter, E Salmon, K Stelmasczczyk, J Yu, A Mysyrowicz, R Sauerbrey, L Woste, and J-P Wolf. Electric events synchronized with laser filaments in thunderclouds. *Optics Express* 16(8); 5757-5763, 2008. doi: 10.1364/OE.16.005757. https://www.osapublishing.org/oe/abstract.cfm?uri=oe-16-8-5757

WP Leemans, AJ Gonsalves, H-S Mao, K Nakamura, C Benedetti, CB Schroeder, CS Tóth, J Daniels, DE Mittelberger, SS Bulanov, J-L Vay, CGR Geddes, and E Esarey. Multi-GeV Electron Beams from Capillary-Discharge-Guided Subpetawatt Laser Pulses in the Self-Trapping Regime. *Physical Review Letters* 113; 245002, 2014. doi: 10.1103/PhysRevLett.113.245002. http://journals.aps.org/prl/abstract/10.1103/PhysRevLett.113.245002

D McNally. Investing in the Army's future. www.army.mil The official homepage of the United States Army. Accessed January 16, 2016. http://www.army.mil/article/132953/Investing_in_the_Army_s_future/

A Mole. One kilogram interstellar colony mission. *Journal of the British Interplanetary Society* 66; 381-387, 2013. http://www.jbis.org.uk/paper.php?p=2013.66.381

EA Nanni, WR Huang, K-H Hong, K Ravi, A Fallahi, G Moriena, RJD Miller, and FX Kärtner. Terahertz-driven linear electron acceleration. *Nature Communications* 6; 8486, 2015. doi: 10.1038/ncomms9486. http://www.nature.com/ncomms/2015/151006/ncomms9486/full/ncomms9486.html

EA Peralta, K Soong, RJ England, ER Colby, Z Wu, B Montazeri, C McGuinness, J McNeur, KJ Leedle, D Walz, EB Sozer, B Cowand, B Schwartz, G Travish, and RL Byer. Demonstration of electron acceleration in a laser-driven microstructure. *Nature* 503; 91-94, 2013. doi: 10.1038/nature12664. http://www.nature.com/nature/journal/v503/n7474/full/nature12664.html

P Polynkin, M Kolesik, JV Moloney, GA Silviloglou, and DN Christodoulides. Curved plasma channel generation using ultraintense Airy beams. *Science* 324(5924); 229-232, 2009. doi: 10.1126/science.1169544. http://www.sciencemag.org/content/324/5924/229.short

M Pomerleau. Air force looking to weaponized ultrashort pulse lasers. *Defense Systems; Knowledge Technologies and Net-Enabled Warfare*. March 6, 2015. Accessed December 25, 2015. https://defensesystems.com/Articles/2015/03/06/Air-Force-ultrashort-pulse-laser-weapon.aspx?Page=1

PG O'Shea, TA Butler, MT Lynch, KF McKenna, MB Pongratz, and TJ Zaugg. A Linear accelerator in space – the BEAM experiment aboard rocket. *Proceedings of the Linear Accelerator conference*, Albuquerque, New Mexico, pg. 739-742, 1990.

JD Sadler, R Nathvani, P Oleśkiewicz, LA Ceurvorst, N Ratan, MF Kasim, RMGM Trines, R Bingham, and PA Norreys. Compression of X-ray Free Electron Laser Pulses to Attosecond Duration. *Scientific Reports* 5: 16755, 2015. doi:10.1038/srep16755. http://www.nature.com/articles/srep16755

M Scheller, N Born, W Cheng, and P Polynkin. Channeling the electrical breakdown of air by optically heated plasma filaments. *Optica* 1(2); 125-128, 2014. doi: 10.1364/optica.1.000125. https://www.osapublishing.org/optica/abstract.cfm?uri=optica-1-2-125

DC Seward. Ball lightning events explained as self-stable spinning high-density plasma toroids or atmospheric spheromaks. *IEEE Access* 2; 1530159, 2014. doi: 10.1109/ACCESS.2014.2308476. https://www.researchgate.net/publication/260603204_Ball_Lightning_Events_Explained_as_Self-Stable_Spinning_High-Density_Plasma_Toroids_or_Atmospheric_Spheromaks

W Shatner and C Walter. *Star Trek: I'm Working on That*. New York: Pocket Books, 2004.

GA Siviloglou, J Broaky, A Dogariu, and DN Christodoulides. Observation of accelerating Airy beams. *Phsycial Review Letters* 99; 213901, 2007. doi: 10.1103/PhysRevLett.99.213901. http://journals.aps.org/prl/abstract/10.1103/PhysRevLett.99.213901

R Sternbach, and M Okuda. *Star Trek Voyager Technical Manual, V1.0. Internal Paramount Picture document*. Published by Paramount pictures, 1994. Accessed December 28, 2015. http://misc.axiom.ky/AR%20Stuff/Akira%20Files/Akira%201.0%20Backup/Akira%201.0%20Backup/Star%20Trek%20Voyager%20Technical%20Manual.pdf

J Sun, E Timurdogan, A Yaacobi, ES Hosseini, and MR Watts. Large-scale nanophotonic phased array. *Nature* 493(7431); 195-199, 2013. doi: 1038/nature11727. http://www.nature.com/nature/journal/v493/n7431/full/nature11727.html

J Sun, E shah Hosseini, A Yaacobi, DB Cole, G Leake, D Coolbaugh, and MR Watts. Two dimensional apodized silicon photonic phased arrays. *Optics Letters* 39(2) 367-370, 2014. doi:10.1364/OL/39/000367. https://www.osapublishing.org/ol/abstract.cfm?uri=ol-39-2-367

IA Vartanyants, A Singer, AP Mancuso, O Yefanov, et al . Cohenernce properties of individual femtosecond pulses of an X-ray free-electron laser. *Physical Review Letters* 107; 144801, 2011. doi: 10.1103/PhysRevLett.107.144801. http://journals.aps.org/prl/abstract/10.1103/PhysRevLett.107.144801

SE Whitfield and G Roddenberry. *The Making of Star Trek*. New York: Ballantine Books, 1968.

P Zhang, S Wang, Y Liu, C Lu, Z Chen, and X Zhang. Plasmonic Airy beams with dynamically controlled trajectories. *Optics Letters* 36(16); 3191-3193, 2011. doi: 10.1364/OL.36.003191. https://www.osapublishing.org/ol/abstract.cfm?uri=ol-36-16-3191#Abstract

DARPA press release, May 21, 2015a. http://www.darpa.mil/news-events/2015-05-21

DARPA press release, May 21, 2015b. http://www.darpa.mil/news-events/2015-05-21-2

GA-ASI press release, April 8, 2015. http://www.ga-asi.com/gen-3-high-energy-laser-completes-beam-quality-evaluation

Chapter 2
Cloaking Devices: Vanishing Into Thin Space

I've served on a half a dozen ships, and none of them've had cloaking devices except the Defiant. Now that we're not using it, I feel naked.

—Miles O'Brien
DS9: Broken Link

2.1 Introduction

Invisibility has been a mainstay of fantasy and science fiction stories for decades. H.G. Wells' *The Invisible Man* had chemistry and biology at the heart of the story of a man rendered permanently invisible by the chemical, monocane. Griffin, the scientist in Wells' story of science and morality, developed a method to change the way human flesh was affected by light; it neither reflected nor absorbed any part of the visible spectrum. Wells was prescient in many of his stories; but even he couldn't imagine just how close he was to early 21st century science's ability to make things invisible.

What could be cooler than moving around without anyone else knowing you are there. You could get away with anything; a license to act badly. Teachers often try to get students to think about the moral dilemma attached to invisibility, they ask students what they would and wouldn't be willing to do if they were invisible. For example, Wells' antagonist Griffin was driven mad by his invisibility and was finally beaten to death by a mob after carrying out a murder spree. His lesson was that invisibility may not be all it's cracked up to be—and Gene Roddenberry was well aware of the moral tightrope one might walk if unseen.

Roddenberry believed in the idea of heroes and fair play. That's why the starships and fighters of the Federation of Planets didn't have cloaking devices. Roddenberry stated in several interviews that, "heroes don't sneak around." To his way of thinking, jumping out of the dark to hit an opponent when they weren't expecting it was not something that those who commanded the moral high ground would do. However, Gene and the myriad writers of the *Star Trek* episodes and movies were more than willing to give some species this technology, it was just too

M.E. Lasbury, *The Realization of Star Trek Technologies*,
DOI 10.1007/978-3-319-40914-6_2

cool a toy to leave out of the franchise. The Klingons, the Romulans, the Borg, the Suliban, the Cardassians—it would be easy to think that only evil empires developed cloaking if the Klingons hadn't later decided to join the good guys. Even the Federation dabbled in cloaking from time to time, either as a means to level a playing field (*DS9*: *The Search, parts I and II*), or sometimes as a tool used by rogue Federation officials as they slipped down that slippery moral slope (*TNG*: *The Pegasus*).

In the Harry Potter tales or the J.R.R. Tolkien epic *The Lord of The Rings*, cloaking and/or invisibility was left to the world of magic or mysticism. This eliminated the need for an explanation of how the cloaks worked. Interestingly, in both instances, the cloaking devices were actually cloaks; I wonder if this was the genesis of the term "cloaking device?" On the other end of the scale, *Star Trek*, like H.G. Wells' invisible man, placed cloaking in the realm of science, using engineering or chemistry to manipulate light. The cloaks of *Star Trek* are engineering marvels, requiring tremendous amounts of energy to render a ship invisible to most, if not all, electromagnetic radiation sensors. In the first season of the original series the cloak is described as an energy field projected through the deflector array, explaining why many ships could not cloak and raise shields simultaneously. In the cloak's earliest appearance in *Star Trek* (*TOS*: *Balance of Terror*), Spock speculates that *if* the enormous power demands could be met, "It is possible that a ship could become invisible." It was a only theoretical technology right up to the time that the Federation first encountered it in a Romulan vessel.

Because cloaking is not a Federation tool, the captains of the various Enterprise or Voyager vessels do not delve deep into an explanation of how they work. The entire franchise is thin on the technical details of the cloaks, it is never revealed to the viewer whether they are particulate or wave energy, or how that energy is generated. It may be that the information was never crucial to a story line, so the writers did not broach the subject, or perhaps a plausible explanation was not readily available. This is in direct contrast to 21st century Earth, where the internet is rife with comparisons of shielding devices to *Star Trek* cloaks. Some "cloaks" use mirrors or cameras and projectors, although these can hardly be considered paradigm-shifting advances, they are more like optical tricks. On the other hand, there are developing technologies that can truly be called cloaks; just imagine a covering material for the Enterprise that could bend light around the star ship or even slow or stop light completely. You don't have to imagine these, they're being improved on even as you read this, but can anyone be trusted with them?

2.2 Ways to Stay Hidden

There are many ways to remain unseen, ask any customer who needs help at a department store. Hiding under a rock or in a cave isn't really becoming invisible, although it gets the job done. The problem with hiding something is that you can't use it. You can duck behind or under something so that light rays do not bounce off you to someone else's camera or eyes, but you have to come out from hiding to do anything, and then they know you're there. It would be much preferred if you could have your equipment and yourself in a strategic position, and yet still not be seen; hiding in plain sight as it were. To go undetected in the open is much more difficult, one must; (1) make the target look like something else, or (2) make it truly invisible. Disguises make you look like something else while camouflage makes you blend into the background, and yet these primitive, passive types of hiding in plain sight often work just fine.

Star Trek loved disguises and costumes; in many cases, Spock's ears are the main problem (*Star Trek IV: The Voyage Home*), while in others the replicator supplies the away team with Fedoras and tommy guns to help them fit in (*TOS: A Piece of the Action*). The shape shifters that play a role in several series and movies (*ENT: Broken Bow; TOS Whom Gods Destroy; TNG Journey's End; DS9: Things Past; Star Trek VI: The Undiscovered Country*) are basically disguising themselves as well, using neural and physiologic powers to look like something or someone else. They are using their talents to blend in by they are taking on the characteristics of an individual, not an area or environment, so they are still disguises, not camouflage and certainly not cloaks.

Camouflage isn't exactly cloaking either, but it lives in the same neighborhood. The differences between the two lie in the method of making something invisible in its environment, and whether the same technique can still work if the background changes. Seamless adjustment of camouflage with a changing observer, background, or light source position is tough. Shadows can easily give a camouflaged target away, and this is especially true for targets with sharp edges. Cloaking on the other hand, is the most technologically advanced form of camouflage in that can account for changes in environment that traditional camouflage cannot. In between these two extremes lies active camouflage, something well represented in *Star Trek* and a technique making its presence known on Earth as well.

2.2.1 Active Camouflage in Star Trek *and on Earth*

Active camouflage is much more techie then painting a tank green or a sniper wearing a ghillie suit that looks like grass. Active camouflage is beginning to develop some features that can hide things in plain sight by responding as the background changes or by helping the target take on a different disguise in real time. It's still not cloaking, although it's getting closer. It is called active because it requires an energy input, be it electrical, mechanical, or biochemical energy. In nature, many animals can change their color, with the cephalopods (squid, octopods, cuttlefish) being the kings of active camouflage (Box 2.1). The energy requirement is constant, for their skin is always being held in one pattern or moved to another, but the energy demand is not too great since they are equally willing to rest on different substrates that require vastly different camouflage patterning (Allen et al. 2010). Perhaps the largest power drain associated with their camouflage system was the evolutionary energy it took to develop those huge eyes. They are much more developed than any other invertebrate animal because good sight and an advanced neural system are crucial to detecting the background and then matching it.

Box 2.1 Camouflage in Nature

Disguises—The crab spider of South America feeds mostly on turtle ants. In order to infiltrate the colony and find more victims, the spider wears the corpse of its last victim on its back so that they other ants will believe the spider is just another ant. A slightly different mechanism, but with the same effect is used by decorator crabs. They stick plants or mostly sedentary animals to their shells to disguise them as they hunt.

Coloration—Birds are often dark on top and light on their bellies. Hunters looking up see light color that blends into the sky while predatory birds diving from above see dark colors that blend in with the ground. Many fish use this technique in reverse.

Mimicry—Leaf insects and stick insects have varied appearances from species to species, depending on the types of plants they live on. Evolution has picked their color as well as their surface texture. The rears of some insects look like a face, so a predator doesn't know which end to attack, while many moths and butterflies have huge eye patterns on their wings. Scarlet king snakes mimic the coloration of coral snakes to make other animals believe they are venomous.

Exact mimicry is not always the best option. Zebra stripes don't mimic their environment; they don't live in a field of football referees. Yet the black and white stripes work for them because the lines look like tall grass, they break up the zebra's outline, and their main predator, the lion, is colorblind. Also, many zebras standing close together will meld stripes together, so a predator will see them as one huge striped mass, it can't pick out a single zebra to attack.

Active (energy requiring) camouflage—Many cephalopods (octopods, cuttlefish, and squids) have layers of different cells in their skin. One layer (chromatophores) can pump more pigment in and out of cells at a moment's notice, another layer (iridiophores) acts as tiny mirrors that can be angled at different directions to reflect light in different ways. A third layer has thousands of small muscles to help the cephalopod change the texture of its skin (Fishman et al. 2015). Some cephalopods have another layer of cells in the epidermis called photophores that produce light via a chemical reaction. The photophores on their bellies produce light to mimic the light filtering down from the surface.

The importance of natural models of active camouflage has not been lost on science. By studying life everywhere on the planet, models to solve engineering problems can be found and mimicked, and predictions about life on other planets can be made. On Earth we call scientists who do this work exobiologists (exo-meaning beyond earth). *Star Trek* is aware of this and wrote the development of active camouflage into the franchise. In the *Voyager* series, it is the exobiologists Magnus and Erin Hansen who develop bio-dampeners for active camouflage. These are energy requiring and environment sensing systems that project a field around a human's body to mimic that of a Borg drone (*VOY: Dark Frontier*). They give off the proper signals and those signals change as the Borg adapts while at the same time hiding any biologic signals given off by the human body. A crew member wearing the bio-dampener could move around in the Borg Cube for hours without being detected as a foreign body.

Back here on Earth, many governments are interested in active camouflage. One example of an active camouflage method that will soon be translated to the battlefield is called Adaptiv. This battle system helps make the point that active camouflage requires a sensing system, like the eyes of a cephalopod, that passive camouflage does not. Adaptiv is a camouflage system that detects and projects an infrared signal, not visible light; therefore, it has sensors that detect the heat signature of the environment surrounding the target to be camouflaged (Fig. 2.1). Hexagonal panels mounted to sides of a tank or personnel carrier generate heat to both match the background and to disguise the heat signature of the target. Because there are nearly 1000 plates per side, patterns of heat generation can be programmed into the system, with some panels being heated, some cooled and some left at ambient temperature. Depending on the pattern selected from the system's "library," the heat signature can be camouflaged to appear as a truck, or a station wagon, or to mimic the foliage behind it. The first prototypes of Adaptiv were demonstrated in 2011 by BAE Systems (London, UK) after they were commissioned to develop just such a system by the Swedish Defence Material Administration, although they have yet to be deployed in the field.

Active camouflage points out three issues that will need to be addressed in cloaking. One, cloaking needs to allow the target to view the outside area—a

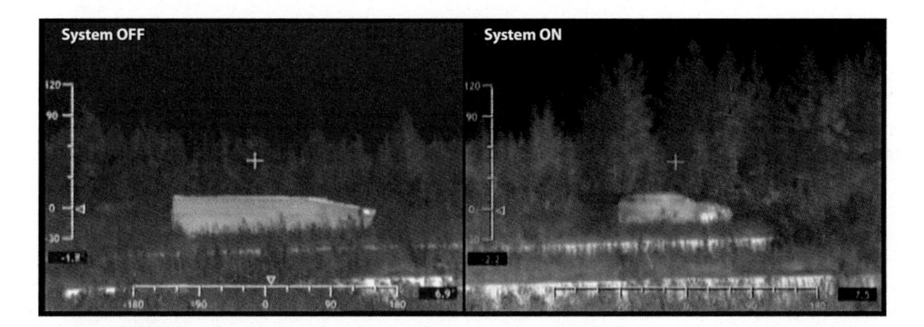

Fig. 2.1 In the *left image*, the Adaptiv System of active camouflage has not been activated, so the heat radiated by the vehicle gives away its presence and shape in a thermal image. The *right image* shows a preprogrammed pattern of heating and cooling of the side panels to make a portion of the vehicle look like an SUV, while the rest of the vehicle matches the heat signature of the background and blends in. (*Image credit* BAE Systems plc)

blinded piece of equipment is of no use. Two, the target signal needs to blend into the environment; a completely eliminated signal will look like a black hole and will be investigated. And three, the cloak needs to respond to changes and make the target hidden from observers no matter where they are or how they might change position. Several advances in optical camouflage are trying to address these issues, but remember—they aren't true cloaking either.

2.2.2 *Mirrors and Rochester Cloaks*

Certain types of optical camouflage have been hailed in the media as cloaking devices; a typical media ploy to stir up interest and readership. They are optical tricks that one can play with to simulate a true cloak and they *are* fun—unless you really need to protect yourself from a moving enemy, then they are worse than useless. All fall short of true cloaking because they require static and specific positioning of both the observer and the target. Take, for instance, two sets of mirrors that reflect light from behind an object first out to the side and then to the front. This light is then reflected back to a mirror positioned directly in front of the object to be hidden so that the light rays from behind again travel forward in their original position and direction. It works well as long as you place the mirrors precisely, hide the side mirrors well, and forbid the observer or target object to move. This is a magician's trick, not a technology.

A more sophisticated trick uses lenses instead of mirrors. Developed at the University of Rochester in 2014 by Joseph S. Choi and John C. Howell, the Rochester Cloak uses the focusing of light to create a visual dead space where an object might be hidden (Fig. 2.2). The distance needed to focus all the rays traveling through a lens to a point (the focal point) is called the focal length. In this

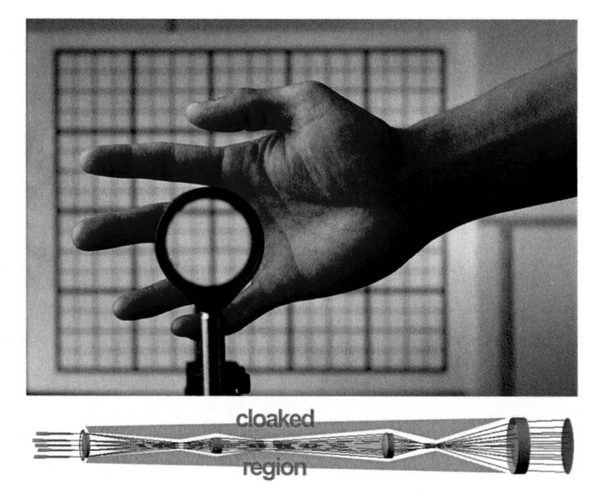

Fig. 2.2 The Rochester cloak uses a series of lens to focus reflected waves so that the region they occupy is smaller than if they traveled in a straight line. Around the center of the beam is an area that is hidden (orange region) as long as the observer views the background from no more than 15° away from the central axis of the lenses. Part of the hand seems to disappear, as long as the fingers do not interrupt the central beam. That is why he has his fingers spread. [*Image credit* (Choi and Howell 2014)]

optical trick, a series of lens of specific focal lengths are set in line on a rail, with very specific distances between each of the lenses, as determined by the focal lengths and the difference between those focal lengths (Choi and Howell 2014). Basically, light bouncing off objects in the background will be focused by the lenses with the rays of the beam compressed into a smaller cone of space as they approach the focal point, held in smaller volume between the middle lenses, and then expanded back to a larger size.

An object placed outside the constricted rays but inside the circumference of the lenses will seem to disappear if and only if the observer is looking straight through the line of all four lenses (or within 15° of that central axis). Nothing to be cloaked can impinge on the constricted cone of light being transmitted from behind the device; this is why demonstrations are often carried out with the human hand. The gap between fingers can be used to allow passage of the constricted cone of light rays, making it appear as if the entire space is cloaked. As with the mirror example above, light can be moved around an object if one adheres to the stiff restrictions on the placement of the object and the observer in order for the effect to be observed. A true cloak has no such limitations.

2.2.3 Stealth Technologies as Cloaks

Stealth technology, as used in the military, can be seen as the union of primitive deflector techniques (Chap. 4) and primitive cloaking techniques to produce an advanced active camouflage by redirecting radar signals. Radar (**RA**dio **D**etection **A**nd **R**anging) is a means by which radio waves (microwaves really) are emitted from a source, bounce off of a target, and are collected back at the source. The direction, timing, and modulation of returned radio waves provide clues as to the shape (identity), location, and movement of a target. The purpose of stealth technology is to make an airplane, ship, or other vessel less visible to radar by greatly reducing backscatter of signal from emitter back to sensor. When radar waves hit a surface, they bounce off in many directions (scattering), with only some of them bouncing back to the source to be detected. If surfaces can be designed to reduce the number of waves bouncing directly back to the collector, either by absorbing them on the surface or directing them in directions other than pure reflection, then the vessel's signal profile will be reduced.

The two most common ways to reduce a ship's radar cross section (how detectable it is) are by making the angles where surfaces meet more acute and by reducing the number of rounded surfaces that would bounce at least some signal back toward the source. Very flat surfaces tend to reflect the radio waves forward (forward scatter) or sideways (side scatter). The sharp angles also help turn the waves away from their original vector. In contrast to this, a rounded surface like a traditional plane fuselage is likely to scatter waves at all angles, including straight back to the detector. In addition to using geometry to reduce backscatter, there are paints that can absorb some of the radio waves and prevent their return to the detector. The most often used paint contains small iron or ferrite spheres covered in quartz as an insulator. The spheres are suspended in a two-part epoxy liquid paint. After application a magnetic field is applied which spreads out the spheres evenly along the field lines, improving their performance. Microwave radar waves striking the surface are absorbed by the quartz insulation, converted to heat, and dissipated across the entire surface. More recently developed paints include multiwall carbon nanotubes (Box 3.4) instead of iron miniballs. The nanotubes are very good at absorbing energy in the microwave range, so they return an even smaller signal back to the detector.

More advanced stealth technologies are under investigation but have not been deployed yet. These would include a plasma (ionized gas, see Box 4.2) stealth shield, much like the plasma deflectors discussed in Sect. 4.4.3 and 4.5.4 for protection of astronauts against cosmic radiation, although here the plasma would absorb and deflect radar energy impinging on the plane's surfaces. This is very high tech, yet perhaps the most *Star Trek*-like stealth technology is the split ring resonator. Printed on a sheet of dielectric material, small copper C-rings have an opening that can be affixed to the surfaces of planes or ships (Abdalla and Hu 2012). The split rings allow microwave energy to enter, with the gap distance tuned to become a capacitor. Each gap width is specific for a different frequency of microwave energy, so varying the size of the rings randomly can make it active over several frequencies. Since the energy becomes stored in the capacitor-like split ring, none (or at least significantly less) is available to reflect off the surface and return to the detector. The use of a conductor with a dielectric to produce this material with unique electromagnetic qualities makes the split ring resonator a type of *metamaterial*, which will become very important as we talk below about designing true cloaks.

2.3 Cloaking the *Star Trek* Way

The optical tricks and radar evaders described in the previous sections are predicated on the idea that light travels in a straight line, it might bounce at an angle when reflected or bend (refract) as it enters a new medium, but it always travels in a straight line. The question is—does it have to? What if we could make light obey commands to bend around an object, so that no light bounces off the object at all? For true cloaks, the aim is to allow waves of EM energy to interact with a device so that the wave front impinging on a target from any direction travels on, uninterrupted by the object being hidden—the cloaked object simply doesn't disturb EM field in any way. This is a much more difficult goal. True cloaking is a sort of stealth camouflage taken to its most amazing end. Instead of hiding the object by scattering a projected beam so that it won't find its way back to a sensor, true cloaking erases an object from the perceivable universe while allowing all interactions of other objects to remain observable—one can see without being seen. This is how cloaks behave in *Star Trek*. A cloaked ship can watch, plan and ready itself, while it rests reasonably assured that nothing can detect its presence. The Klingons are especially good at making use of cloaks, although they didn't develop them originally.

Box 2.2 Basic Geometric Optics

If you look at a green glass bottle of soda, you will witness many of the characteristics of optics:

Reflection—You look closely and you can see yourself.

Absorption—If you leave it in the sunlight, it will warm up because some light is held by the glass and the energy is converted to heat.

Transmission—The bottle is a solid object, but you can see how much soda is left, so some light waves pass through the glass.

Refraction—The level of soda inside is altered slight from where it actually is because light passing through the glass is bent slightly as it enters and bent back as it leaves. Different materials have different indices of refraction and the value can be altered by temperature, wavelength of the incident light, and density of the medium. The reason for the bend is due to something called phase velocity, where the speed of the peaks of the light wave will seem to travel slower in more refractive materials. To maintain a constant speed of light, it must bend to accommodate the peaks slowing down. More refractive materials have slower phase velocities. Diamond bends light a lot (has a high refractive index). This is why diamonds sparkle so much. Water has a low index of refraction; this is why the stick you fiddle with in the water looks slightly bent where it enters.

Scattering—You can see the bottle, so light waves strike the bottle, bounce off in several directions and some of those scattered rays travel to your eyes. The green light is reflected and scattered more, while other frequencies are absorbed. This is why the bottle appears green. Different materials absorb different wavelengths of light, longer wavelengths are the reds, and the wavelengths get shorter and higher energy as you travel from red through orange, yellow, green, blue, indigo and violet (ROY G. BIV). The same thing happens outside the visible range of electromagnetism, but we don't see it.

For reflected and scattered light, angle of incidence equals angle of reflection. If the surface isn't perfectly smooth, then some of the light will be reflected at many angles; this is scattering. You can see a searchlight beam directed into the sky at night because some of the rays (waves) are scattered by particles in the air and even the air molecules themselves; some rays are scattered toward your eyes and you see the beam. Scattering by molecules (things smaller than the wavelength of the incident light) directs rays in all directions.

The Klingons received cloaking technology from the Romulans during their brief alliance, so all the early cloaks probably operate on the same principle, an energy field projected through the deflector array (*TOS: The Enterprise Incident*).

Romulan cloaks are made of a tetryon compositor and a projection matrix—neither of these is explained fully, although tetryons are supposedly a type of subspace particle (*DS9: The Die is Cast*). Fortunately, space station Deep Space Nine can detect tetryons particles, so Romulan cloaks probably aren't much good in the 24th century. This may be why Romulans develop cloaks with different actions, with at least one based a prismatic mechanism (*TOS: The Enterprise Incident*). In the *Enterprise* series, the Suliban receive cloaking technology as a gift from an unknown benefactor in the 28th century (*ENT: Broken Bow*). It works differently from the Romulan cloaks, as it is based on a particulate substance, as witnessed by engineer Trip Tucker spilling some of the Suliban particles on his hand and arm, rendering them invisible for some time (*ENT: The Communicator*).

Using any of these cloaking techniques, planets (*TNG: When the Bough Breaks*), ships, minefields (*ENT: Minefield*; *DS9: Call to Arms*), even individuals (*Star Trek: Insurrection*) can be hidden from prying eyes. They also happen to be undetectable to heat, microwave, gamma ray, or most other wave or particle radiation emission sensors; true cloaks work over the entire EM spectrum. It is true that some cloaks have weaknesses that can be exploited where some exotic particle or energy trace can be detected, like the "quantum beacon" used to locate a cloaked a minefield (*ENT: Shockwave*). These weaknesses set up a back and forth cloaking/detection race that is so familiar to the arms races we have seen in the recent past. In this case, the Klingon ships show up in short order with a new cloak on which the quantum beacon is ineffective (*ENT: Minefield*). As Spock says, "Military secrets are always the most fleeting." (*TOS: The Enterprise Incident*).

2.4 Cloaking by Bending Light

Commander Spock notes in *Balance of Terror* that while it might be theoretically possible to selectively bend light around an object, it would take a tremendous amount of energy. Well, he was wrong on both counts. Cloaking is more than theoretically possible; cloaks that can obscure objects from specific portions of the EM spectrum exist today. And while broad band cloaks may require an input of energy, some recently demonstrated cloaks are completely passive; light strikes them and they, and the object they are surrounding, just disappear. And while Spock ponders the possibilities of cloaks in the 23rd century, cloaking actually has its scientific genesis in the 20th century mind of Albert Einstein, by way of two British geniuses—James Clerk Maxwell (1831–1879) of Scotland and Sir John Pendry (1943–) of Imperial College, London. Using the knowledge of optics that these giants gave us, the cloaking techniques that exist as of 2015 fall into two broad categories—EM wave bending and EM wave scattering suppression. Since Spock contemplated light bending, we will first look at how Einstein, Maxwell, and Pendry have taught us to move light around an object.

2.4.1 *Transformation Optics*

Einstein considered space, light, gravity, and time as interconnected. He proposed that if you distort space with a large mass (like a sheet being deformed by a ball placed on its surface), and then shine a light beam through that distorted space, it will bend around the mass instead of traveling in a straight line. Maxwell had earlier developed the mathematical equations for the behavior of electromagnetic waves in free space, and Einstein used these to formalize the behavior of electromagnetism in his curved space. This part of his theory of general relativity was proven correct in 1919 (Sect. 4.2.4). Jump forward to the 1990s when Dr. Pendry wonders about the implications of the reverse situation. If light bends in distorted space, what would space do in the presence of distorted (bent) light?

The fact that electromagnetic radiation travels in straight lines makes it tough to bend light around a body, allowing an observer to see what is directly behind it. However, light does bend as it passes from one medium to another, like air to water or air to glass. When light hits an object, it may bounce off in many directions. Sometimes the light is reflected off of an object's surface and returns back in the direction from which it came, while other rays may be scattered at some angle from the surface. If light can pass through the medium, then it may be bent at a certain angle (called refracted) based on the material it is entering and the wavelength of the light itself (Fig. 2.3). However, in all these cases (reflection, scattering,

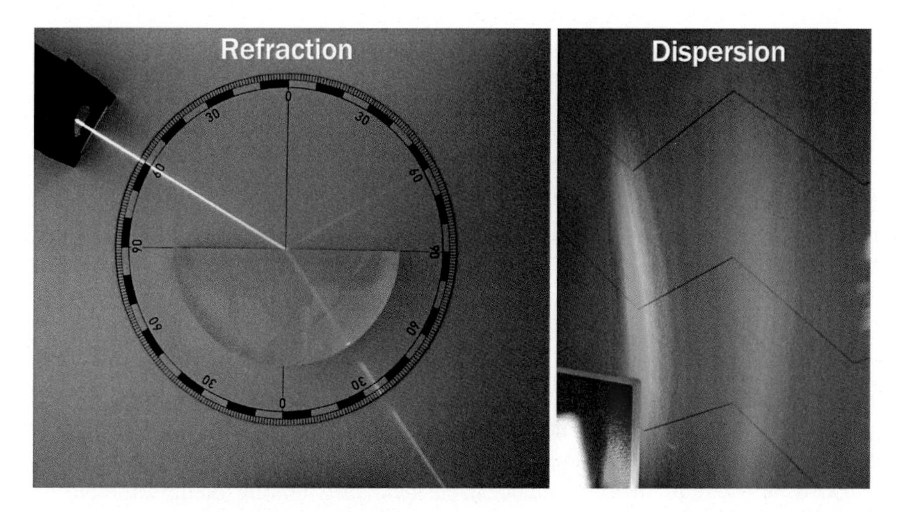

Fig. 2.3 Light shown through the plastic of a protractor bends due to the plastic's index of refraction. The angle of incidence equals the angle of reflection for reflected rays, but the transmitted waves bend by some 25°. The index of refraction is dependent on the wavelength of light (amongst other things), so white light containing all colors will be refracted to different degrees (dispersed), spreading the individual frequencies (*colors*) out over a distance. (*Image credit for refraction* By Zátonyi Sándor (ifj.) Fizped (Own work) CC BY-SA 3.0 (http://creativecommons.org/licenses/by-sa/3.0), via Wikimedia Commons, the image has been merged and labeled) (*Image credit for dispersion* By no machine-readable author provided. Andreas Rejbrand assumed (based on copyright claims). [GFDL (http://www.gnu.org/copyleft/fdl.html)], via Wikimedia Commons)

transmission and refraction) the light wave is still traveling in a straight line. It is the straight line movement of light that allows us to locate objects in our field of vision. Light from a source strikes the object, and some of those rays bounce off and travel to our eye. If they didn't travel in a straight line, how would we know where the object was?

Consider the case where a hot blacktop road heats the air directly above it. Temperature affects index of refraction (how much it will bend light as it enters the new medium) so the different layers of air above and beside the road will bend light to different degrees. As the light waves that bounce of a palm tree in the distance and travel through volumes of air with different temperatures, they are bent, then bent more, and then bent back. The result is a wavy image that can't quite be placed at one specific location. Philip Ball's 2015 book, *Invisible: The Dangerous Allure of The Unseen*, explains nicely how light bends (refracts) when entering a different medium because it is looking for the shortest time of passage. Light in a vacuum travels at 299,792 km/sec (186,000 miles/sec), but in other media one aspect of light (phase velocity) slows a bit. In order to traverse the new medium in the same time as it would in a vacuum, it must take a shorter path, so it bends. For each material the degree of bend is slightly different, therefore every transparent substance has a unique index of refraction, which is also influenced by temperature, density and the wavelength of the light.

What Dr. Pendry realized was that the bending of light to distort space is mathematically equal to changing the index of refraction over a distance. Theoretically, one could bend light to distort space just as easily as distorted space can bend light. His conclusion: if you bent the light sufficiently, you could move it around an object just as water moves around a rock in a stream, rejoining the rest of the water on the other side as if the rock had not been there. The bending of light would be the moral equivalent of folding space into a tiny volume so the light could pass right by unscattered. Maxwell's equations could be modified to tell you how much to change the index and at what positions. This wasn't just a mental exercise for Dr. Pendry, he told me that an attempt to make something disappear was a, "grand challenge; something to capture the imagination of the public."

This was the birth of the new field of transformation optics, where "optics" describes the study of how light moves through a medium. The problem was that the equations called for a material with unique properties of bending light, properties that don't exist in many (or any) natural substances. Instead of slowing down light and therefore causing it to bend a certain degree, a cloak might require a material to speed up light at some spots and slow it down in others. Speeding up and slowing down light in this instance is a matter of looking at the different components of the light. The speed of a single part of the wave, like all the peaks, is called the phase velocity, while the propagation of the light wave as a whole is called the group velocity (Box 2.3). The movement of light into and out of a medium alters the phase velocity, and this is why it bends to take a shorter path. By taking the shorter path, the group velocity can be maintained at the universal constant speed of light. Or is it a constant? We shall see later that this constant isn't so constant, and might be used as a cloaking technique (Sect. 2.6.2).

Natural materials have indices of refraction between 1.0002 for air and a whopping 4.05 for germanium. Diamond has a high index of refraction (2.417) while liquid helium hardly bends light at all (1.025). What science needed for bending light and distorting space was a material that could be designed to have arbitrary indices of refraction at different points in space, and together these would manipulate the light direction and speed as it traveled around an object. Importantly, the material of the cloak would have to have a different index of refraction for rays of light that strike it from different directions, a property called anisotropy. You've seen the effects of anisotropy many times (Fig. 2.4). Birefringent butterfly wings that change color depending on your position relative to them and automobile paints that shine purple to green or blue are examples of anisotropy. Since each frequency of the EM spectrum reacts differently to a material's index of refraction, white light and sunlight will be dispersed (scattered) differently because it is made up of many

Fig. 2.4 Calcite crystals are anisotropic. Light travels through in different ways based on the angle of the material to the light. The anisotropy produces a double refraction, where a ray of light will be split and take two different paths, each with a different index of refraction. Notice how the position of the doubled letters changes as the orientation of the crystal changes. (*Image credit*: D. Walker. *Examples of the animation of macro and microscopy subjects using sequential jpeg images.* Micscape, May, 2004)

frequencies. Looking from one direction scatters it in a particular way so that purple waves are reflected to your eye, but look from another angle, and the anisotropy determines that green waves will be scattered.

Working with Dr. Pendry on this idea, and doing his own work on changing indices of refraction, was Dr. David Smith of San Diego. Dr. Smith explained to me that the importance of anisotropy in transformation optics is hard to visualize; suffice it to say that it is crucial because the math says so. In transforming space, some areas get squeezed and others stretched changing the relative direction from which light is striking the material. Therefore, being able to modulate light from different directions in the distorted space becomes necessary (Smith, personal communication). Dr. Smith had found a way to make extremely small copper wires behave exotically with respect to microwaves by giving them specific shapes and diameters. A chance meeting between he and Pendry led to a discussion of printing small copper wires on a surface to produce a material that could do even more—change the material's index of refraction. Because light is electromagnetic radiation, it has both an electric field and a magnetic field. The embedded copper wires changed the *permittivity* of the material (how much it responds to the electrical field of light), but a C shaped ring of the wire also altered the material's magnetic *permeability* (the degree to which a substance responds to light's magnetic field). Together, the permittivity and permeability determine a material's index of refraction, so manmade materials with specifically designed permittivities and permeabilities can alter light's direction and speed in predictable ways.

2.4.2 Gradient Index Metamaterial Cloaks and Their Limitations

These manmade materials were already known by this time and had been named *metamaterials*, although they had not yet been used for this novel purpose. In 2006, Smith (Schurig et al. 2006) and (Pendry et al. 2006) both published papers wherein they demonstrated similar methods to hide a volume of space from a narrow band of microwaves. The invisibility device itself, and anything held within the center of its concentric circles of metamaterial, looked as though it wasn't there when a beam of microwaves was sent across its path and sensed by a detector on the other side—a true cloaking device. This was a seminal moment for Dr. Smith, who had grown up a *Star Trek* fan and had a scene about cloaks from the original series playing constantly in his head as he contemplated the possibilities of cloaking (Smith, personal communication). Actualizing a piece of *Star Trek* technology was a very big deal to this particular trekkie.

For the transformation optics cloaks of 2006, each of ten concentric circles of the metamaterial bent the light a specific amount so that it would curve around the object and then back to its original path (Fig. 2.5). The equations of transformation optics designated what change in index was needed for each ring, creating a gradient of

Fig. 2.5 This is one of the first gradient index cloaks to hide space from EM radiation. Light passes around the cloak through the various layers of metamaterial that have unique abilities to bend light. The combination of the copper wire and translucent material in each repeating unit cell determines the index of refraction, and the shape of the unit cells changes from ring to ring; this produces the correct overall bend. (*Photograph credit* Dr. David Schurig)

indices of refractions. Therefore, this is called a gradient index cloak, or more generally, a transformation optics cloak. The light traveling through the cloak was seemingly sped up and then slowed down by changing refraction, yet they arrived at the front of the cloak at the same time as all the waves that did not travel through the cloak (were above or below it or off to one side or the other). The mathematical distortion of space to a point meant that it was just as if all the waves traveled in a straight path to the observer. This maintenance of a common wave front is important for a cloak; only in this way will the background remain uninterrupted and no edges be detected where some waves went through the cloak and others didn't. Press releases heralded the age of invisibility cloak, despite the obvious problems that need to be overcome—and there are many. Scientists and the military use extended parts of the EM spectrum for emission and detection, but for most of us, we know something is there because we see it with our eyes, making use of only visible light. Because of this bias, the public wants a cloak in the visible light range, yet no current gradient index cloak can work in visible light, most work in the infrared or microwave range.

For materials with large crystalline structure, light takes an "average" of the material's optical properties and develops a behavior based on this. For subwavelength materials, like metamaterials, the light wavelength is actually larger than the unit cell (sometimes by an factor of 1000), and this keeps the waves from "seeing" the material. Light doesn't scatter off a metamaterial with the complexity that it does off natural materials, and this simplicity makes the light easier to manipulate, as long as the wavelengths are relatively long. Visible light waves are about 20,000 times shorter that microwaves used for radar and most gradient cloaks. To be able to manipulate visible light as needed for cloaking, the unit cells will have to have much smaller, and this is harder to manufacture. This is a problem, but it isn't the only problem with moving cloaks into the visible range. Metamaterials also need to keep the electrons moving with certain properties. As the wavelength gets shorter, the electrons in the metamaterial must speed up more and more to maintain the same movements. More speed from the electrons means more distortion, and this makes the cloak itself visible and less effective for hiding objects (Pendry, personal communication).

Another problem to solve before transformation optics cloaks will achieve *Star Trek*-like invisibility is that index gradient cloaks have the unfortunate side effect of blinding the cloaked object from the outside world. The cloaks work by bending light around the object instead of letting it reach the surface and be reflected or scattered. Just like how no light reaches your retina and you see nothing when you close your eyes, a gradient index cloaked target's sensors receive no incoming light; therefore, the crew can't see anything going on around the ship. How will the commander know when it is safe to decloak or where or when to fire? Starships in *Star Trek* can still see space and the objects in it while cloaked, so this is a severe limitation that must be overcome.

Finally, there is the issue of effective limited bandwidth that must be solved. Each wavelength of light has a specific phase velocity change as it enters a new medium (aka—a different index of refraction). Metamaterials are designed with very specific indices of refraction, so if you shine light of many different wavelengths on a cloak, they bend to different degrees (called dispersion) and do not all follow the path described by the equations for that particular cloak. A rainbow is a good example of this dispersion of wavelengths. As white light passes through a droplet of water, the different colors of light (based on their wavelength) are refracted differently. As the waves pass through many droplets, the dispersion increases, so that you eventually see the different colors distinctly, each based on how far they refract, and each refraction based on the wavelength, with red light refracting most and violet light refracting least. The order always ends up the same; red, orange, yellow, green, blue indigo, violet. The same thing happens across the entire EM spectrum, we just don't perceive it with our eyes. Therefore, gradient index cloaks are necessarily functional over a very narrow bandwidth to reduce dispersion. If a wider band is used, some of the waves will disperse and cause a shadow, making the cloak itself will be detectable. Science fiction cloaks seem to work over very wide bands; second generation Romulan cloaks hide ships or other objects from *all* EM waves. Dr. Smith has a way to go before he realizes a *Star Trek* cloak that matches that of the Romulans.

Box 2.3 Cloaks Of All Sorts

Anything that travels as a wave, pressure, sound maybe even heat, can theoretically be cloaked. Examples include:

Acoustic cloaks—Sound travels as mechanical waves and travels well in water, and is the basis of sonar detection of ships and submarines. A cloak developed at Duke University might uses metamaterials on the hull that modulate acoustic properties and have precisely sized holes drilled in them based on wavelengths of sound to be masked. The cloak returns a signal of exactly the same type that was sent, suggesting that no metallic object reflected the wave (Zigoneanu et al. 2014). There are possible uses for the acoustic cloak in air as well, such as reducing echoes and giving people in the cheap seats a decent chance to hear the music at the symphony.

Earthquake cloaks—Two different methods being tested to reduce the physical and financial damage of earthquakes. One method places rubber stressed in particular directions *around but not touching* the foundation, and this stress deflects seismic waves away from the building (Parnell 2012). A second type of earthquake cloak modifies the soil around the building to deflect and absorb seismic waves. To do this, scientists are developing seismic metamaterials that alter the soil density and soil's elastic modulus, just like gradient index or metascreen metamaterials alter optical properties. In early tests, simply drilling strategic holes in the soil reduced the force of seismic waves encountered by a target by 80 % (Brule et al. 2014).

Mechanical cloaks—It may be possible to cloak the pressure you feel from an object when you touch it. Some types of metamaterials exhibit unique physical properties that redirect pressure to a flat surface around any object that fitted inside the cloak (Buckmann et al. 2014). Nanometer sized cones in the cloak have tips that bend and meet with precisely designed forces generated. Using an inversion of the mathematical equations for optical properties, solids can be designed to be stiff in one direction, but shear easily in another, or vice versa.

Temporal Cloaks—Metamaterials can be used to hide events in time as well. A beam of light can be sped up in a metamaterial so that it reaches a point before an event takes place. The light behind it can be slowed down, so that by the time it reaches the same point, the event has already occurred. The sped up portion can then be slowed down and the slowed down waves can be sped up, so the wave front arrives at an observers eyes without disruption, yet lacking a recording of the event that occurred in the middle. This technique has been demonstrated (Fridman et al. 2012) and may first play a role in secure data transmission (Lukens et al. 2014).

Make no mistake—the gradient index instruments are true cloaks; objects are being hidden from light waves. What is more, they will only get better with time. For example, transformation optics equations yield results that are hard to reproduce in materials, so approximations must be made. This reduces the effectiveness of the cloak or limits the directionality and dimension in which it will work; fortunately, the number of approximations is being reduced all the time. Dr. Smith and his colleague Nathan Landy published a demonstration of a microwave cloak (Landy and Smith 2013) that used no approximations to the transformation optics equations and yielded an effective cloak over a larger range of wavelengths. However, to widen the range they had to limit the number of dimensions (X/Y) that it would work in and limit the source light to coming from one direction. The math involved in creating a wide band cloak that functions with light from any direction and can cloak a spherical space (X, Y, and Z dimensions) is harder to reproduce without making shortcuts and materials to produce such a cloak are harder to manufacture. Once again, nature comes to the rescue. Natural materials are giving scientists clues how to build transparent cloaks and cloaks that work with visible light.

2.4.3 *Natural Calcite Carpet Cloaks*

In general, making cloaks work in three dimensions means that one must account for light scattering in more directions. Limiting the dimensions in which the cloak acts simplifies the math and reduces the approximations. The first metamaterial microwave cloaks were produced as flat surface and then curved to produce a cylinder, giving a cloak in the X and Y dimensions only. This is much easier than trying to produce the unit geometry on a spherical form, although the object would not be cloaked from microwaves that shone on it from below or above the plane of the cylinder. The math of a spherical cloak produces a wire shape—a hard area to hide, but calculating the compression of space by a cylindrical cloak reduces the volume inside to a single point. Even easier to produce and predict than a cylinder is a flat sheet cloak, a so-called carpet cloak (Li and Pendry 2008). Natural crystals have been manipulated to produce decent carpet cloaks, and some even work in the visible range, an impossibility for present index gradient cloaks because of the smaller units cells required.

Natural calcite (calcium carbonate, $CaCO_3$) crystals are an example of a cloaking material functional in the visible spectrum. Small prisms of the crystal are attached to one another as a flexible, two-dimensional surface. Anything placed under the cloak, even a ball under a carpet that deforms the surface, will have its reflected light refracted in many arbitrary directions. The light passes through the cloak, bounces of the target object and some of that light bounces back through the crystalline cloak again. All the combined reflection and refraction means that little of the light reflected from ball will return to the eye, and even less will be returned in a manner that shows the shape of the object. Instead, the light of the background will overwhelm all other patterns and the cloak will blend in perfectly. Granted, the object must be laid on a mirrored surface to ensure that all the refracted light will be bounced through the prisms, but it allows for the target under the carpet cloak to disappear. This is one way to achieve cloaking in the visible range, and even though it comes with limitations, it is actively being pursued by the military.

Meshed crystal prisms are not the only way that calcite can be used as a carpet cloak. If you have two larger crystal prisms lined up back to back, the natural optical effects of the crystals can break up the light enough to hide an object placed in a void beneath them. Despite calcite being transparent, the light is broken up as it passes through the crystal because it is anisotropic and double refracting (birefringent, see Fig. 2.4). The carbonate ions are arranged in flat sheets, so light perpendicular will be affected differently than light that enters parallel to the sheets.

Calcite breaks an entering light beam into two rays that have different indices of refraction and therefore take different paths. One ray, called the ordinary, is refracted according to the index of refraction for calcite. The other ray, the extraordinary, is refracted at an angle that depends on the angle at which the light strikes the crystal relative to the flat carbonate sheets. Turning a calcite crystal as it lies over printed image will produce two images, with one of the images moving around the other as the crystal is rotated. The moving image results from the extraordinary ray.

By placing calcite crystals back to back, the combination has four paths and speeds during each path through the crystal. The overall effect is similar to the flat surface described above made from many small flat crystals, although the light must be shone directly at the crystals. These dimensional optical glass cloaks have been demonstrated in the visual range by B. Zhang at MIT in 2011 (Zhang et al. 2011) and S. Zhang at the University of Birmingham (Chen et al. 2011) in recent years, hiding anything beneath them as long as they are narrower than the width of the crystals themselves. Of course, the cloaks are visible and cask shadows unless the light is polarized and the entire setup is under water (Fig. 2.6), so their use in cloaking starships in their current form is limited to say the least.

Fig. 2.6 The calcite crystals are joined back to back, each with a wedge removed at the bottom to form a central trough for hiding objects. The light rays reflected from the pink tube are bent and re-bent by the double refracting calcite. Notice that the entire apparatus is in a fish tank of water to eliminate shadows and the light is shone directly on the calcite crystal. However, that portion of the pink tube under the calcite blanket does disappear from view. (Photograph courtesy of B. Zhang and G. Barbastathis)

2.4.4 Natural Guanine Crystal Carpet Cloaks

The skin of some fish act as natural cloaks, helping them blend into the color of the water around them by perfectly reflecting the light back to the observer. Most reflectors are imperfect, they tend to polarize the light as it bounces off. The polarized light has more rays traveling in one certain direction, instead of scattering off the surface equally in all directions. The higher number of rays traveling in one direction creates a more intense light that we see as glare. Fisherman wear polarizing sunglasses to block the horizontally polarized rays and glare of the sunlight off the water, just as photographers use polarizing filters to eliminate glare caused by polarized reflected light.

For fish, reflecting the light helps conceal them, but the glare created from polarization would give them away. To combat this problem, some silvery skinned fish, like herring and sardines, have evolved a crystalline cloaking mechanism that gives high reflectivity without polarization. Engineers and researchers in optics usually overcome polarization in optical communication devices by using sophisticated filters and surfaces, with the unfortunate side effect that these lower the efficiency of the reflectors, thereby increasing the noise in the signal. By mimicking the fish system, these optical devices can be made much more efficient.

The fish don't care about optical communication, they are more worried about cloaking themselves. The system they have evolved uses multiple layers of hexagonally shaped guanine crystals to help them blend in with the water. Guanine is one of the chemical bases used in DNA and RNA. Here their function is structural, not informational; many of the small guanine crystals join together in a lattice formation to create crystals with interesting optical properties. It has been known for some time that sardine skin contains multiple layers of these crystals, with scientists assuming that the result of sunlight striking them would be full polarization—equal scattering in all directions and no glare, but also a loss of reflectivity. The truth, as reported in a recent study from the University of Bristol in the UK, is that the glare is prevented while high reflectivity of the surface is maintained (Jordan et al. 2012).

The experiments showed that the layers contain two different types of guanine crystals, each with their own optical properties, and it is the mix of the two crystals that gives the high reflectivity, no glare characteristic, similar to how the mixing of the dielectric and anisotropic solids in metamaterials achieve unique permittivities and permeabilities. However, unlike gradient index cloaks that bring the light from behind an object to the front, the carpet cloak of fish skin reflects the light that was in front of the cloaked object. If the fish swims across the front of a colorful reef, a predator would see them as a shadow against the background. This may be why this evolutionary strategy has evolved in fish that mostly in open water and are attacked from below. Looking up, the predator is much more likely to lose the fish against the uniform background.

A similar guanine system is responsible for the invisibility of a class of small crustaceans called Sapphirina copepods. These marvelous looking animals can be iridescent in color, from fluorescent blue to gold or orange, yet by turning their body slightly they can disappear from view in an instant. The hexagonal layers of guanine crystals are very similar to those seen in fish scales, but their arrangement here dictates that when the animal turns 45° to the sun, all the reflected light shifts into the UV ranges and they disappear from sight (Gur et al. 2015). Amazingly, it was shown a few years ago that magnetic fields can alter the guanine crystal optical properties, changing the wavelength of the reflected light. By increasing the magnetic field strength from two to ten tesla, the fish scales tested changed from white-green to dark blue or violet (Iwasaka et al. 2012). This shift from lower to higher frequency light suggests that an even stronger field could shift the light into the ultraviolet spectrum. This suggests a mechanism to control a carpet cloak of guanine crystals, allowing a target object to go from iridescent to invisible on command. Metamaterial index cloaks can neither be tuned on command nor work in the visible spectrum, so there may be something to be said for following nature when designing these first generation cloaking devices.

2.5 Cloaking by Destroying Light

Bending light around an object or breaking light up so much that a cloaked object disappears in the reflections/refractions are not the only ways of cloaking that are being studied. Another class of metamaterial cloaks attack the problem from a different angle, and may be able to overcome some of the cloaking limits described above. For these cloaks, the metamaterial generates an optical signal equal and opposite to any light scattered by the target object and cancels it out. It is as if the cloaked object doesn't exist because no light reflected from it ever reaches your eye.

2.5.1 Plasmonic Cloaks

First demonstrated in 2012 (Rainwater et al. 2012), the "plasmonic" or "light-canceling" cloak is a device that is thinner than the typical gradient index cloak. Plasmonic cloaks are made from a single layer of an engineered compound that is placed around the target object with just a narrow gap between the two. The gap is essential to the functioning of the cloak, because unlike the gradient index cloak that keeps light away from what is to be hidden, the plasmonic cloak requires that incident light be scattered off the target so it can be canceled. The types and wavelengths of light scattered from a target object depend on the material it is made from and the structure of its surface. Therefore, to cancel out only the light scattered from the target, a plasmonic cloak must be designed with both the target and the wavelength of the incident light in mind. This is different from the transformation optics cloaks that can hide whatever object is placed in the cloaked region.

To achieve their goal, light-canceling cloak designs use metamaterials. However, the term metamaterial is a catchall term according to Sir John Pendry; it really encompasses many different types of materials that have unique optical properties. Plasmonic cloaking metamaterials have very different properties than those of transformation optics metamaterials, but one thing they all have in common is that they can be fabricated to have very specific optical characteristics at any given point. Which property you wish to manipulate then determines the constituent materials and geometry of the given metamaterial. Plasmonic cloaks utilize metamaterials that have very low permittivity properties; they react strongly to the electric field generated by light waves. Their electrons move around more within their molecular structure when light strikes them (high electron flux) and this has a polarization effect.

Taken together, the cloak and target work to radiate a light wave with a very specific wavelength and phase, which is designed to be the negative of the wave scattered by the cloaked object. The scattered light from the cloak is exactly that light which is 180° out of phase with the object being cloaked, so the interference between the waves scattered from the object and the waves scattered from the target causes both to be canceled. This don't just render the cloaked object invisible, it renders the cloak invisible as well. This makes a plasmonic cloak very much like Harry Potter's cape; you can see it when Harry isn't wearing it, but both the cape and Harry disappear when he puts it on. In the case of a cloaking device, if you remove either the cloak or the target inside it, then both become visible.

The cancellation of the scattered light explains why one doesn't detect the cloak or the target, yet the light coming from behind the cloak is unaccounted for. It would seem that a shadow would form, giving away the cloak. Dr. Andrea Alu of the University of Texas at Austin explained to me that the light coming from behind the cloaked object is restored by the plasmonic cloak because some of the incident light from behind can travel through the cloak unimpeded. Remember, light must be able to get to the target object within the cloak for it to work, therefore, much of the light from behind will pass straight through to the observer. If the target itself doesn't transmit, some light from behind will trickle around the surface of the object as charges called *surface plasmons*, although the name of the cloak comes from plasmons generated in the metamaterial, not the target. The surface plasmons help reestablish a front on the other side along with those that did not interact with the object or the cloak. The canceling of only scattered waves and the transmission of unimpeded waves means that you can see the background, just like with the transformation optics cloaks.

2.5.2 Mantle Cloaks

One year after introducing the plasmonic cloak, Dr. Alu and colleagues demonstrated a mantle cloak based on the similar principles (Soric et al. 2013). This cloak was much thinner (166 μm, 0.0065 in.), about as thick as a sheet of paper. They

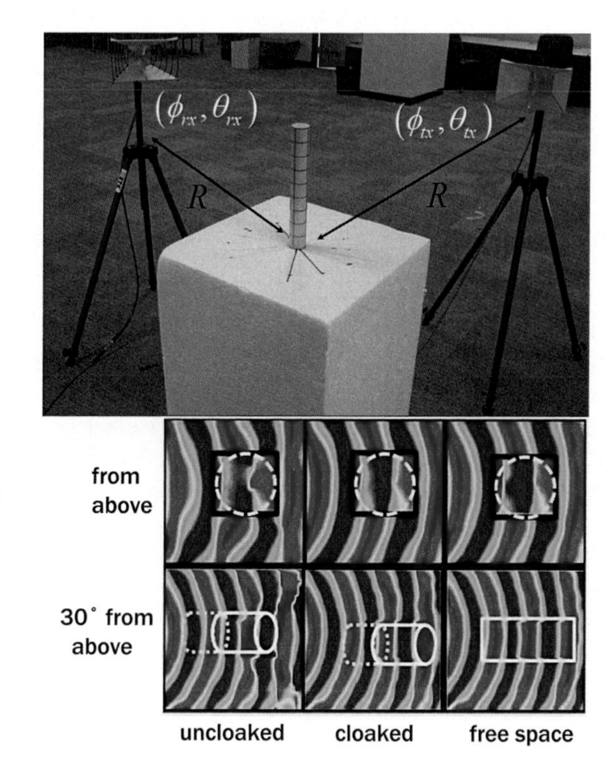

Fig. 2.7 This plasmonic metascreen cloak is only 166 μm (0.0065 in.) thick, yet can destroy reflected and scattered rays in the microwave region. The cylinder target is wrapped with the cloak in the image above, and microwave emitters and detectors are directed straight through it or from 30° above the plane. The lower images shows how the wave front is distorted in the uncloaked situations, but the cloaked versions more closely resemble the gentle curve of the wave front in free space. (*Image credit* (Soric et al. 2013) under CC BY 3.0 license (https://creativecommons.org/licenses/by/3.0/us/legalcode), portions of three figures have been merged)

termed this a metascreen, since it is mostly a transparent sheet with nanometer (one millionth of a millimeter) thick copper wires embedded in it. The advantage to this design is that the thinner the cloak or mantle, the broader the range of wavelengths it can cancel out. The metasurface, as it is sometimes called, radiates out of phase waves based on slightly different properties of the material as compared to the plasmonic cloak. Namely, it resists a flow of electrons after an incident light wave strikes it. That resistance causes the light wave to be radiated instead of propagating on the material as a surface plasmon. However, as with the plasmonic cloak, the impedance (resistance to flow) varies with the frequency of light, so these cloaks are effective over only certain bands of EM, not all bands. Alu's plasmonic and mantle cloaks were both designed to cloak within the microwave region (Fig. 2.7).

The idea of mantle cloaks has taken off in the past couple of years. Zhang and colleagues at the US Department of Energy's Lawrence Berkley National Laboratory call their metascreen a "skin cloak" (Ni et al. 2015). At only 80 nm (3.15×10^{-5} in.) thick, this cloak could easily be used to cover any three dimensional object. In Zhang's demonstration it was used to hide a space about the size of a few human cells, but the authors say it could one day be scaled up to cover macro-objects, perhaps even a Klingon or Romulan ship. Their current version is effective at the 730 nm wavelength (infrared light) only, right on the edge of the visible range. This is a major step forward as compared to other metamaterial cloaks.

Even beyond the possibility of visible light cloaks, the principal advantage to cloaking with metascreens or plasmonic cloaks is that the cloaked object is not blind when it is hidden. Light strikes the cloaked object and can be detected by sensors on the object—whether they be the eyes of the crew or EM sensors on the ship. Either way, the captain can direct the ship more satisfactorily since he/she can see what is going on around them. Also an advantage, specific transmissions can be sent from the cloaked object, as long as they are not in the functional spectrum of the cloak, they will be broadcast just fine. Finally, the size of mantle and plasmonic cloaks is much more manageable. The width of a gradient index cloak is often double that of the object being cloaked, whereas with the metascreen it can be a thin coating held just off the surface of the ship or other target object.

On the downside, Dr. Alu laments the fact that the properties used make mantle cloaks are very sensitive to frequency. The size of the possible target goes down as the wavelength gets shorter. Since visible light has a much shorter wavelength than radio waves (on the order of a hundred million times shorter), the size of objects that could be cloaked from visible light using current mantle cloaks is in the microscopic range. This is why the skin cloak of Ni and Zhang described above was effective only for an object the size of a few cells. However, Dr. Zhang is of the opinion that the cloak could be scaled up for larger objects, since it uses nanometer sized gold nanoantennae that can be switched on or off to change their polarization, a method slightly different than that of Alu.

2.5.3 *Spoofing and Illusion Coatings with Mantle Cloaks*

Jiang and Werner of Penn State University (Jiang and Werner 2014) recently demonstrated another mantle cloak with a peculiar twist. It is a single metasurface layer similar to Alu's, but is effective on larger objects. The cloak was designed to be functional against radio waves, and it will hide objects from those wavelengths, yet it's best use may come in another area. Often, military or communication vehicles with tops that house many different antennae have interference and jamming problems because the signals of one antenna disrupt those of another. The authors of the Penn State study found a way to modulate their skin cloak's optical properties so that the hidden object would appear to be silicon instead of metal or metal instead of glass. This technique is called "spoofing," an engineering riff on the software technique of masking a website or email address as something else. In spoof cloaking, one object is cloaked in a way to make sensors detect a false signal. This is one of the areas in which Dr. Pendry sees real promise for metamaterials, and is possible because the transformation optics math that he derived allows one to design metamaterials that can manipulate the light that travels to an object, not just bend it. With a gradient cloak one can make an object look larger or smaller or change the ways it's edges are perceived, and a metascreen can change the signature to make it appear to be another structure, including hiding one antenna from another.

Of course, *Star Trek* led the way here as well. Chief O'Brien once configured the navigational deflector (Sect. 4.2.1.1) to project a false Federation warp drive signal onto a weapon platform's generator (*DS9: Tears of the Prophets*), fooling it into firing on its own power source. This is a form of spoofing, and although O'Brien uses the deflector instead of the cloak there isn't too much of a difference. In real life, spoofing with thin mantle cloaks sometimes called "illusion coatings" could very well have significant value in the military, when troop positions may be given away by sensors trying to locate a radio-emitting antennae. Beyond the military, being able to make one antenna invisible to others will keep communications clear and allow for more co-sited devices. Perhaps most impressive, spoofing with metamaterial cloaks can make an object appear to be where it is not—a spoof in location. This is not the only way to cloak an object by changing its apparent position; we will see below that this might be done by slowing or stopping light.

2.6 Future Cloaking Possibilities

Star Trek has cloaks that hide ships and other large objects from every frequency of EM radiation. Current real life cloaks cloak small objects over small bands of the EM spectrum, although the progress described above suggests that this won't be the case for long. Not only will the capabilities of cloaks be expanded by incorporating diverse synthetic and perhaps natural materials, the uses for cloaks will grow as technology calls for shielding signals in optical computing and even in aviation or automobiles. Imagine having a 360° view from your driver's seat because the roof and sides of your car are cloaked from your point of view. Sir John Pendry sees cloaks "galloping off in all directions" in the next few decades, including thermal cloaks to protect electronics from damaging heat. He is, however, less optimistic about earthquake cloaks (Box 2.3), at least until the mathematics is better demonstrated.

2.6.1 Wide Band and Active Cloaks

Expansion of the effective bandwidths is probably the most prominent issue for realizing a cloak like those used in *Star Trek*. Dr. David Smith at Duke is of the opinion that the problems of narrow bandwidth will be solved over the next half-century and some sort of wide band cloak will be available, either by mastering more parameters within the mathematics or by producing cloaks that use energy to produce canceling waves. Much research is being conducted on metasurfaces that can be tuned to different frequencies, based on the needs at the time. These frequency-selective metamaterials are active cloaks, energy is used to alter the characteristics of the metascreen, giving it a different impedance, permittivity, or polarization (depending on the techniques used) and therefore changing the frequencies over which it is effective. This is in fact how the skin cloak of Ni et al. (2015) works. The gold nanoantennae are switched on with a small current that changes their polarization and makes them functional over the entire visible spectrum, a much broader bandwidth as compared to passive cloaks.

Dr. Alu is of the opinion that active cloaks are the only possible way to increase bandwidth significantly. He recently proposed an active cloak with a battery pack that would be effective over a very large range of radio frequencies (Chen et al. 2013). However, even with this advancement, Dr. Alu believes that cloaks in the visible range will not be realized anytime in the near future. Dr. Smith agrees, he doubts that a cloak that is effective against all EM frequencies will be possible for decades, and will likely require active technology. Perhaps Spock was right all along, cloaks might be possible, but they will be a terrific drain of energy. Then again, it may not be necessary to cloak over all wavelengths, planes don't need to be cloaked against visible light to avoid radar, and many fly too high to be spotted from the ground. Cloaking for specific frequencies may be sufficient.

2.6.2 *Slow Light and Positional Cloaks*

One of the most unbelievable things about transformation optics is its ability to slow down light. The speed of light is nearly constant; it travels only 0.03 % slower in air than it does a vacuum, although it slows by 3 % in glass—this is the basis for the index of refraction of a material. Yet there are ways to slow light down without moving it from one medium to another. Slow light could yield an amazing cloaking device; by the time the image of your ship arrives at that Borg vessel that is trying to assimilate you, you won't be there anymore. This is not unlike how light traveling extremely long distances gives a view of events in the past at that location. A transformation optics cloak might achieve the same thing by making the light take a bit longer to get to an observer.

One interesting way of slowing or even stopping light is by forcing a material that normally absorbs light to become transparent over a very small band of frequencies. This electromagnetic induced transparency (EIT) uses two lasers and takes advantage of the multiple quantum states allowed in a material's electrons (called superpositioning, Sect. 8.3.4). If a substance absorbs light over a certain range of wavelengths *and* a material's optical characteristics meet a set of very specific parameters, then a laser (the coupling laser) tuned to a near resonant frequency with one of the electron quantum states can make the material transparent to a very narrow range of frequencies within the absorbed range—a "spectral window" or "spectral hole." If a second laser (the probe) is shown on that material in the same area and is tuned to the frequency of the spectral window, it will be transmitted through the substance just like it was transparent, like looking through a brick wall (Fig. 2.8).

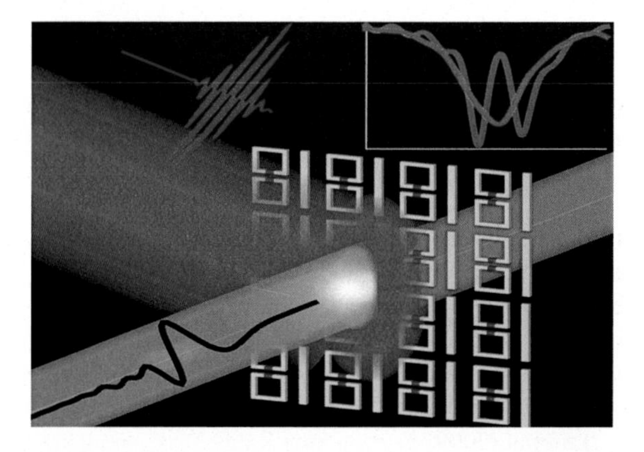

Fig. 2.8 Electromagnetic induced transparency uses two lasers. The *red* coupling laser, at one specific frequency is absorbed by the material (*yellow pattern*). The transmission spectrum at top right dips to zero showing it is not transmitted. The *blue colored* probe laser is a different set of frequencies; a few frequencies are transmitted through the opaque material as long as the coupling laser is on. This is indicated by the spike in transmittance in the *green line* in the graph at *top right*. If the coupling laser is shut off while the probe laser is on, the light will be trapped inside the opaque material until the coupling laser is turned on again. It can stop and start light. (*Image credit* Los Alamos National Laboratory)

This is interesting enough on its own, yet it gets weirder. When the coupling laser interacts with the material's quantum states, it very rapidly changes the index of refraction within that volume of material. An extremely fast change in refraction creates a large amount of dispersion of light, with parts of the ray passing through the material at each point along its change in refractive index so each is refracted to a different degree. A large range of refractions is called dispersion, like our example of the rainbow above where the different frequencies (colors) have different indices of refraction in water drops (see Fig. 2.2). When a narrow range of frequencies (the spectral window) undergoes extreme dispersion, the group velocity of the light slows down considerably, sometimes to just thousandths or millionths of the normal speed of light. Using techniques like this, light has been slowed down to just 17 m/s (38 mph) (Hau et al. 1999) and in other cases has been stopped completely using EIT in an ultracold cloud of strontium metal atoms (Ginsberg et al. 2007).

In a similar set of experiments carried out at the University of Darmstedt in Germany, scientists actually stopped the probe laser light in the substrate material. Since the coupling is what made the substance transparent to the probe laser frequency, turning the coupling laser off stopped the probe light inside the opaque solid. They had imprinted the probe laser signal with a picture of three horizontal bars, so in essence, this information was trapped in the solid as stopped light. A full minute later, the researchers turned the coupling laser back on and the probe laser light restarted and traveled through the rest of the material (Heinze et al. 2013). In its current state, this technology uses very small amounts of very cold gas molecules (called a Bose-Einstein condensate) and manipulates light over an extremely low number of frequencies for fractions of a second and across distances of less than an inch. The mechanics of the system would ostensibly work over larger distances and wavelengths as well, although the number of materials with which to work is limited because the characteristics of the quantum states to produce a spectral window are very specific.

Since this mechanism of slow light production utilizes the dispersing properties of the material, it is called material dispersion, slightly different than waveguide dispersion techniques that use guided pathways for the waves as a means of inducing dispersion and slowing light. Metamaterials can be designed as waveguides that are specific geometric shapes and generate either negative permittivities and/or negative permeabilities that create group velocity changes and huge dispersive effects based on the wavelength of the light. These metamaterials have been used to slow down (Lu et al. 2009) or even stop light (Tsakmakidis et al. 2007) in transparent materials embedded with nanowires of copper. In a waveguide, the total dispersive effect is roughly an addition of the material dispersion and the waveguide dispersion, but some materials for optical communication devices can be produced where these two parameters cancel each other out and have zero dispersion.

Waveguide dispersion has also been achieved with a photonic crystal, which is basically several metamaterial layers put together to form a waveguide (Alagappan and Ping 2015) and also with micro-resonator metamaterials that act on light as if they were 10 cm (4 in.) thick even though they are flexible sheets only 1/1000th as thick (Hokmabadi et al. 2015). It is important to realize that all these metamaterial slow light generators are passive; it is the geometry of their constituents that produces the dispersion that slows light. This is direct contrast to the huge energy input needed to produce Bose-Einstein condensate material dispersion devices. However, waveguide dispersion generators also come with some of the limitations of metamaterial cloaks, including the difficulty in reducing unit cells to subwavelength sizes when moving from microwave or terahertz radiation up into infrared and visible light and the limited frequencies over which these cloaks are functional.

Luckily, even if the slowing of light to create a positional cloak proves difficult, the concept of slow light may help other cloaks work better. One problem with current metamaterial cloaks is that a moving cloak can't keep up with the changes to the background scenery and the light coming to the observer gets blurred. This has to do with the limited ability to speed up the phase velocity of all wavelengths when bending light around a cloaked object. By introducing a small optical device that the undergraduate developer at the University of St. Andrews calls an "invisible sphere" into a cloak, it can slow the light down in all frequencies (colors) so that the when the wave front is sped up again, the speed required does not surpass the limits of the material (Perczel et al. 2011). Therefore, all the colors of the wave front travel together and the observer sees a crisp image even if it is changing rapidly.

2.7 Conclusion

Disguises, camouflage, stealth technology—mankind often goes out of his way not to be seen. Whole technologies and commercial markets are based on the sole idea of keeping things hidden. While cloaking is the next step in this line, there are differences between cloaking and the other methods of hiding that make it both better at its job and harder to accomplish. It is clear that cloaking is more than just camouflaging or disguising an object to look like something else. The point is not to have the object just blend into the background, but to be undetectable while preserving the environmental signals that would normally be hidden by the object's presence. More than stealth, cloaking doesn't just prevent a signal from being detected, it allows for the detection of normal signals, the situation to be expected if the object wasn't there at all.

Camouflage is often limited by positional issues. The camouflaged object blends in from one direction, yet the same camouflaged object may stick out like a sore thumb from another direction. Birds and fish are colored differently on their backs and bellies just to avoid this problem. This makes them less visible from below and above, although changes in the background may still point them out to their predators. True cloaking renders an object invisible from all directions at all times, no matter the movement of the object, the observer, or the background. Unfortunately, this strength of cloaking is also one of the reasons it is so difficult to achieve. Current cloaking methods are limited in frequency range, size, dimensions in which they are functional, and mobility. Metamaterials have opened up the field of cloaking, but the transformation optics discipline is so young that most cloaks are only functional over very narrow frequency ranges. As of now, the expansion of bandwidth destroys very optical characteristics that render an object cloaked. Likewise, expanding the number of dimensions (X, Y, Z) in which the cloak is active makes the mathematics and production of the metamaterial much more difficult. Only a few current metamaterial cloaks are attachable to the object being cloaked, so for the majority, the cloak is immovable and the object must remain unmoving as well.

While these limitations are real and difficult to overcome, one must realize that the progress that has been made in just the past two years has been amazing. The expansion of cloaking types from gradient index to the thinner mantle cloaks has made cloaking moving objects easier, although the need to take the target object's optical characteristics into consideration introduces another issue that must be dealt with. Likewise, the advent of slow light technology portends the development of positional cloaks, yet have limitations of their own. As of now, positional cloaks use ultracold atoms, the choice of atoms is limited, and the distances over which light can be slowed are much too small to be useful in making a starship disappear.

Overall, the take home message is that cloaking is very real, very young, and very misunderstood. Most of the people who work in these fields cringe at the mention of *Star Trek* or Harry Potter cloaks. The possible applications for cloaking beyond movie tricks and military tactics are numerous and important. The same technology that *could* hide a starship from detection *will* improve wireless communication, open up the field of optical computing, and improve information security through communication devices. Space exploration will be aided by metamaterial thermoelectric and heat dissipation technologies as well as cheaper and lighter communication instruments that are based on the materials being developed now. Medical imaging, landmine detection, cheap and less intrusive airport scanners, even allowing a surgeon to see around his hands to get a batter view of field—the uses of cloaking technologies are limited only by the imagination of the engineers. Even if Gene Roddenberry refused to let his heroes hide, perhaps he could see that humanity may be able to use cloaking technology and still hold on to their moral center.

References

MA Abdalla, and Z Hu. On the study of development of V band metamaterial radar absorber. *Advanced Electromagnetics* 1(3); 94-98, 2012. doi:10.7716/aem.v1i3.25. http://aemjournal.org/index.php/AEM/article/view/25

G Alagappan, and CE Ping. Broadband slow light in one dimensional logically combined photonic crystals. *Nanoscale* 7(4); 1333, 2015. doi:10.1039/C4NR05810K. http://pubs.rsc.org/en/Content/ArticleLanding/2015/NR/C4NR05810K#!divAbstract

JJ Allen, LM Mathger, A Barbosa, KC Buresch, E Sogin, J Schwartz, C Chubb, and RT Hanlon. Cuttlefish dynamic camouflage: repsonses to substrated choice and integration of multiple cues. *Proc R Soc B* 277(1684); 1031-1039, 2010. doi:10.1098/rspb.2009.1694. http://rspb.royalsocietypublishing.org/content/277/1684/1031

S Brule, EH Javelaud, S Enoch, and S Guenneau. Experiments on seismic metamaterials: molding surface waves. *Physical Review Letters* 112; 133901, 2014. doi:10.1103/PhysRevLett.112.133901. http://journals.aps.org/prl/abstract/10.1103/PhysRevLett.112.133901

T Buckmann, M Thiel, M Kadic, R Schittny, and M Wegener. An elasto-mechanical unfeelability cloak made from pentamode metamaterials. *Nature Communications* 5; 4130, 2014. doi:10.1038/ncomms5130. http://www.nature.com/ncomms/2014/140619/ncomms5130/full/ncomms5130.html

P-Y Chen, C Argyropoulos, and A Alu. Broadening the Cloaking Bandwidth with Non-Foster Metasurfaces. *Physical Review Letters* 111; 233001, 2013. doi:10.1103/PhysRevLett.111.233001. http://journals.aps.org/prl/abstract/10.1103/PhysRevLett.111.233001

X Chen, Y Luo, J Zhang, K Jiang, JB Pendry, and S Zhang. Macroscopic invisibility cloaking of visible light. *Nature Communications* 2; 176, 2011. doi:10.1038/ncomms1176. http://www.nature.com/ncomms/journal/v2/n2/full/ncomms1176.html

JS Choi, and JC Howell. Paraxial Ray Optics Cloaking. *Optics Express* 22(24); 29465-29478, 2014. doi:10.1364/OE.22.0293465. https://www.osapublishing.org/oe/fulltext.cfm?uri=oe-22-24-29465&id=304785

A Fishman, J Rssiter and M Homer. Hiding the squid: patterns in artificial cephalopod skin. *Journal of the Royal Society Interface* 12(108); 20150281, 2015. doi:10.1098/rsid. 2015.0281. http://rsif.royalsocietypublishing.org/content/12/108/20150281

M Fridman, A Farsi, Y Olawachi, and AL Gaeta. Demonstration of temporal cloaking. *Nature* 481; 62-65, 2012. doi:10.1038/nature10695. http://www.nature.com/nature/journal/v481/n7379/full/nature10695.html

NS Ginsberg, SR Garner, and LV Hau. Coherent control of optical information with matter wave dynamics. *Nature* 445; 623-626, 2007. doi:10.1038/nature05493. http://www.nature.com/nature/journal/v445/n7128/abs/nature05493.html

D Gur, B Leshem, M Pierantoni, V Farstey, D Oron, S Weiner, and L Addadi. Structural basis for the brilliant colors of the sapphirinid copepods. *J Am Chem Soc* 137(26); 8408-8411, 2015. doi: 10.1021/jacs.5b05289. http://pubs.acs.org/doi/abs/10.1021/jacs.5b05289

LV Hau, SE Harris, Z Dutton, and CH Behroozi. Light speed reduction to 17 metres per second in an ultracold atomic gas. *Nature* 97; 594-598, 1999. doi:10.1038/17561. http://www.nature.com/nature/journal/v397/n6720/abs/397594a0.html

G Heinze, C Hubrich, and T Halfmann. Stopped light and image storage by electromagnetic induced transparency up to the regime of one minute. *Physical Review Letters* 111(3); 033601, 2013. doi:10.1103/PhysRevLett.111.033601. http://journals.aps.org/prl/abstract/10.1103/PhysRevLett.111.033601

MP Hokmabadi, J-H Kim, aE Rivera, P Kung, and SM Kim. Impact of substrate and bright resonances on group velocity in metamaterial without dark resonator. *Scientific Reports* 5; 14373, 2015. doi:10.1038/srep14373. http://www.nature.com/articles/srep14373

M Iwasaka, Y Miyashita, M Kudo, S Kurita, and N Owada. Effect of 10-T magnetic fields on structural colors in guanine crystals of fish scales. *J Appl Phys* 111(7); 07B316, 2012. doi:10.1063/1.3675206. http://scitation.aip.org/content/aip/journal/jap/111/7/10.1063/1.3675206

ZH Jiang and DH Werner. Quasi-Three-Dimensional Angle-Tolerant Electromagnetic Illusion Using Ultrathin Metasurface Coatings. *Advanced Functional Materials* 24(48); 7728-7736, 2014. doi:10.1002/adfm.201401561. http://onlinelibrary.wiley.com/doi/10.1002/adfm.201401561/abstract

TM Jordan, JC Partridge, and NW Roberts. Non-polarizing broadband multilayer reflectors in fish. *Nature Photonics* 6; 759-763, 2012. doi:10.1038/nphoton.2012.260. http://www.nature.com/nphoton/journal/v6/n11/full/nphoton.2012.260.html

N Landy and DR Smith. A full-parameter unidirectional metamaterial cloak for microwaves. *Nature Materials* 12; 25-28, 2013. doi:10.1038/nmat3476. http://www.nature.com/nmat/journal/v12/n1/full/nmat3476.html

J Li and JB Pendry. Hiding under the carpet: a new strategy for cloaking. *Physical Review Letters* 101; 203901, 2008. doi:10.1103/PhysRevLett.101.203901. http://journals.aps.org/prl/abstract/10.1103/PhysRevLett.101.203901

WT Lu, S Savo, BDF Casse, and S Sridhar. Slow microwave waveguide made of negative permeability metamaterials. *Microwave and Optical Technology Letters* 51(11); 2705-2709, 2009. doi:10.1002/mop.24727. http://onlinelibrary.wiley.com/doi/10.1002/mop.24727/abstract

JM Lukens, AJ Metcalf, DE Leaird, and AM Weiner. Temporal cloaking for data suppression and retrieval. *Optica* 1(6); 372-375, 2014. doi:10.1364/OPTICA.1.000372. https://www.osapublishing.org/optica/abstract.cfm?uri=optica-1-6-372

X Ni, ZJ Wong, M Mrejen, Y Wang, and X Zhang. An ultrathin invisibility skin cloak for visible light. *Science* 349(6254); 1310-1314, 2015. doi:10.1126/science.aac9411. http://science.sciencemag.org/content/349/6254/1310.abstract

W Parnell. Nonlinear pre-stress for cloaking from antiplane elastic waves. *Proceedings of the Royal Society A.* 468; 563-580, 2012. doi:10.1098/rspa.2011.0477. http://rspa.royalsocietypublishing.org/content/early/2011/10/19/rspa.2011.0477

JB Pendry, D Schurig and DR Smith. Controlling electromagnetic fields. *Science* 312; 1780-1782, 2006. doi:10.1126/science.1125907. http://www.ece.utah.edu/~dschurig/Site/Recognition_files/1780.pdf

J Perczel, T Tyc, and U Leonhardt . Invisibility cloaking without superluminal propagation. *New Journal of Physics* 13 (8); 083007, 2011. doi:10.1088/1367-2630/13/8/083007. http://iopscience.iop.org/article/10.1088/1367-2630/13/8/083007/meta

D Rainwater, A Kerkhoff, K Melin, JC Soric, G Moreno, and A Alu. Experimental Verificationn of three-dimensional plasmonic cloaking in free-space. *New Journal of Physics* 14; 013054, 2012. doi:10.1088/1367-2630/14/1/013054. http://iopscience.iop.org/article/10.1088/1367-2630/14/1/013054#metrics

D Schurig, JJ Mock, BJ Justice, SA Cimmer, JB Pendry, AF Starr and DR Smith. Metamaterial electromagnetic cloak at microwave frequencies. *Science* 314; 977-980, 2006. doi:10.1126/science.1133628. http://science.sciencemag.org/content/314/5801/977.abstract

JC Soric, PY Chen, A Kerkhoff, D Rainwater, K Melin, and A Alu. Demonstration of an ultralow profile cloak for scattering suppression of a finite-length rod in free space. *New Journal of Physics* 15; 033037, 2013. doi:10.1088/1367-2630/15/3/033037. http://iopscience.iop.org/article/10.1088/1367-2630/15/3/033037/meta

KL Tsakmakidis, O Hess, and AD Boardman. Trapped rainbow storage of light in metamaterials. *Nature* 450(7168); 397-401, 2007. doi:10.1038/nature06285. http://www.nature.com/nature/journal/v450/n7168/abs/nature06285.html

B Zhang, Y Luo, X Liu, and G Barbastathis. Macroscopic invisibility cloak for visible light. *Physical Review Letters* 106; 033901, 2011. doi:10.1103/PhysRevLett.106.033901. http://journals.aps.org/prl/abstract/10.1103/PhysRevLett.106.033901

L Zigoneanu, B-l Popa, and SA Cummer. Three dimensional broadband omnidirectional acoustic ground cloak. *Nature Materials* 13; 352-355, 2014. doi:10.1038/nmat3901. http://www.nature.com/nmat/journal/v13/n4/full/nmat3901.html

Chapter 3
The Replicator: Maybe You *Can* Have Everything

The replicators on decks four through nine are producing nothing but cat food.

—Commander Riker
TNG: A Fistful of Datas

3.1 Introduction

Money makes the world go around—does it do the same for the Galaxy? Within the *Star Trek* community there is discussion as to whether the Federation of Planets is a currency-free society and the degree to which money and trade play a role in daily life. Jonathan Frakes (Commander William T. Riker) said in several interviews that series creator Gene Roddenberry told him before he read for the part, "In the 24th century there will be no hunger, there will be no greed, and all the children will know how to read." Roddenberry insisted that the Federation is a post-scarcity society with no need for money. People work only for the betterment of themselves and all other beings (*Star Trek: First Contact*). However, some of the writers may not have gotten that memo.

In the second season, Kirk asks Spock if he knows how much money Starfleet has invested in him (*TOS: The Apple*). In the Next Generation pilot, Beverly Crusher selects some fabric and asks it be charged to her account (*TNG: Encounter at Farpoint*). In Star Trek books, Federation credits are used to produce starships, with a Constitution class ship costing several billion credits (*Star Trek: The Kobayashi Maru,* #47). Yet Captain Picard still contends that the accumulation of wealth is no longer a driving force in the lives of Federation citizens. Does he mean wealth as in money or wealth as in possessions? If Roddenberry meant no hard currency but still allowed for Federation credits, then this is a distinction without a difference; the credits are still a symbol of value that is traded for goods or services.

The Federation is not a closed society and has many dealings with species and planets that thrive in a capitalist economy, so maybe the Federation has a monetary system just for use with other societies. For instance, the Ferengi have a culture focused almost exclusively one accumulation of traditional wealth through business. *Star* Trek

M.E. Lasbury, *The Realization of Star Trek Technologies*,
DOI 10.1007/978-3-319-40914-6_3

has the opinion of 20th century Earth—Kirk annoys Dr. Gillian Taylor when he can't pay for their lunch after the Enterprise goes back in time to 1980s San Francisco (*Star Trek IV: The Voyage Home*). She says sarcastically they probably don't use money in the 23rd century, and all he can say is, "Well, we don't!"

Economic story lines exist in all five series, with several alien cultures possessing their own currencies. The Klingons have the Darsek, the Cardassians use the Lek, yet it is the pursuit of gold-pressed latinum by several cultures that is most often discussed. The Ferengi admired the humans' historic Wall Street (*VOY: 11:59*), with Quark telling Rom that greed is the noblest of all emotions (*DS9: Prophet Motive*). On station Deep Space Nine, Jake Sisko's lament that he can't bid on a Willie Mays baseball card does not bring sympathy from his Ferengi playmate Nog, "It's not my fault that your species decided to abandon currency-based economics in favor of some philosophy of self-enhancement." (*DS9: In The Cards*).

What could make it possible for the Federation abandon currency-based economics? Pure altruism is one possibility, but more likely it's the work of the replicator. If a society has machines to produce all the food and goods it asks for, then the value of any one particular item disappears; there is no need for money. Of course, there are still services that could be bartered. Money isn't required for barter of services, and if you can replicate all your physical needs, then you could afford to contribute your skills to society. So, if this is how the Federation system works, why does Mr. Scott brag that he just bought a boat (*Star Trek: The Undiscovered Country*)? There are internal inconsistencies here.

Then again, individual Federation citizens aren't required to use replicators. A small band of people who refuse to "buy into" a cashless society would need an alternate economic platform. Captain Picard's father, Robert, is just that sort of man. He doesn't allow a replicator in his house (*TNG: The Family*), and Riker is of a similar mind when it comes to food—he prefers cooking his own omelets (*TNG: Time Squared*). On the other hand, it sometimes appears that we humans have already given up cooking food in favor of processed food-like products, and this may be preparing us for the day when just about any product or food will be built for you before your eyes. *Star Trek* started with food replication and progressed to making hard goods while our science is approaching replication in the opposite direction. Goods manufacture has been achieved, but cheap, healthy food fabrication for the masses lags behind. Perhaps we aren't as altruistic as the Federation.

3.2 *Star Trek* Replicators

Replicators in *Star Trek* begin as a way to make food (*ENT: Fight or Flight*) using proteins (for *ENT*) or organic molecules of an ill described nature (*TOS, TNG, DS9, and VOY*). This "protein resequencer" rearranges amino acids in proteins to produce other proteins (with some exceptions) for Captain Archer and his crew, while several decades later, Spock, Kirk and the Constitution class Enterprise crew can make any food (*TOS: Assignment Earth*) in a food replicator.

The original series takes place just as standard and industrial replicators for hard goods are appearing, with major advances in replicator technology after we enter the 24th century. By the time of Captains Picard, Sisko and Janeway, replicators are ubiquitous, producing products, food, and even some living tissues. The writers use them or refer to replicators in almost every episode, indicating that they are an important part of 24th century life. Therefore, the franchise gives the audience ample examples of how they are used, how they function, and what their limitations are. A discussion of these issues is important for determining if we can, or should, mimic them on Earth.

3.2.1 Replicators Are Both a Necessity and a Convenience

In a situation where satisfying physical needs are only a replicator away, leisure time would undoubtedly be increased. Shopping might be one activity where a replicator could be convenient. Picard's Enterprise has replicator centers with several build platforms each—like having an entire Galleria in one room. Worf and Data use it to shop for a wedding gift for Miles O'Brien and Keiko O'Brien (*TNG: Data's Day*). Similarly, Deep Station Nine has a replimat, a version of the 1950s automat with replicators, where one can select from a wide range of foods for one's dining pleasure (*DS9: Emissary*). One of the ultimate conveniences for a starship stuck in space for years is the holodeck. Its programs can whisk a bored or burned out crew member to the beach, or anywhere else they can think of for any kind of adventure they might desire. The holodeck uses replicator technology; while many of the environmental objects in the program are holographic, things that the participants might touch are replicated and then dematerialized when not needed any longer (*VOY: The Cloud; Dark Frontier*).

However, replicators are more than just a time saver or expediency device, they are an integral part of functioning of Constitution and Galaxy class starships. Chief O'Brien points out one crucial function when marvels at how a historical starship could have made a long journey, "I just don't see how this ship could have made the trip. They didn't even have replicators back then. They would've had to store their air supply and there's only enough room on board for a few week's worth." (*DS9: Explorers*) Not just their air comes from replicators, so do clothes, although you have to specify the exact size. The Emergency medical hologram on *Voyager* complains that Naomi has grown 5 cm in just the previous three weeks and he is constantly calling on the replicator for larger clothes for her (*VOY: Mortal* Coil). However, Starfleet uniforms are another matter—they can't be replicated since the patterns are not allowed to be inputted into the replicator databases (*TNG: The Game; DS9: For The Cause*). If the crew could make their own uniforms, the ship would be full of admirals and no one would ever wear a red tunic.

Other limits are placed on the replicators as well. Obvious weapons like phasers cannot be fabricated, but dual function items present a problem. A screwdriver can be used as a weapon, so the captain will disable the replicators of crewmembers or prisoners confined to quarters. For example, Dr. Bashir can't get breakfast because the Starfleet internal affairs officer switches off his replicator after Bashir is accused of being a Dominion spy (*DS9: Inquisition*). Replicators will not produce poisons or toxins either (*VOY: Death Wish*) because the replicator computer will not accept instructions (patterns) for obviously dangerous items. However, if a user wants to enter the instructions for a novel item to the database, it can be replicated … with a few exceptions. Power source dilithium crystals must have a pattern that can't be saved because they are only be obtained from the mines that are spread out across the galaxy.

In a strange twist of fate, the same replicator that removes economic motive is also the reason that one currency still has value—latinum can't be replicated. The structure of latinum is so complex and so dense that the replicators can't handle the pattern. Gold may be worthless because it can be replicated, although it does have one use with respect to latinum. Gold is suspended the liquid latinum to keep produce a solid; bars are easier to use for business transactions and to store assets than containers of liquid (*DS9: Who Mourns for Mom?*).

3.2.2 Replicator Function

A replicator assembles an object from individual atoms based on a pattern that is stored in the replicator's computer. It uses some of the same technology as the transporter (Chap. 8) to have objects materialize on the build platform, with some important differences. Teletransported objects have their atoms converted to energy (perhaps) and then reconverted to matter at the destination. Therefore, there is a 1:1 correspondence of building materials; for every atom of the dematerialized item there is one for the rematerialized form. In contrast, a replicated product is fabricated from a saved pattern—perhaps from an object that had once been transported (*TNG: Lonely Among Us*), or from a set of instructions entered into the database specifically for that fabrication. Therefore, replicated objects don't have a built in cache of atoms from which to be assembled. This lack of matter for replication can be overcome in a few of ways; use recycled atoms or use atoms from a stockpile.

The growth in recycling on Earth in the 1990s is reflected in the replicator's ability to reuse atoms. Dirty dishes, waste materials, that dead Romulan, or broken ship parts can be placed in the replicator and they disappear, broken down to their atomic constituents (*TNG: Lonely Among Us*, *DS9: Hard Time*, *VOY: Year of Hell*). Atoms from the recycled objects can be used to make something that is always needed, like air, or can be stored and reused when needed. This is certainly ingenious, but it is a bit unappetizing to think that Geordi's late night cheeseburger might have been an old sock just minutes before. Even worse, recycling also includes the electrolytic fractioning of wastewater to reclaim the water and the organics (TNG Technical Manual, p. 154). Yes, that does mean that they make new products from human waste.

Fortunately, the atoms for food can also come from stores of raw materials pumped into holding tanks in the hull each time the ship is at a starbase for service/repairs, and the same is true for atoms to make nonfood items. Raw materials are pumped into hold tanks called "headends" which are connected by to the replicator terminals for materialization of the finished products that are requested. The replicators can use these raw materials because patterns are stored at the molecular, not the quantum level, so pure energy can't be used to replicate items. If the database of all possible objects contained information at the quantum level, the memory storage would exceed that of the entire fleet (TNG Technical Guide, p. 90). Luckily, it's not necessary to have every electron spin and energy level to make a good 3D chess set.

In the spirit of recycling, it is likely that both raw materials and recycled atoms are used for fabrication; this would make raw material supplies last longer during long voyages. The raw and recycled materials are dematerialized from the headends or other storage containers and through something called the "quantum geometry transformational matrix field." This poorly explained device modifies the matter stream to fit the selected pattern as it travels as energy through the conduits but before it rematerializes at the replicator terminal. Nowhere in the canon does it say whether the dematerialized raw atoms are rematerialized as the same atoms, or whether they are rearranged at a subatomic/quantum level by the quantum geometry transformational matrix field to produce the right types of atoms and the right numbers—maybe that is asking too much.

3.2.3 Replication of Organic Materials

The technical manuals do say that more information is needed for replication of organics (food, biomedicines, living tissue) than for inorganic objects, and that specialized replicator terminals are housed in the galleys and sickbay for this purpose. The raw materials are molecules containing carbon, hydrogen nitrogen, oxygen and phosphorous (plus trace elements) fashioned into a powder. Though chemically bonded, the powder is the simplest form possible and takes the least rearrangement by the matrix field to produce food or the like. Higher resolution is needed to ensure sure the products are safe and because the higher resolution makes the food taste more natural.

Replicators include biofilters (*DS9: Babel*) to make sure pathogens are not included in food or medicine and for sickbay replicators that manufacture anatomical replacements and some tissues. This points out a possible internal inconsistency. Replicators can't be used for living organisms because the lack of quantum level information in pattern induces radiation-like damage (TNG Technical manual, p. 91). But at the same time, replicators have biofilters to keep living organisms out of replicated items. Perhaps the justification is that radiation-like replicator damage might just as easily produce superbugs by inducing mutations as kill them.

There does exist a specialized Federation replicator for living tissue, the Genitronic. It replicated Lt. Worf's crushed spinal column—without the damage of course, to successfully treat his paralysis because this replicator can pattern at the quantum level (*TNG: Ethics*). However, some non-Federation species have more sophisticated standard replicators that can reproduce whole organism, as when Picard impostor was generated by the Allegiance species, right down to his memories (*TNG: Allegiance*). Obviously the Allegiance doesn't worry about conserving power and computer storage space; they don't lower the resolution on their replicators. The Allegiance, the Ferengi, the Federation—they all have replicators, and we humans are working to join that group.

3.3 Current Replicators

Humans have entered the digital manufacturing age, where coded instructions can be converted directly to large objects, and science and engineering are working toward manufacturing on a molecular level for food, biologics and commercial products. Computer aided manufacture (CAM) has been around since the last quarter of the 20th century; just enter a set of instructions and it will make the part. Although this sounds just like replication it is a different process. CAM methods mimic traditional manufacture by cutting or carving final products from large sheets or blocks, or pouring excess material into molds and then finishing the rough product. Twenty-first century science is now learning how to build products from the bottom up, bit by bit. This is much more like *Star Trek* replicators.

Box 3.1 3D Printing Methodologies

Stereolithography: The oldest method and only uses plastics. The products must be smoothed by hand, so it used mainly for hobbyists and prototypes. Within a pool of liquid plastic, the build platform is covered by a one 0.1 mm (0.004 in) thick layer of the liquid. A laser moves in X/Y directions according to the plan to cure the plastic in specific places. The platform lowers 0.1 mm and another layer of liquid flows over the cured first layer. It is fast (hours to a day or so) and inexpensive.

Powder bed or polyjet matrix: A moveable head dispenses binder (glue) onto a layer of powdered build material. UV light or a laser fuses the powder to the binder and to the layer below. The platform is lowered and more build powder is laid down. If the printer head dispenses build material instead of binder, then the method is called polyjet matrix. Up to 14 different build materials can be combined in the same build, depending on the number of dispensers on the printhead.

Fused deposition modeling (FDM): One of the most popular methods; the only technique that can use high performance thermoplastic. Support structures for undercuts or cavities are printed out of water-soluble material and dissolved after production. The build material is heated to the melt point in the moveable head and deposited in the proper positions. It cools and hardens as it fuses with the previous layer.

Laminated object manufacturing (LOM): This method is used by artists or architects. The build material (paper, plastic, metal) comes on thin rolls with adhesive on one side. The rollers heat a sheet as it is laid down. This adheres it to the previous layer. A laser then cuts the outline and crosshatches any areas to be removed. If paper is used, the finished product looks like laminated wood, and is usually lacquered or varnished.

Selective laser sintering or melting (SLS/SLM): These methods are often used for aerospace parts. In sintering, a laser fuses the metal powder with the layer below, but does not reach the melting temperature. The energy of the laser compresses them with pressure into a solid. SLM uses a stronger laser so that the material reaches the melting temperature. In some cases, an electron beam is used instead of a laser. The unfused powder acts as a support, so no supports are needed in the design. Many different materials can be used with these methods, including nylon, glass, ceramics and metals (silver, aluminum, steel, chromium, titanium).

3.3.1 Additive Manufacturing Is 3D Printing

Invented in the 1980s, three-dimensional printing techniques have been the darlings of the media since about 2010. The media loves to call them "replicators," and the commercial industry has adopted this terminology. One machine from MakerBot (New York, NY) is actually called "The Replicator," although it takes several hours or days for the object to materialize and it can only use one material. MakerBot calls the Replicator a 3D printer, but "additive manufacturing" being a more accurate term. The two terms overlap in that the various 3D printing methods are additive in the sense that they join new layers to previously fabricated layers. They build products higher and higher one layer at a time, although some methods do use a vast excess of build material, with a portion of it being fused together and the rest subtracted away. This is more like traditional manufacturing where a block of material is cut away, much like how Michelangelo sculpted his David. However, all 3D printing techniques build objects a layer at a time.

In their current forms, 3D printers are more like plotters for printing large blueprints. The head moves in the X and Y directions to place the laser or material delivery nozzle in the correct location, as opposed to the single platform of the *Star Trek* replicator on which things just appear. A printer head delivers a small amount of spot of build material or binder to a specific spot as designated by the computer instructions, or energy in the form of laser or UV light is shone upon a layer of build material. In either case, that small spot is cured to make it solid, and the head then moves to each adjacent spot until the entire layer is completed according to the pattern.

Depending on the method used and the size of the project, a build can take from hours to days, and this doesn't include the time needed to design and input the pattern. The pattern is a software program, not unlike the card programs used to order foods from the earliest food synthesizers on *Star Trek*. Later food replicators and standard replicators as well are able to accept spoken commands via speech recognitions systems (Sect. 6.3.2). In a limited way, we can now ask for our products as well. Yahoo! Japan recently developed a 3D printer that can respond to speech commands for a limited number of objects. The technology was developed for visually impaired children and was demonstrated in several schools, but can be applied more broadly. The child asks the unit for a toy or object, and the computer searches the internet for its 3D blueprint or something similar. It takes about fifteen minutes for the device to produce the toy by a method called stereolithography (Pearson 2013).

3.3.1.1 How 3D Printing Works

The etymology of the name for the first and still most common form of 3D printing, stereolithography (SLA), gives one an idea of just what is going on in the process. It means "writer of solid stone." The writing is the pattern, the stone is the build material (actually plastic), and the solid means that it is fused or solidified in the process. Taken together, it is using a written pattern to fuse material together into a solid. First, the pattern is created with a computer aided design (CAD) program, which reflects the outer profile of the object and includes any cavities or internal structures. Then the CAD plan is converted to standard tessellation language (STL), a compilation of 2D layers.

The STL instructions are uploaded to the printer and the first layer is begun. The build material is placed on the build platform. In techniques like SLA, a pool of liquid or powdered build material surrounds the platform, but only to a depth of one build layer, perhaps 0.10 mm. In techniques like selective laser sintering (SLS), a pool of powdered build material is held separate from the platform and rollers spread a thin layer of polymer or metal powder onto the build platform for fabrication of each layer. In both cases, after a layer is completed, the platform is lowered by the depth of one layer and the next layer is fabricated directly on top. See Box 3.1 for details on the common 3D printing techniques.

Once the polymer, metal, alloy or other build material is in place for a given layer, only some of it is cured to become solid. The areas cured are determined by the computer pattern for that layer. Several mechanisms are possible, based on the properties of the build material. UV or visible light can cure resins, heat may cure liquid plastic, lasers fuse metal powders by melting them or by compressing them (as in SLS), or a binder (glue) can be laid down on the build material in a specific pattern and then cured by UV or laser light. The build platform is lowered and the process is repeated until completed (Fig. 3.1).

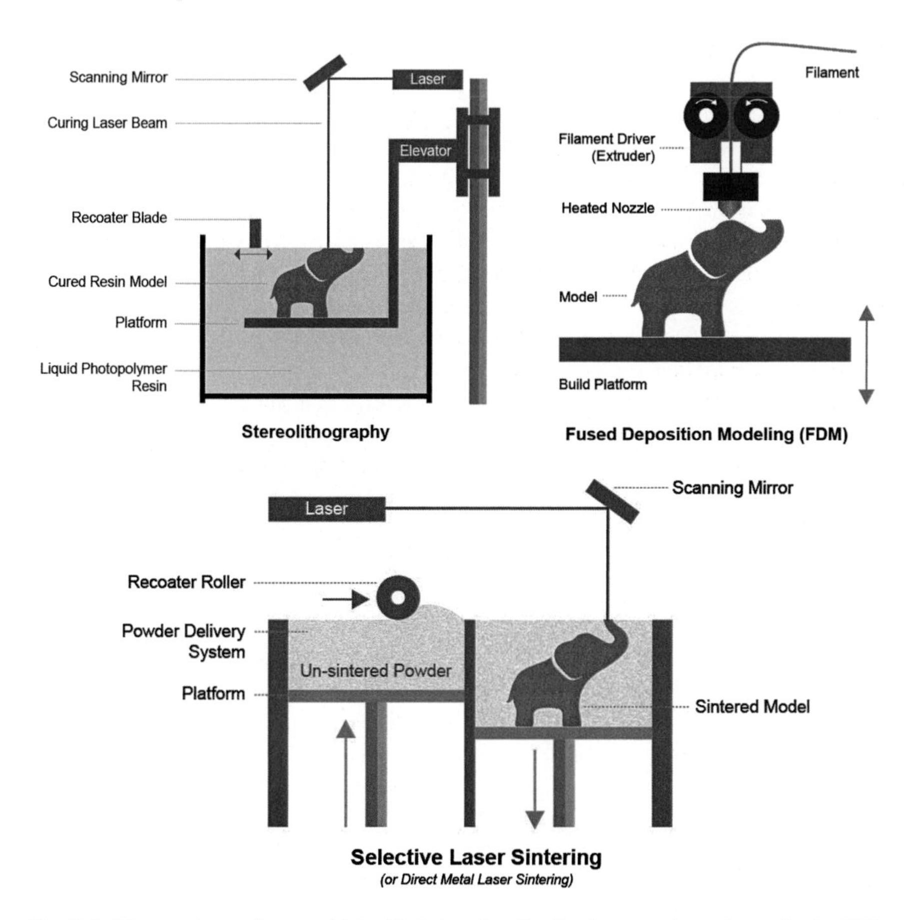

Fig. 3.1 These cartoons do a good job of helping visualize the three most popular methods of 3D printing. In stereolithography (STL), the platform is lowered and the object sits in a bath of unpolymerized build material. In fused deposition modeling (FDM), the build material is heated and builds the object as either the platform lowers or the extruder rises. In select laser sintering (SLS), the powdered build material is rolled over the build platform with each layer, with the unfused material supporting any undercuts or otherwise unsupported portions of the object. (Images courtesy of Mark Jaster, www.printspace3d.com)

Fused deposition modeling (FDM) is a reverse process; the build material is extruded from a printer cartridge(s) above the platform. The cartridge moves in three dimensions, it moves in the X/Y direction to lay down build material for one layer, and then it moves up in the Z direction for the next layer, etc. In some types of FDM, the build material is melted prior to deposition and cools to solid form after it is extruded. In other types of FDM, the build material is laid down cold and is cured by laser, UV, or heat. FDM is versatile because several build materials can be used, with one extruder on the print head for each. For instance, undercut areas can be supported with a water-soluble build material from one extruder that can then be dissolved once the object is completed. The total time required to complete an object may be from hours to days depending on the shape, size, and limitations of the technique used.

Research into additive manufacturing materials has picked up in the past five years, where early projects could only be made in low density plastics, many techniques can now use metals and even ceramics or glass. One of the biggest advances has been the ability to use high performance or engineering grade thermoplastics in the FDM systems. Thermoplastic products can be used directly in engines and motors without additional finishing or machining. They can be used in situations with strong forces and high temperatures instead of only for prototypes or finished products that undergo little stress. NASA has already tested FDM thermoplastics to make rocket parts; their tests indicate that 3D printed parts rival those fabricated in traditional ways for strength and durability. (NASA press release, August 27, 2015). Even more amazing, additive manufacturing won't just have an effect on how we get to space—NASA is already 3D printing *in* space.

3.3.1.2 3D Printing in Space

Printing parts in space is very *Star Trek*. There would be no need for packing extra parts, and if something really unexpected happened, you could print a solution on the spot. Imagine how routine the Apollo 13 mission would have been if the astronauts had been able to print a service module and a CO_2 scrubber—well, now they could. The first 3D printed parts fabricated on the International Space Station (ISS) materialized in November of 2014 (NASA press release, November 25, 2014). The printed wrenches used FDM with plastic that becomes malleable after only a small temperature increase. Space printing will have to use extruded build material, as pools or thin layers of liquid or powder will not stay put in microgravity. It turns out that microgravity is different than printing on Earth. The first efforts resulted in parts that couldn't be separated from the build platform, because microgravity enhances the effect of the binder/build material fusion. Things stick to the platform more doggedly, but this also produces parts that are stronger.

Many copies were made on the ISS and brought back to Earth to assess all effects of microgravity on the process, since 3D manufacture is going to be a big part of extended space travel. As the next step in this process, Made In Space (Mountain View, CA) has been contracted by NASA to develop Zero-G printing. They recently (August, 2015) developed and tested a 3D printer that can work not just inside the ISS, but in the vacuum of space itself (Made In Space press release, August 10, 2015). Parts for the outside of the ISS could be made on the spot. Quincy Bean, the lead investigator on the Zero-G printing project said in an interview, "I grew up watching *Star Trek* and *Star Wars*.......... what excites me most about 3D printing in space is that if we really want to have increased human presence in space, we really have to have the capability of building what we need in space, instead of having to launch spares over and over again." (Oberhaus 2015)

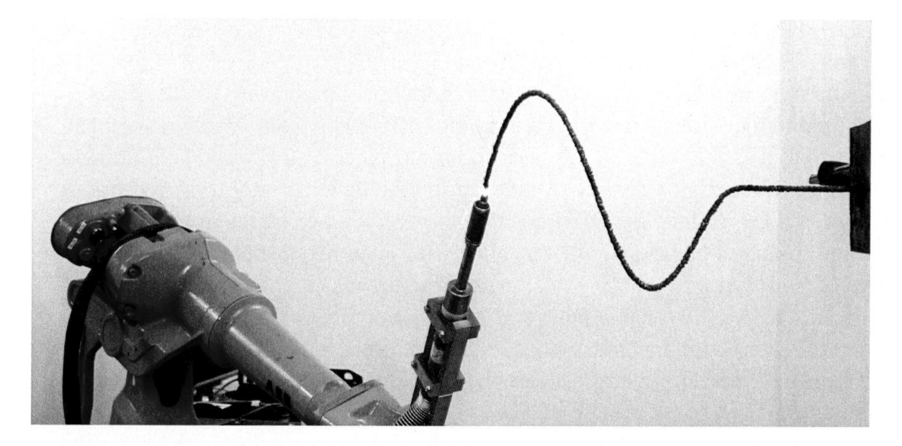

Fig. 3.2 The robotic printer from MX3D can fuse metal or resin using a dispenser and heat source (or welder) on a robotic arm. It can build in three dimensions, to any size the robot arm can reach, with the robot able to build its own track to make it freely moveable. The system can be used horizontally or vertically and can build with gravity or against it. (*Image credit* MX3D/Joris Laarman Lab, Amsterdam, The Netherlands)

The MX3D system (MX3D, Amsterdam, Netherlands) might also be useful in space someday. It can print metal or resin in free space (Fig. 3.2). The system uses a dispenser to extrude small amounts of metal, a robotic arm to place the dispenser at any particular point in 3D space, and a welder to fuse metal. Lines of material can be fabricated in any direction and attached to any vertical or horizontal surface. The company is using MX3D to print a pedestrian bridge across a canal in Amsterdam. The entire bridge is printed in place by a pair of robotic arms that slide along the rails of the bridge as it is built. Imagine a smart robot sliding across the hull of the ISS or a moon station to build modules or scaffolds in free space.

3.3.1.3 Electronic Beam Melting and Glassy Metals

Materials science is working hand in hand with 3D printing companies to improve the capabilities of additive manufacturing. Electronic beam melting (EBM), is a new method of fusing build material that is similar to laser sintering or laser melting techniques except that an electron beam is used to fused the build powder (metal or resin) into a solid. EBM is particularly good at producing and printing with a type of build material called "glassy steel" or "glassy metal." (Koptyug et al. 2013). The material is a metal because it has made from metal atoms, while it is like glass because the metal atoms do not assume a rigid crystalline structure, they remain amorphous, as in glass. Glassy metals are solids that are stronger than their crystalline counterparts. They are more flexible, noncorrosive, lighter, and exhibit low metal fatigue; basically everything you would want out of a material that will hold your life in the balance as you hurdle through space—but no, they aren't transparent like the aluminum Scotty traded for plexiglass to make whale tanks for the Enterprise (*Star Trek IV: The Voyage Home*).

Amorphous steel is an alloy made with atoms of different metals that are significantly different sizes. Like pebbles packing the voids in a jar of rocks, the small metal atoms fit between the larger ones filling up more of the space and giving a liquid alloy with a high viscosity. A pure crystalline lattice cannot form because of the smaller atoms and the higher viscosity prevents the large atoms from moving to their most ordered positions. As the alloy cools, atoms assume an amorphous structure, providing no grain boundaries and producing very hard surfaces, corrosion resistance, high wear tolerance and low thermal conductivity. Yet the glassy metals are more flexible and less brittle than traditional glasses or ceramics.

Glassy steel for 3D printing uses a yttrium additive that makes it 3–4× harder than traditional steels and can be used in nontraditional applications because it is not magnetic at room temperature. The yttrium slows the crystal growth so that the metal cools before falling into a typical lattice structure (Lu et al. 2004). Likewise, glassy titanium alloy containing small metal atoms is stronger than regular titanium while being as flexible as human bone and a very good joint replacement material (Sugiyama et al. 2009). These strong, resilient, yet lightweight metals will be important for building in space, just as space will be important in learning more about the alloys. The microgravity of the ISS allows metal alloys to be studied without holding them in containers that could trigger crystal formation.

3.3.1.4 Quasicrystalline 3D Printing

Nobel laureate Wolfgang Pauli once said, "God made the bulk; surfaces were invented by the devil." (Schroeder 1991, p. 230) Additive manufacturing in space is a good example of the problems that can come from surfaces. The surfaces of 3D printed objects can have unfused build material or solvent on their surfaces. Even cleaning the surface will not get rid of all the particulate, and more is lost over time. In a microgravity environment, the unfused build material can float behind panels and disrupt electronics (Science Daily summary from ESA, June, 25, 2014). This is why NASA and the ESA build rovers and modules in clean rooms.

Box 3.2 Quasicrystals

As the name implies, quasicrystals (QC) act only somewhat like traditional crystal structures. Traditional crystals are rigid, highly ordered, repeating units of a single type that give strong materials. In three, four, or six fold symmetry, all space is ordered and the unit is repeated, so these are allowable symmetries for crystal growth. Regular pentagons cannot be assembled without a void occurring, so five-fold symmetry isn't allowed for traditional crystals. In contrast, QCs can have any fold symmetry in any number directions. They have a finite number of repeating units that work together to fill all available space and build a crystal. This makes them a completely new class of solids.

Daniel Shechtman of the Technion-Israel Institute of Technology in Haifa won the Nobel Prize for chemistry in 2011 for describing QCs. His results were initially rejected because no one thought a nonperiodic set of repeating units could self assemble and remain non-repeating. Other scientists thought that knowledge of the entire crystal would be needed, a difficult task since the units have only local knowledge. Now they know that a single repeating unit with overlap can place itself in the growing chain properly to maintain the quasiperiodicity (Olberhaus 2015).

The overlapping rule gives a maximum density and the lowest energy for the most stable configurations. Their seemingly impossible structure gives QCs some characteristic properties. They diffract light differently than normal crystals, allowing for absorption—quasi-cloaking as it were (Chap. 2). The hull of a starship coated in QCs would be strong, yet appear black in space. The slightly askew periodicity means that QCs may conduct electricity in one direction but resist a flow of electrons in several others.

QC containing materials are light while remaining strong; the crystals reinforce metal structures while the repeating units allow for some elasticity without brittleness. QC coatings can be as slippery as Teflon without the tendency to flake or chip. At the same time, QCs have very poor thermal conductivity and can be used as insulators. These properties are good for cookware and for 3D fabrication of aerospace materials.

Natural QCs were found in Russia in 2009 (Bindi et al. 2009) and again in 2015 (Bindi et al. 2015). Interestingly, both natural QC examples are from meteorites that are billions of years old and from billions miles away. The symmetry of the story is nice—QCs identified from outer space just as science is learning how to use QCs to take us further into space.

Quasicrystalline materials provide an answer to this problem in 3D printing by providing harder, smoother surfaces on printed objects. Quasicrystals (QCs) are solids that do not have a *conventional* crystalline structure, yet they don't have amorphous structures like glassy metals either. QCs are a completely new type of solid matter first observed in the 1980s. The unique structure of quasicrystalline metals results in dense surfaces that are very smooth, due to higher orders of symmetry filling all available space in the crystal with atoms. Instead of having a normal repeating period for the units as crystals normally do, QCs have more multiple repeating units that can fit together in an ordered structure (Fig. 3.3). The crystal is ordered, without a strict repeating pattern—it's a quasicrystal because it has quasi-periodicity. Box 3.2 details the unique structure of their crystals.

QC solids have unique properties based on their non-repeating and dense structure. Their structure predicts that they would be conductors in one direction

Fig. 3.3 Quasicrystals are a bit like these mosaic tile tessellations found in Marrakech, Morocco. The pattern repeats, but not simply, and the space cannot be filled by one of shape, different shapes must be pieced together. The different tiles are analogous to crystals of different symmetries in quasicrystals. This occurs very rarely in nature. (*Image credit* By Ian Alexander [Own work] (CC BY-SA 3.0 [http://creativecommons.org/licenses/by-sa/3.0], via Wikimedia Commons, image cropped)

and dielectrics in another or absorb light in some directions and reflect it in others. The testing of a decagonal quasicrystal for several properties confirmed this anisotropy. The crystal was electrically conductive in the direction of 10 fold symmetry, yet when the plane was turned any of three ways in the quasicrystalline planes, the solid showed insulating behavior (Bobnar et al. 2012). The same held true for thermal conductivity and diamagnetism. These properties could become important for production of transformers and inductors, semiconductors and in thermoelectric systems.

Most important for additive manufacturing in space, QC build materials give 3D printed parts with very hard, smooth surfaces; surfaces of which Wolfgang Pauli could be proud. Laser sintering and laser melting are especially good for QC incorporation in surface coatings and metals or polymers for reinforcement because they permit slow cooling and easier non-periodic crystal formation. QCs can be produced through traditional means, but they are much more expensive when produced this way and are hard to place in specific arrangements for reinforcement or coating. QCs characteristics and properties are better exploited by 3D printing than by traditional manufacturing techniques, and this creates a good partnership for use of QC materials in space.

3.3.2 4D Printing

One step beyond three is four, and 4D printing is another step toward production of a replicator that appeared in just the last year or two. The fourth dimension is time, and 4D printing produces products that can change structure and shape over time. This gives printed material functionalities that haven't been possible before, and

Fig. 3.4 The cube configuration is printed by traditional 3D printing techniques, although the resin use to print the various parts is different based in which way they need to change shape and the amount of environmental change (temperature, etc.) that must occur for that change to be induced. When heat is applied, the pattern folds up in proper order to give a cube that locks itself in that configuration. (*Image credit* 4D Printed Self-Folding Cube, Self-Assembly Lab, MIT + Stratasys Ltd + Autodesk Inc.)

takes 3D printing from static objects to machines that can react to conditions and perform work.

Products printed in 4D are shape shifters, but not like Garth of Izar who makes Spock choose between two identical Kirks (*TOS: Whom Gods Destroy*). Four-D printed objects have their possible shape changes programmed into their instructions; more like the DNA thief that needs to incorporate DNA instructions from its target in order to change what it looks like (*VOY: Vis a Vis*). They are printed from shape memory polymers (SMP) that have actuating fibers incorporated into their liquid state. Some are activated by heat, some by water, others by mechanical forces. When the polymers encounter these changes in environment, they contract or stretch and the parts change shape. The SMPs are generated with different amounts of the actuating material in the polymer to correspond to how much change is needed to activate them, and they are printed in specific orientations to make the shape change occur in the desired direction (Fig. 3.4). These parameters, along with other factors, such as speed and degree of change are written into the printing instructions (Ge et al. 2013).

When printed, the polymers will assume their temporary (deformed) shape. The actuating stimulus precipitates a change to their permanent shape, by folding, twisting, curling or stretching. The point where SMPs of different properties meet becomes a joint; the change in shape of one compared to the other can create an edge. For example, some flat, printed sheets will fold and lock themselves into cubic boxes when exposed to heat. The key is in the design. The timing of the folds is a function of different amounts or types of actuators and work to avoid self-collisions during the shape changes. The top of the cube needs to fold after the side, so it needs either less of the actuating material or an actuator that requires more heat exposure. One vertical side might need to fold in a different direction, so the SMPs need to extruded in a different orientation during the build. If the patterns are properly designed and executed, almost any shape can be formed and/or change in function be induced (Ge et al. 2013).

For example, a functional valve was printed recently using SMPs at a university in Australia. The valve, as printed, allows water to pass through, but if the water is hot, the valve closes automatically. No additional mechanical parts are needed, both the original function and change in function are programmed into the design (Bakarich et al. 2015). For this particular application, four different extruder heads were required, each distributing a different SMP material. The researchers made the four SMPs by varying the proportions of just two different formulations. More amazing, a 2014 project at Harvard University included a 4D-printed capacitor with a contact sensor and a 4D-printed mechanical switch in a self-assembling lamp containing layers of copper as electrodes and internal wiring. The lamp lights up and adjusts its brightness when touched after the addition of only a bulb and external wiring to a power source (Shin et al. 2014).

3.4 Current 3D Printing of Organics

Products fabricated with 3D and 4D techniques will pay off when humans begin to live and work in space in greater numbers. But inanimate, structurally engineered products are just one part of the additive manufacturing revolution, and may not be the field *most* affected by 3D printing. Medicine may very well be the largest recipient of 3D printing advances, followed closely by food. And it makes sense that the ability feed or treat people will be instrumental in human conquest of space.

3.4.1 Printing Cells and Tissues

Bioprinting is the term used to describe laying down cells in an ordered pattern to mimic tissues or organs. Biochemicals and various cell types are fed *very* gently through extruder heads and are laid down on and in a hydrogel binder. The cells themselves don't immediately adhere to one another when laid down by the printer, so the binder keeps them in place and happy until they learn to form cellular junctions with their neighbors. The most difficult aspect of bioprinting is developing the proper bio-ink, the combination of cells, growth factors, hormones, nutrients and hydrogel binder. In regular 3D printing, the build material is more or less standard; many different things can be made from it. Bioinks must be developed for each individual application, and the cells must be grown just for that build and to match the specific patient (if to be used in a patient).

The bio-ink must be thin enough to flow easily, yet thick enough to build up in three dimensions. The binder must harden in response to temperature or UV light before the next layer is deposited, but it can't dry out the cells. Even though there is a two-hour window to get the printed tissue back into growth media, the hardening can't be so fast or rigid as to damage the cells. One way to manage hardening and shape is to extrude bioink gelatins at one temperature into slightly cooler gelatin supports. Building up a few layers is no problem once the bioink is formulated properly, yet organs can't be printed yet for a couple of reasons. One problem is that larger structures such as organs collapse under their own weight or become pressure damaged without a proper scaffold. In your body, fibrous tissues provide the scaffold.

To solve this problem, scaffolds are printed first, and the cells are printed directly on to the scaffolds in a second run. This method is gaining traction in the medical community as materials and design techniques improve. The design of the scaffold might be based on imaging techniques of a patient if the shape and fit are crucial. The unique anatomy of each patient means that every build for a scaffold (for a bone defect, a trachea, etc.) is a one off fabrication. You need not store a pattern for building the same scaffold again—it won't fit anyone else. Therefore, it is a more economical process to print the scaffold than to make a 3D model based on a CAT scan or MRI, mold it, and cast a scaffold that will be used only once.

The scaffold must be porous to allow infiltration of blood vessels and tissue cells. Poragens, specific materials that are dissolved away from the scaffold to leave the final porous structure, are included in bioink when printed. Cartilage printed this way maintains populations of chondrocytes (cells that lay down and live within cartilage) that survive and grow for long periods of time and will work for implantation (Izadifar et al. 2015). As an example of how these techniques are being used, a baby was born in 2012 with a trachea that wouldn't stay open. Small spasms would cause the trachea to collapse and make it impossible for baby Kaiba to breathe. Groups at the Universities of Michigan and Illinois were testing the ability of a sleeve placed around the trachea to hold it open. Unfortunately, a static sleeve or a stent inside the trachea would not grow with the baby and would not expand and contract with each breath. What was needed was a biocompatible scaffold that would support the natural trachea and then disappear on its own by the time Kaiba had grown to the point that his trachea was strong enough to stay open on its own (Fessenden 2013).

The medical groups used an indirect 3D printing technique. A model of Kaiba's trachea was printed in synthetic polymer, using CAT scan images of the Kaiba's own trachea. Then the sleeve was designed to fit around the model. This was cast as a mold and the sleeve was produced in the mold from a degradable polymer, the same material used for resorbable sutures. It took a couple of efforts to get a mold and then a cast that would fit perfectly (Zopf et al. 2013), and they experimented with the technique on a couple of piglets with the same defect before applying to the FDA for an emergency approval to do surgery an sew one into place around baby Kaiba's trachea. Happily, the boy has been happy and healthy in the time since the surgery with no further breathing incidents. Kaiba's trachea has reconstructed itself and the splint will now degrade over time. What's amazing is that with proper equipment on a spacecraft, this entire procedure could have been carried out on the way to Mars or beyond, far from any hospital on Earth.

3.4.1.1 Printing Blood Vessels

The above examples all have one thing in common; the printed medical products don't contain many blood vessels. Developing an adequate vasculature is the second issue that prevents 3D printing of organs and large tissues as of now. Without blood vessels, some of which are so narrow that red blood cells must squeeze through them one at a time (100–200 μm diameter), tissue cells do not receive the oxygen, glucose, and hormones they need, and cannot get rid of their waste products. In general, cells must have a capillary system no more than 10 μm distant in order to adequately diffuse oxygen and nutrients to the cell and carbon dioxide and other wastes to the blood vessel. Despite this massive number of very small capillaries, blood vessels come in many shapes and sizes, as detailed in Box 3.3, and this makes printing them even more difficult.

Box 3.3 Blood Vessels Aren't Just Pipes
For a tissue or organ to survive, it must have a blood supply, arteries and veins are not just tubes through which blood travels. They is harder to bioprint because they must have a tubular structure with several cell types and be housed within a sold organ.

Arteries: elastic, and react to increased pressure when the heart pumps so that the pressure can be regulated. Each artery has three distinct layers made from endothelial cells, elastic fibers, smooth muscle cells, and fibrous tissue and cells.

Large arteries have longitudinal and circumferential muscle layers to control their volume when the heart pumps, while small arteries only have circumferential muscle fibers.

Arterioles: smaller than arteries. They have thinner walls because the pressure is lower, but still include innervation and muscle to control blood flow and pressure.

Capillaries: The walls of some capillaries are only one cell thick, and they are leaky to allow flow of fluid in and out of tissues. This is where oxygen exchange occurs. Capillary systems are redundant in several ways; certain pathways can be opened or closed to maintain flow and pressure. The capillary system must supply every cell that needs nutrition.

Veins: thin walled vessels with no muscle. Blood is pushed through the veins by the beating heart and the movement of large muscles that surround them; walking helps pump blood back to your heart. Veins have one-way valves to prevent backflow of blood when the heart is between beats and to counter the effects of gravity.

Printing vasculature must account for every cell type and every structure, and must be vast enough to supply all the cells of the printed tissue. Implanted printed tissues can't wait for vessels to grow in from the surrounding area, so bioprinting must supply complete or nearly complete vascular systems. The print design needs to include positions to tie the printed arteries and veins into the existing blood supply of the intact tissue.

Various techniques are being studied to produce vasculature in 3D printed tissues. Researchers at Carnegie Mellon University in Pittsburgh use a method coined "freeform reversible embedding of suspended hydrogels" (FRESH) to print large vascular meshes and even whole hearts (miniaturized versions at least). A gelatin scaffold is used to support proteins, collagens and cells as they are laid down. When these take hold and bind together using growth factors and mild electrical stimulation in a "bioreactor," the gelatin is melted away at body temperature and different vasculature cell types are both injected into the lumen and layered over the collagen scaffold tube (Hinton et al. 2015). This works well for individual vessels, but an organ may have thousands of them.

A similar system has been developed by groups working together in Australia and the US. They print a network of fibers and then coat them in vascular cells and a protein matrix. After these are stable, they pull the fibers from inside the constructs and are left with intact vessels of selected diameters (Bertassoni et al. 2014). Other groups use "fugitive ink." While this polymer bio-ink is a gel at temperatures for printing, it turns into a liquid when it is *cooled* slightly. Tissue bio-ink with tissue cells is printed at the same time a network of fugitive ink is laid down to approximate a vasculature network. When the hydrogel solidifies and cools, the fugitive ink runs out, leaving a system of tubules. The network is flushed with blood vessel precursor cells and growth factors that help the different layers of the vessels develop. A greater degree of complexity for the vasculature is possible with this technique (Kolesky et al. 2014) as compared to the others and is receiving more attention (Fig. 3.5). Once the scaffolding and vasculature problems are solved and standardized, printing of organ replacements will be closer to reality, something that even *Star Trek* can't manage.

The newest system avoids the fuss of printing blood vessels by leaving pores in the printed structure that can provide oxygen to cells while natural blood vessels grown into the printed tissue. The Integrated Tissues-Organ Printer (ITOP) combines an imaged design of specific patient's needed structure with a biodegradable substructure interwoven with printed cells that holds the living tissue in place and provide the pores for oxygen diffusion (Kang et al. 2016). Implanted ITOP examples of bone, cartilage and muscle have been well accepted in mice and rats and a human ear structure suitable for transplant has been produced. It is hoped that for external structures like ears, the oxygen available through the pores will keep the tissue alive until a growth factor-stimulated blood supply can grow into the construct.

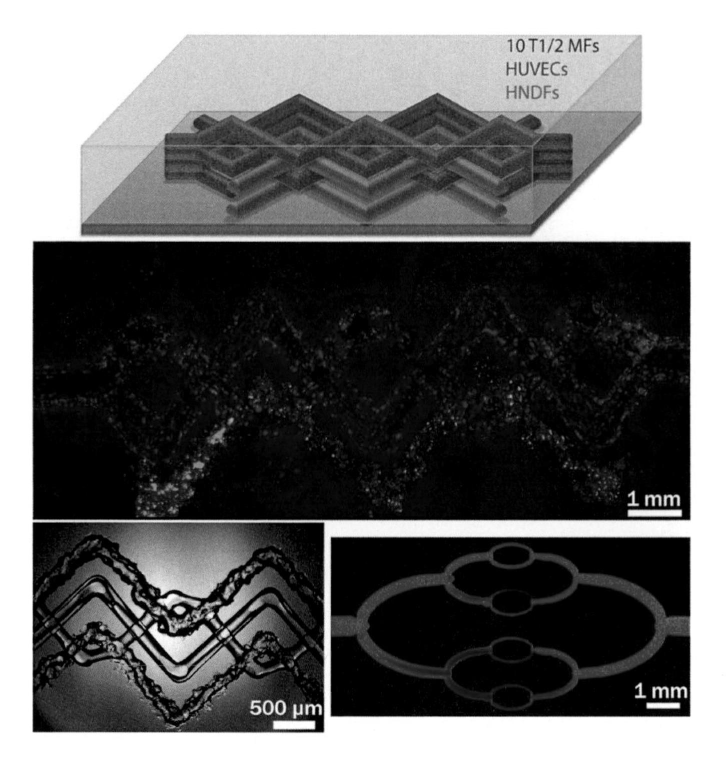

Fig. 3.5 Its hard to print a blood supply for tissue. The *top* image shows how tissue fibroblasts in gel (*blue*), dermal fibroblasts in gel (*green*) and fugitive ink to produce vessels (*red*) can be printed in three dimensions. The *middle* image shows the fibroblasts cells (*blue* and *green*) after they start growing and dividing in the gel matrix. The *red dots* are growing blood vessel cells they were introduced into the vessels after the fugitive ink was evacuated. The *bottom left* image shows the growing constructs with the vessels being smooth as to not damage red blood cells. The *bottom right* image is a two-dimensional vessel construct filled with *red* dye to illustrate how printed vessels can form a network to deliver oxygen via artificial capillaries (the thinnest vessels in the middle). (*Image credit* [Kolesky et al. 2014], Copyright Wiley-VCH Verlag GmbH & Co. KGaA. Reproduced with permission)

3.4.2 Printing Food

We can't print full organs or organisms yet, but neither can *Star Trek* without lots of help. However, the Federation is way ahead of us in terms of materializing food. Replicated food is such a given in 24th century space that it is rarely a subject of any thought. Only when Voyager is lost in the Delta Quadrant does replicating food become an issue, with replicator rations instituted to save both power and raw materials (*VOY: The Cloud*). Most space travelers seem to think there is no difference between grown and replicated food. Some actually prefer the artificial sustenance (*VOY: The Killing Game*), with few characters favoring the real deal (*TNG: The Wounded*). NASA once used freeze-dried food, then food in a tube, while now the ISS

Fig. 3.6 Chloé Rutzerveld has designed a food printer of the new wave. The Edible Growth printer prints a case of carbohydrate into which is printed a slurry of seeds and mushrooms pores. Over time, the sprouts and mushrooms grow and ferment, giving the substrate more depth of flavor. The process leads to fresher food for people and fewer miles that food will be shipped. (Image courtesy of Chloé Rutzerveld and the Edible Growth Project)

astronauts eat very much like they would on Earth. Unfortunately, when longer missions are undertaken, the good food arriving from Earth on every restocking mission will come to an end. Therefore, NASA has funded a small business grant to look into the possibility of printing food for journeys to Mars and beyond.

In the 1990s, inkjet printers were co-opted for use with cakes and chocolates. Rice paper was loaded into the tray, and the ink in the printer head was replaced with food dyes. Picture and text could be printed and laid over foods to create visual masterpieces, although this was hardly printing food, more like printing on food. A decade later, FDM 3D printers were adapted to use melted chocolate or sugar for printing designs as decorations on cakes. These example seem trivial, but make no mistake, printed food is a serious issue. Better printing techniques will help with food sustainability in areas of drought and will put nutritious food in everyone's hands. When issues of unpalatable texture and production and storage of raw materials are overcome, users will have the ability to produce a food with arbitrary amounts of vitamins, salts, proteins, carbohydrates, etc. Fewer hormones and antibiotics will be needed because there will be less waste, so there will be no need for extreme productivity (Fig. 3.6).

In a small step toward this goal, Nature Machines has a product called the Foodini (available 2016, third quarter) that can print burgers, pizza, etc. using fresh foods in steel capsules. It extrudes natural foods as layers that are then cooked in a conventional fashion; spaghetti with sauce and spinach quiche are two of the choices. If the eggs are scrambled in the capsule, quiche will be easy but it impossible to make a deviled egg, so the versatility of this technology is limited. The Foodini isn't actually a food printer, standard food printers now extrude pastes or doughs that can be cooked for crispness, yet food engineers are a long way from printing a carrot stick. The closest thing to a commercial printer is the Smoothifoods, used in German eldercare facilities. For patients who have trouble chewing solid foods, mashed vegetables or fruits bound together with edible glue are much preferred to food purees; the texture better mimics regular food. Eating is about texture, flavor and visuals; maximizing these is harder than printing plastic. While it's true that you eat first with your eyes, who cares what a replacement part looks like.

Flavor and aroma are important for food as well, and NASA is looking to improve these in printed foods. The Systems and Materials Research Corporation in Texas has developed a system for NASA that uses freeze-dried, powdered ingredients in print head capsules that can be mixed with water and printed. The printed material is enhanced by print heads that spray liquid flavoring, odorants, and nutrients. The first item printed by the prototype was a cheese pizza, where the dough layer was cooked as it was printed, followed by reconstituted tomato powder reconstituted with oil and water and some protein flakes that mimic cheese. Someday, biosensors worn by the astronauts will be able to sense the state of each user and have the printer match their nutritional needs using sprayed on minerals and vitamins.

Futurist Thomas Frey believes that 3D food printing is going to change our relationship with food. He wrote in 2011, "The labels we use today to describe our diets, labels such as vegetarian, Kosher, gluten-free, vegan, lactose-free, will be replaced with new terminology ...... there are no such things as a pig molecules, or a fish molecules, or a wheat molecules. We have other types of molecules that make up plants and animals, but on the molecular level there is no such thing as vegetarian and non-vegetarian molecules." (Frey 2011) This idea of thinking of the things we assemble at the molecular level is prescient. The future of fabrication will involve individual atoms and molecules.

3.5 The Future of Replicators

Ray Kurzweil is a futurist with a twist—he doesn't just predict the future, he makes it. Kurzweil was *the* key figure in inventing the first optical character recognition system to read multiple fonts, the text-to-speech synthesizer and the text-to-speech reading machine, all of which will be important for realizing a true universal translator (Chap. 6). The CCD flatbed scanner and music synthesizers that better mimic instruments were Kurzweil projects. Now he is Director of Google Engineering with an eye to creating computers that understand and produce natural language. All are amazing feats and goals, but it's his predictions that seem to get more publicity.

Amongst many other prophecies, Kurzweil sees a coming virtual reality revolution in the next ten years, and true AI by 2030. In association with AI, he is predicting that nanoscale robots (nanobots) will be traveling through our bodies, replicating by themselves, fixing pathologies in our tissues and allowing us to connect our brains to computers. Kurzweil also predicts that within a century molecular assemblers will fabricate all our products, starting at the smallest subunits (Google I/O presentation 2014). While Neil Gershenfeld of MIT predicts macro-sized personal fabricators for everyone in a couple of decades, molecular assemblers will be the next big step toward true replicators; the ability to assemble objects one atom at a time.

3.5.1 Nanobots

A group scientists and futurists hail the coming age of the nanobot. These molecule-sized, self-programmed and self-replicating machines will build the products we need from atoms and perhaps swim through our bodies curing diseases and fixing defects. Small advances have been made toward these goals, but most scientists now doubt that they are will come about in our lifetimes. Ray Kurzweil is not one of the doubters. He predicts that by 2030 nanobots will attach our brains to the internet. Based on small strands of self-replicating DNA, his envisioned nanobots will latch on to individual neurons after passing through the blood and into our brains, and will translate our electrical activity into a machine language that can communicate with the internet. However, Paul Cherukuri, scientist and manager of the James Tour laboratory in the Richard E. Smalley Institute Nanoscale Science and Technology at Rice University, sees self-replicating nanobot technology as a very long-term goal, if achievable at all. And he ought to know, the Tour lab is building the most advanced nanoscale machines in the world (Sect. 3.5.2.2).

True self-replication of nanobots would require them to be autonomous; the instructions for both their job(s) and for their self-replication would be contained within the nanobots themselves. It is more likely that the next generation of nanobots (which will also be the first generation), will replicate based on a system called exponential assembly. This method uses an outside control to assemble the nanobots from a massive number of preproduced nanobot parts. Laid out in parallel, the product of one assembly step in one nanobot would trigger the same assembly step in at least two other bots. One will build one, and those two will build four, and those four will build eight; the exponential nature will produce millions of bots in a short time, given that the energy and raw materials remain available. Considering that there are about 5×10^{21} atoms in a single drop of water, it is going to take millions of nanobots to assemble any product that human might be able to use.

3.5.2 Molecular Assembly

Before either self-replication or exponential assembly are mastered, engineered nanomachines will likely work under direct programming to help science with fabrication tasks. Imagine small grapplers and rollers moving atoms or molecules around, placing them in specific spots and inducing the chemical bonds that will build larger structures on a nanoscale construction site. Unfortunately, nanomachine contributions to construction will remain small until science can ramp up the production of the nanomachines themselves. It would take a long time to build a house if the crew was just one guy with a hammer.

Nanomachines or chemical processes building at the level of atoms or molecules is referred to as "molecular assembly" or "molecular manufacturing." A true molecular assembler will fabricate any stable structure in three dimensions that does not specifically break the laws of physics and chemistry. This is not a new idea. Richard Feynman said in a 1959 speech to the American Physical Society (Feynman 1960), "The principles of physics, as far as I can see, do not speak against the possibility of maneuvering things atom by atom." Engineers, physicists, and biologists can already observe, control and move atoms in a precise manner, it's just that they must do it by hand using macroscopic tools. Useful products won't be possible until the scale of the tools decreases and the rate of production increases. It may take decades, but will be worth the wait.

The advantages of molecular manufacture are many. Atomically engineered materials can have strength, low weight, durability, corrosive resistance, unique electrical and magnetic properties, and will have zero manufacture defects. The products made by molecular assembly will have characteristics that engineers can't get from traditionally manufactured items. Nanoscale fabrication can be accomplished by top-down or bottom-up techniques; however, only the bottom-up method will let engineers control material properties through true molecular assembly. When Data sculpts a musical treble clef from clay, he is using a top-down approach (*TNG: The Masks*). He has a lump of clay that he manipulates with his hands using the pattern in his brain to remove clay he does not need. The replicator is a bottom-up approach; atomic or molecular scale components are assembled together into ordered structures using the natural bonding proscribed by physics and chemistry.

Current technologies can manipulate atoms to form 2D films or 3D structures of precise arrangement, with most using macromachinery and relying on excesses of atoms; these are top-down processes. Chemical vapor deposition for production of graphene or carbon nanotubes and the stamping out of nanoscale structures from rolls of film (roll to roll processing) may use microfabrication, but they are also top-down since they require tools for cutting, milling, shaping materials to achieve the desired order. The goal is to produce structures from the atoms up, without outside control or instructions. While current bottom-up techniques are crude, costly, time-consuming, and labor intensive, they will eventually allow for production of large numbers of objects in parallel and will be cheap compared to traditional fabrication.

3.5.2.1 Microscopy for Molecular Assembly

Builders still need to see what they are building during this period when science is just learning how to manipulate atoms. Once engineers learn reliable techniques and protocols for manipulating atoms, direct (or indirect) visualization will be less important, but for the near future it is crucial. Therefore, microscopes that can easily image on the atomic level are important tools for observing, building, and assessing atomic placements. Scanning tunneling microscopy (STM), atomic force microscopy (AFM), field ion microscopy (FIM), and other types of scanning probe microscopic techniques are all available now, and these may soon be joined by superlens microscopes using metamaterials to resolve images below the level of the atom.

The most useful of these techniques also allow the scientists to move them atoms in three dimensions while they observe. Several methods use a very sharp tipped probe, perhaps only one atom wide at the point. The probe interacts with the surface of a material, although the nature of the interaction is different from technique to technique. All the tip-based methods are versions of scanning probe microscopy, although some measure the interaction force between the tip and the atoms of the surface (AFM, Fig. 3.7), and others measure an electrical current between the tip of the material and a detector (FIM). In the case of AFM, the interactions produce a movement of the tip in the Z-direction, drawing it closer or driving it farther from the surface, so an image of the surface can be obtained at the atomic level. Depending on the change in tip conditions and distance, the position and type of each atom can be identified. This is how they image the atoms, and research in the 1990s showed that they can also perform atomic manipulations.

AFM utilizes the natural van der Waals force to stick an individual atom to the end of the tip, scoot it to a new location, and then raise the tip until the force between the surface and the atom is greater than that between the tip and the atom (Custance et al. 2009). STM is another manipulation technique, it can vary the electrical forces at the tip to pick up atoms and move them (Stroscio et al. 2006). Using STM, atoms have been moved around on conducting, semiconducting, and recently on insulating surfaces (Kawai et al. 2014) opening the door to producing atomic level logic gates and storage devices. Even more amazing, scanning with an electron beam can alter the positions of atoms *beneath* the surface, akin to building the needle inside the haystack (Jesse et al. 2015). The desired internal structures grow in same crystalline lattice as the parent material to ensure that electrical properties are maintained. The hope is that this will one day lead to three-dimensional nanochips within solid substances.

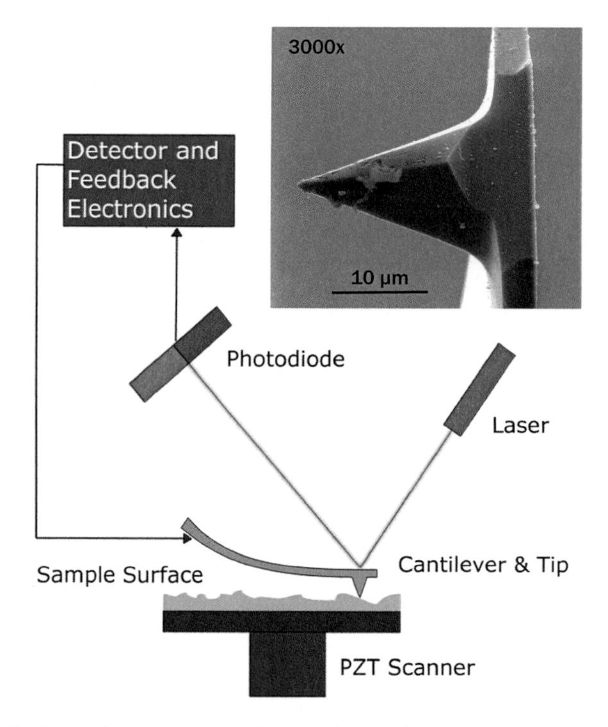

Fig. 3.7 Atomic force microscopy uses a sharp tip brought into close proximity to a surface. The forces that play between the materials of the surface and the tip will move it up and down, based on the size and identity of the individual atoms that make up the surface, as these determine the forces between the surface and cantilever tip. The cantilever is nearly one atom wide at the tip and its movement is detected by changes in laser beam position after reflection from the cantilever's face. (*Image credit* By AFM_(used)_cantilever_in_Scanning_Electron_Microscope,_magnification_3000x.GIF: SecretDisc derivative work: Materialscientist (CC BY-SA 3.0 [http://creativecommons.org/licenses/by-sa/3.0] via Wikimedia Commons)

3.5.2.2 Nanotools for Molecular Assembly

It is a peculiarity of molecular assembly that before we can use machines to efficiently build things atom by atom, the machines themselves will need to be built atom by atom. Microscopic atomic manipulation has been used to create atomic smiley faces, tiny advertising signs for laboratories, and even a short atomic animated movie as proof of concept for molecular assembly, but to do things on a larger scale, science is going to need many tiny tools.

To begin the process of inventing nanoscale tools and machines, some groups working with small molecules are using natural chemical bonding to produce useful tools. In 2005, the James Tour lab at Rice stunned the world with a ~250 atom, single molecule car on which the buckyball wheels actually turned when the car was pushed (Shirai et al. 2005). The car itself is only 3×4 nm ($1.18 \times 10^{-7} \times 1.57 \times 10^{-7}$ in); if a nanocar traffic jam left them bumper to bumper, it would

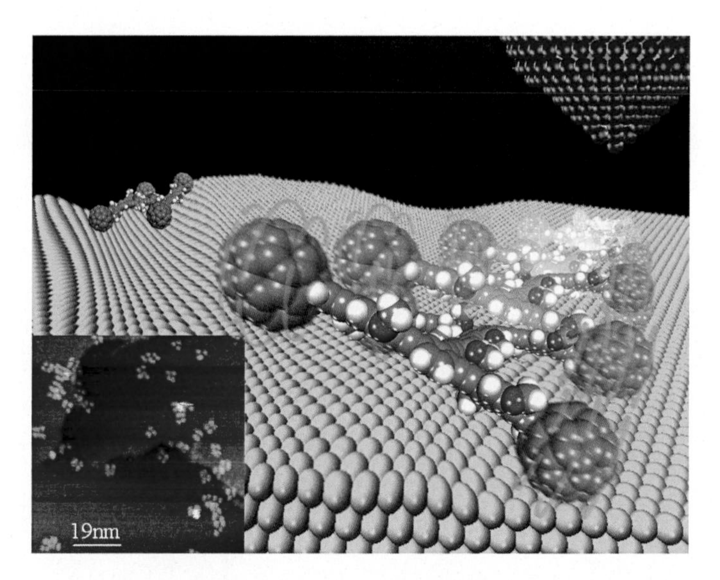

Fig. 3.8 This molecule-sized car is made of just two hundred or so atoms and is powered by light. The chemical bonds in the wheels force turns as they move from double to single bonds and back. The *inset* shows a photomicrograph of many cars to give scale. (Images courtesy of the Richard E. Smalley Institute for Nanoscale Science and Technology—See more at: http://news.rice.edu/contact-us/#sthash.0uxBSrNY.dpuf and Rice University)

take 30,000 of them to cross the diameter of a human hair. In 2006, they powered their car (Morin et al. 2006) by adding molecular motors that had been developed in a Danish laboratory. UV light is the driving force of the motor; it converts a double bond to a single bond in each of the wheels (Fig. 3.8). The conformation change in the wheel structures causes each to turn a quarter turn. The bonds want to return to the lower energy state, so they turn through another quarter rotation to reestablish the double bond and then the pattern repeats itself. The turning continues as long as the UV light stays on (Browne and Feringa 2006).

The Dutch developers of the motor demonstrated their own motorized nanocar in 2011, but their was powered by electricity. The energy was supplied through an electrified scanning probe microscope tip that touched the top of the car. The electrons tunnel their way through the car to the copper road it was driving on, again stimulating double bond breaks and rotations to re-establish the bonds (Kudernac et al. 2011). On the down side, the car had to be recharged after each half turn and it could only travel forward. This lack of a reverse gear also hampers the Tour group's nanocar, as well as their latest development—the nanosubamarine.

The submarine is composed of 244 atoms in a single molecule and has two UV light-driven outboard motors. Each submarine is the product of twenty highly sequenced chemical reactions, making them a bottom-up fabrication (Fig. 3.9). Every rotation moves the submarine forward 18 nm (7.08×10^{-7} in). This may not seem like much, until you realize that the motors spin a million times per second.

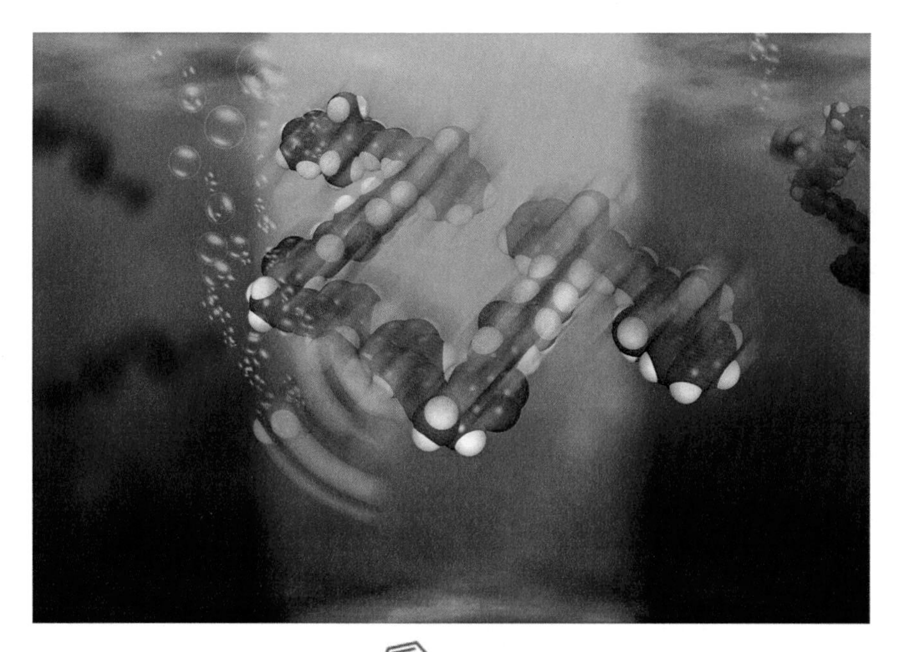

Fig. 3.9 The nanosubmarine of the James Tour laboratory uses the same type motor as the nanocar, and like the car, there is no reverse gear. The *red* portion of the molecular diagram represents the motor that spins at over a million revolutions per second. (Images courtesy of the Richard E. Smalley Institute for Nanoscale Science and Technology—See more at: http://news.rice.edu/contact-us/#sthash.0uxBSrNY.dpuf and Rice University)

That equates to just under an 18 mm (0.7 in) of forward progress every second, which is blistering by atomic standards, fully 26× faster than the molecule would diffuse in water (García-López et al. 2015). The fastest bacteria only manage about 0.2 mm/sec (0.008 inch/sec) because at small scales the bumping of molecules has a bigger effect and water behaves more like molasses. The Tour lab hopes that the nano-vehicles will be adapted to moving atoms for molecular assembly and delivering payloads such as medicines to specific cells. First they will have to learn to make more than one at a time.

Other important nanotools, if not quite as imaginative as submarines half the size of a red blood cell, have also been fabricated. A wrench, just 1.7 nm (6.7×10^{-8} in) wide, is actually a piece of anthrocene coal that can rigidly hold other molecules without flexing. It can help in syntheses and modifications of molecules to alter their properties (Liu et al. 2015). A different group has built a mechanical clutch for transmitting torque in different thermal situations. It works the same way gears transmit force through other gears (Williams et al. 2016) but is built from only 75 particles and is driven by optical laser traps (Sect. 5.3.3.1). An axle can be placed at the center of the clutch to be turned in either direction. Finally, there is a tentacle-like grasping tool that can reach out to hold microscale objects. Elastic tubules powered by small pneumatic actuators can curl themselves into a spiral loop of just 0.2 mm (0.008 in), small enough to capture a single dust mite (Paek et al. 2015). Engineer all these tools together and one can envision a molecule-sized torque wrench that can drive or swim to where it is needed and grab molecules for assembly.

While these are amazing goals, we may be getting ahead of ourselves. As things stand in 2016, atomic manipulation tools are human controlled and are million billion times larger than the building blocks they move. Creating nanomachines today is really a matter of one person, a microscope, some atoms, and a lot of patience. Smaller tools are starting to be developed, and the programming machinery to direct those tools needs to be scaled down many fold as well. Dr. Cheurkuri of the Tour in Houston described to me a newer facet of their work using a technique called laser-induced graphene (LIG). This may one day provide a method for programming instructions directly into nanomachines and creating the link that will bring nanofabrication into the macroworld.

Box 3.4 Graphene and Carbon Nanotubes

Graphene: a single atom thick layer (the only known 2D substance) of crystalized carbon in tessellated hexagon form. It is a one atom thick slice of diamond that is incredibly strong, transparent, and electrically versatile. It has an incredibly high surface area to mass ratio; one gram of graphene would cover five tennis courts.

Electrons have limited directions of travel in a 2D material. The dimensions and honeycomb pattern give graphene ballistic electron mobility. Electrons blast through graphene faster than through any other known material, behaving like they were traveling at the speed of light (relativistically), making graphene a good material for electronics. Unfortunately, an inability to produce large amounts at low cost within narrow tolerances and a large bandgap (energy difference between conductive and non-conductive states) means that graphene hasn't reached its potential.

Carbon nanotubes (CNTs): a sheet of graphene rolled into a cylinder. They can be extremely long (up to mm) compared to their width (as low as 1 nm). There single walled nanotubes (SWNT) or cylinders within cylinders (multiwalled, MWNT). CNTs are stronger, less reactive, and more thermally, chemically and mechanically stable than linear graphene. CNTs have unique properties with respect to conductivity that make them even more useful in electronics. Those properties depend on the direction of the graphene sheet when it is rolled; in X direction, in Y direction or diagonally. The diagonal versions (with names like armchair, chiral, zigzag) can have different twist, which determines if the CNT will be metallic or semiconducting (Fig. 3.10).

CNTs are 200× stronger and 5× more elastic than steel. They have 5× the conductivity and 1000× the current capacity of copper, but only half the density of aluminum. Transistors made from CNT can be spaced only single nanometers apart, less than half the gap of the best silicon versions, and the performance is not compromised by a reduction in size as it is in silicon.

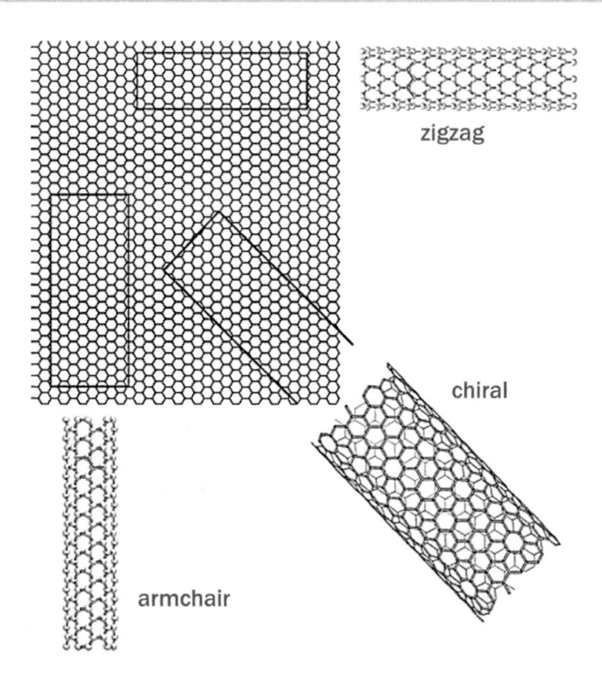

Fig. 3.10 This diagram illustrates how different types of carbon nanotubes can be produced from a single sheet of graphene. Depending on the angle at which the tube is rolled, the electrical properties will vary. In zigzag, some carbon-carbon bonds lie parallel to the axis of the tube and conveys semiconductor electrical properties. The *red line* overlay shows the zigzag pattern. In armchair (pattern shown by *red line* as well), the bonds are perpendicular to the axis of the tube and result in them conducting electricity like a metal. Finally, the chiral nanotube has a corkscrew axis, either left-handed or right-handed and act as semiconductors, with additional energy from light or an electrical field is necessary to make the electrons mobile. (*Image credit* [Fischer 2000])

CNT forests can be "microcombed" to produce even stronger and more conductive structures (Zhang et al. 2015), and new allotropes of carbon such as carbyne (Liu et al. 2013) and diamond nanothread (Fitzgibbons et al. 2015) may be stronger than even conventional CNTs.

3.5.2.3 Laser-Induced Graphene for Molecular Assembly

Graphene is one type of molecular carbon that may be only one atom thick and can form sheets, tubes, or balls. Graphene is electrically flexible, incredibly strong, durable at very small scales and has the potential to change manufacture in some fundamental ways. Dr. Cherukuri described the Tour lab's efforts at producing graphene by shining a powerful laser beam on an inexpensive polymer. The energy of the laser converts the polymer to graphene. By controlling the path of the laser, fine graphene channels can be etched onto the polymer surface and connected to electrodes. They have used this technique to produce stackable (Peng et al. 2015a) or linear (Peng et al. 2015b) microsupercapacitors that store more energy than regular microcapacitors and release that energy much faster. By doping the graphene with metallic salts, the LIG products can also substitute for platinum in fuel cells (Ye et al. 2015). So they may work as internal energy sources for nanomachines.

Carbon nanotubes and graphene, because of their tunable electrical properties (see Box 3.4), have the potential to replace silicon as the material of choice for semiconductors, but are expensive to produce. The LIG process can alleviate this problem while at the same time produce usable nanoelectronics. LIG can etch transistors, logic gates, and integrated circuits, while also allowing for resistors and insulators just by changing the orientation of the carbon crystals—all on a 10–20 nm scale. These may very well become the brains that run molecular assemblers.

3.6 Conclusion

The task of building a *Star Trek* replicator will remain moot until science solves that pesky teleportation problem. Hopefully, advances in fabrication will soon make 3D printers look like clumsy attempts at adapting existing technology and make it look like molecular assembly. This is not to say that 3D printing isn't an important step along that path. Engineering is a matter of massaging technologies and mixing and matching them until a paradigm-shifting advance comes along. Nanobots, or at least semi-autonomous nanomachine tools will represent a fundamental change in how fabrication can proceed.

There are two integrated parts to molecular manufacture, how to place atoms where you want and link them to existing structure, and procuring stockpiles of the atoms themselves. Atoms we have in abundance, gaseous forma of elements can be used as raw materials without problem. Solids will need to be reduced to atoms, and that will take energy; this makes it more a matter of power and labor rather than technology. With regard to assembling the atoms, there will likely be an accelerating cycle of technology. As science becomes better and faster at building nanotools, the products of molecular assembly will increase in complexity and will be produced at a faster rate. Some of these products will be better and more autonomous nanomachine tools, which will then fuel another leap in productivity. Add to this cycle advances in LIG-based circuitry and materials science and molecular assembly will move forward in leaps and bounds. The applications cross all facets of society, from medicine and biology to computing, the arts, food production, and space exploration.

The advances in material science will represent another cycle within molecular assembly. The products that can be made with glassy metals and quasicrystals might not be otherwise possible and will allow for new properties and functions to be exploited, but these are just the beginning phases. Imagine how long ago quasicrystals might have been discovered if science had the ability to play around with crystal lattices and build tessellated flat surfaces simply by trying out different atomic arrangements. In similar fashion, what new forms of solids and surfaces will be devised through atomic manipulation and what unique properties might those materials possess. More kinds of materials will allow for more detailed and productive molecular manufacturing, which will in turn lead to more new materials.

Will everyone eventually have their own molecular assembler and be able to make what they want? Neil Gershenfeld of the MIT's Center for Bits and Atoms started the ball rolling when he convinced the Mayor of Barcelona to institute a challenge that city be self-sufficient in the next decades, producing everything it needs through smart factories and a series of neighborhood Fab Labs stocked with 3D printers and manufacturing equipment of all kinds. With this as a model, Gershenfeld's goal of personal fabricators will work itself out, from city to neighborhood to family to individual, much like how the television spread in the 1950s. Single fabricator devices will first come down to who can afford them, then the price will fall as technology becomes ubiquitous and people will pay mostly for the atom supplies to make things. After science develops a cheap or free way to deliver atoms, the economic motive will be lost and we will approach a post-scarcity society.

At that point, society will need to tackle the question of whether we *want* to remove humans from the production line. During the transition to manufacture on demand from freely available resource, how we will we fell when people in manufacturing start to lose their jobs? Are we ready for a post-scarcity economy and to work hard for the sake of something other than money and will innovation still thrive in a society where it is not directly rewarded? At least Captain Picard will be able to get his, "Tea. Earl Grey. Hot."

References

SE Bakarich, R Gorkin, M in het Panhuis, and GM Spinks. 4D Printing with Mechanically Robust, Thermally Actuating Hydrogels. *Macromolecular Rapid Communications* 36(12); 1211-1217, 2015. doi:10.1002/marc.201500079. http://www.sciencedaily.com/releases/2015/04/150423213500.htm

LE Bertassoni, M Cecconi, V Manoharan, J Hjortnaes, AL Cristino, G Barabaschi, D Demarchi, MR Dokmeci, Y Yang, and A Khademhosseini. Hydrogel bioprinted microchannel networks for vascularization of tissue engineering constructs. *Lab Chip* 14(13): 2202-2211, 2014. doi: 10.1039/c4lc00030g. http://pubs.rsc.org/en/Content/ArticleLanding/2014/LC/C4LC00030G#!divAbstract

L Bindi, P Steinhardt, N Yao, and P Lu. Natural quasicrystals. *Science* 324(5932): 1306-1309, 2009. doi: 10.1126/science.1170827. http://science.sciencemag.org/content/324/5932/1306.abstract

L Bindi, N Yao, C Lin, LS Hollister, CL Andronicos, VV Distler, MP Eddy, A Kosin, V Kryachko, GJ MacPherson, WM Steinhardt, M Yudoskaya, and PJ Steinhardt. Natural quasicrystal with decagonal symmetry. *Scientific Reports* 5; 9111, 2015. doi: 10.1038/srep09111. http://www.nature.com/articles/srep09111

M Bobnar, P Jeglic, M Klanjsek, Z Jaglicic, M Wencka, P Popcevic, J Ivkov, D Stanic, A Smontara, P Gille, and J Dolinsek. Intrinsic anisotropic magnetic, electrical, and thermal transport porperties of *d*-Al-Co-Ni decagonal quasicrystals. *Physcial Review B* 85(2); 024205, 2012. doi: 10.1103/PhysRevB.85.02.024205. http://journals.aps.org/prb/abstract/10.1103/PhysRevB.85.024205

WR Browne, and BL Feringa. Making molecular machines work. *Nature Nanotechnology* 1; 25-35, 2006. doi: 10.1038/nnano.2006.45. http://www.nature.com/nnano/journal/v1/n1/full/nnano.2006.45.html

O Custance, R Perez, and S Morita. Atomic force microscopy as a tool for atom manipulation. *Nature Nanotechnology* 4; 803-810, 2009. doi: 10.1038/nnano.2009.347. http://www.nature.com/nnano/journal/v4/n12/abs/nnano.2009.347.html

M Fessenden. 3-D printed windpipe gives infant breath of life. *Scientific American Online.* May 24, 2013. Accessed 09/12/15. http://www.scientificamerican.com/article/3-d-printed-windpipe/

R Feynman. There's plenty of room at the bottom. *Caltech Engineering and Science* 23(5): 22-36, 1960. http://www.zyvex.com/nanotech/feynman.html

JE Fischer. Storing energy *in* carbon nanotubes. *Chemical Innovation* 30(10); 21-27, 2000. http://pubs.acs.org/subscribe/archive/ci/30/i10/html/10fischer.html

TC Fitzgibbons, M Guthrie, E Xu, VH Crespi, SK Davidowski, GD Cody, N Alem, and JV Badding. Benzene-derived carbon nanothreads. *Nature Materials* 14; 43-47, 2015. doi: 10.1038/nmat4088. http://www.nature.com/nmat/journal/v14/n1/full/nmat4088.html

T Frey. The coming food printer revolution. FuturistSpeaker Blog, October, 17 2011. Accessed November 20, 2015. http://www.futuristspeaker.com/2011/10/the-coming-food-printer-revolution/

V García-López, PT Chiang, F Chen, G Ruan, AA Martí, AB Kolomeisky, G Wang, and JM Tour. Unimolecular Submersible Nanomachines. Synthesis, Actuation, and Monitoring. *Nano Letters*. November 5, 2015. Epub ahead of print doi: 10.1021/acs.nanolett.5b03764. http://pubs.acs.org/doi/abs/10.1021/acs.nanolett.5b03764

Q Ge, HJ Qi, and ML Dunn. Active Materials by four-dimension printing. *Applied Physics Letters* 103; 131901, 2013. doi: 10.1063/1.4819837. http://scitation.aip.org/content/aip/journal/apl/103/13/10.1063/1.4819837

TJ Hinton, Q Jallerat, RN Palchesko, JH Park, MS Grodzicki, HJ Shie, MH Ramadan, AR Hudson, and AW Feinberg. Three-dimensional printing of complex biological structures by freeform reversible embedding of suspended hydrogels. *Sci Adv* 1(9); e1500758, 2015. doi: 10.1126/sciadv.1500758. http://advances.sciencemag.org/content/1/9/e1500758

Z Izadifar, T Chang, AM Kulyk, D Chen, and BF Eames. Analyzing biologicalperformance of 3D-printed, cell-impregnated hybrid constructs for cartilage tissue engineering. *Tissue Eng Part C Methods* Nov. 23, 2015 (Epub ahead of print) doi:10.1089/ten.TEC.2015.0307. http://online.liebertpub.com/doi/abs/10.1089/ten.TEC.2015.0307

S Jesse, Q He, AR Lupini, DN Leonard, MP Oxley, O Ovchinnikov, RR Unocic, A Tselev, M Fuentes-Cabrera, BG Sumpter, SJ Pennycook, SV Kalinin, and AY Borisevich. Atomic-Level Sculpting of Crystalline Oxides: Toward Bulk Nanofabrication with Single Atomic Plane Precision. *Small* 11(44); 5895-5900, 2015. doi: 10.1002/smll.201502048. http://onlinelibrary.wiley.com/doi/10.1002/smll.201502048/abstract

H-W Kang, SJ Lee, IK Ko, C Kengla, JJ Yoo, and A Atala. A 3D bioprinting system to produce human-scale tissue constructs with structural integrity. *Nature Biotechnology* Published online February 15, 2016. doi: 10.1038/nbt.3413. http://www.nature.com/nbt/journal/vaop/ncurrent/full/nbt.3413.html

S Kawai, AS Foster, FF Canova, H Onodera, S Kitamura, and E Meyer. Atom manipulation on an insulating surface at room temperature. *Nature Communications* 5; 4403, 2014. doi: 10.1038/ncomms5403. http://www.nature.com/ncomms/2014/140715/ncomms5403/full/ncomms5403.html

DB Kolesky, RL Truby, AS Gladman, TA Busbee, KA Homan, and JA Lewis. 3D bioprinting of vascularized, heterogeneous cell-laden tissue constructs. *Adv Mater* 26(19); 3124-3130, 2014. doi: 10.1002/adma.201305506. http://onlinelibrary.wiley.com/doi/10.1002/adma.201305506/abstract;jsessionid=569053B40DB17B11847A93900790129D.f02t03

A Koptyug, LE Rannar, M Backstrom, and R Langlet. Bulk metallic glass manufacturing using electron beam melting. In: *Proceedings from Additive Manufacturing & 3D Printing*, Nottingham, UK, July 2013, Nottingham, UK, 2013.

T Kudernac, N Ruangsupapichat, M Parschau, B Maciá, N Katsonis, SR Harutyunyan, KH Ernst, and BL Feringa. Electrically driven directional motion of a four-wheeled molecule on a metal surface. *Nature* 479 (7372); 208, 2011. doi: 10.1038/nature10587. http://www.nature.com/nature/journal/v479/n7372/full/nature10587.html

M Liu, VI Artyukhov, H Lee, F Xu, and BI Yakobson. Carbyne From First Principles: Chain of C atoms, a Nanorod or a Nanorope. *ACS Nano* 7(11); 10075 – 10082, 2013. doi: 10.1021/nn404177r. http://pubs.acs.org/doi/abs/10.1021/nn404177r

X Liu, ZJ Weinert, M Sharafi, C Liao, J Li, and ST Schneebeli. Regulating Molecular Recognition with C-Shaped Strips Attained by Chirality-Assisted Synthesis. *Angewandte Chemie International Edition*, 54(43); 12772-12776, 2015. doi: 10.1002/anie.201506793. http://onlinelibrary.wiley.com/doi/10.1002/anie.201506793/abstract

ZP, Lu, CT Liu, JR Thompson, and WD Porter. Structural amorphous steels. *Physical Review Letters* 92; 245503, 2004. doi: 10.1103/PhysRevLett.92.245503. http://journals.aps.org/prl/abstract/10.1103/PhysRevLett.92.245503

JF Morin, Y Shiarai, and JM Tour. En route to a motorized nanocar. *Org Lett.* 8(8); 1713-6, 2006. doi: 10.1021/ol060445d. http://pubs.acs.org/doi/abs/10.1021/ol060445d

D Oberhaus. Quasicrystals are nature's impossible matter. *Motherboard* May 3, 2015. Accessed 11/04/15. http://motherboard.vice.com/read/quasicrystals-are-natures-impossible-matter

J Paek, I Cho, and J Kim. Microrobotic tentacles with spiral bending capability based on shape-engineered elastomeric microtubes. *Scientific Reports*, 5; 10768, 2015. doi: 10.1038/srep10768. http://www.nature.com/articles/srep10768

K Pearson. Voice recognition search engine connected to 3D printer by Yahoo! Japan. *MakerFlux, The Open Maker Community*, September 19, 2013. Accessed 10/14/15. http://makerflux.com/voice-recognition-search-engine-connected-to-3d-printer-by-yahoo-japan/

Z Peng, J Lin, R Ye, ELG Samuel, and JM Tour. Flexible and stackable laser induced graphene supercapacitors. *Applied Materials and Interfaces* 7(5); 3414-3419, 2015a. doi: 10.1021/am509065d. http://pubs.acs.org/doi/abs/10.1021/am509065d

Z Peng, J Lin, R Ye, JA Mann, D Zakhidov, Y Li, PR Smalley, J Lin, and JM Tour. Flexible boron-doped laser-induced graphene microsupercapacitors. *ACS Nano* 9(6); 5868-5875, 2015b. http://pubs.acs.org/doi/abs/10.1021/acsnano.5b00436

M Schroeder. Fractals, Chaos, *Power Laws: Minutes from an Infinite Paradise*. New York: WH Freeman, 1991.

BH Shin, SM Felton, MT Tolley, and RJ Wood. Self-Assembling Sensors for Printable Machines. *IEEE International Conference on Robotics and Automation (ICRA)*, Hong Kong, China, May 31 – June 7, 2014. https://micro.seas.harvard.edu/papers/ICRA14_Shin.pdf

Y Shirai, AJ Osgood, Y Zhao, KF Kelly, and JM Tour. Directional control in thermally driven single-molecule nanocars. *Nano Lett* 5(11); 2330-4, 2005. doi: 10.1021/nl051915k. http://pubs.acs.org/doi/abs/10.1021/nl051915k

JA Stroscio, F Tavazza, JA Crain, RJ Celotta, and AM Chaka. Electronically induced atom motion in engineered CoCu nanostructures. *Science* 313 (5789); 948-951, 2006. doi: 10.1126/science.1129788. http://science.sciencemag.org/content/313/5789/948

N Sugiyama, HY Xu, T Onoki, Y Hoshikawa, T Watanabe, N Matsushita, X Wang, FX Qin, M Fukuhara, M Tsukamoto, N Abe, Y Komizo, A Inoue, and M Yoshimura. Biocative titante nanomesh layer on Ti-based bulk metallic glass by hydrothermal-electrochemical technique. *Acta Biomaterialia* 5(4); 1367-1373, 2009. doi: 10.1016/j.actbio.2008.10.014. http://europepmc.org/abstract/MED/19022712

I Williams, EC Oğuz, T Speck, P Bartlett, H Löwen, and CP Royall. Transmission of torque at the nanoscale. *Nature Physics* 12; 98-103, 2016. doi: 10.1038/nphys3490. http://www.nature.com/nphys/journal/v12/n1/full/nphys3490.html

R Ye, Z Peng, T Wang, Y Xu, J Zhang, Y Li, LG Nilewski, J Lin, and JM Tour. In situ formation of metal oxide nanocrystals embedded in laser-induced graphene. *ACS Nano* 9(9); 9244-9251, 2015. doi: 10.1021/acsnano.5b04138. http://pubs.acs.org/doi/abs/10.1021/acsnano.5b04138?journalCode=ancac3

L Zhang, X Wang, W Xu, Y Zhang, Q Li, PD Bradford, and Y Zhu. Strong and Conductive Dry Carbon Nanotube Films by Microcombing. *Small*, 11(31); 3830-3836, 2015. doi: 10.1002/smll.201500111. http://onlinelibrary.wiley.com/doi/10.1002/smll.201500111/abstract

DA Zopf, SJ Hollister, ME Nelson, RG Ohye, and GE Green. Bioresorbable airway splint created with a three-dimensional printer. *New Engl J Med* 368; 2043-2045, 2013. doi: 10.1056/NEJMx1206319. http://www.nejm.org/doi/full/10.1056/NEJMc1206319

Online Material

Made In Space Press Release, August 10, 2015. http://www.madeinspace.us/made-in-space-announces-in-vacuum-additive-manufacturing-breakthrough/

NASA Press release, November, 25, 2014. https://www.nasa.gov/content/open-for-business-3-d-printer-creates-first-object-in-space-on-international-space-station

NASA Press release, August 27, 2015. http://www.nasa.gov/press/2013/august/nasa-tests-limits-of-3-d-printing-with-powerful-rocket-engine-check/#.Vl0NWWSrSmM

Science Daily summary from ESA, June, 25, 2014. http://www.sciencedaily.com/releases/2014/06/140625133309.htm

Ray Kurzweils' talk at Google I/O 2014. https://www.google.com/events/io/io14videos/4bdebcad-11da-e311-b297-00155d5066d7

Chapter 4
Deflector Shields: The Best Offense Is a Good Defense

Lasers!? Lasers can't even penetrate our navigation shields. Don't they know that?

—Captain Jean Luc Picard
TNG: The Outrageous Okona

4.1 Introduction

When informed that the shields of the *USS Defiant* were down to 25 % and that one more hit could finish them, Captain Benjamin Sisko of station Deep Space Nine once stated the first rule of fighting. He said, "Then we'll have to make sure we don't get hit." (*DS9: Shattered Mirror*) Indeed, the best offense is sometimes a good defense; you can't kill what you can't hit. Speed of light weapons like phasers are hard to dodge, so it is really a choice of firing first or relying on some defensive technology to keep your ship intact. If you feel you have to fire first, then you better have some big weaponry, and the Enterprise does. While Captain Kirk isn't completely honest when he tells the immortal Flint that the Enterprise's weapons are defensive (*TOS: Requiem for Methuselah*), it is true that their phaser banks and photon torpedoes are almost always fired as a last resort (*TNG: Silicon Avatar*). Unfortunately, an individual or species you have just met can't know that your intentions are peaceful. Perhaps the best approach is to have both a good offense and a good defense.

No matter how much a boxer might bob and weave to avoid an opponent's jabs and roundhouse punches, he is going to get hit. It is just as important to minimize the damage of any landing blow; deflect the majority of the energy away from you and you'll last a lot longer. This is especially true if you find yourself in space. A spacecraft is really nothing more than a bubble of livable environment in an ocean of instant death. The *USS Enterprise* and other starships are marvels of fictional (so far) engineering, even though they're basically just a way to carry around a bit of Earth (air, temperature, gravity) as the crew travels from one planet to another. The big drawback—one hit with a phaser or photon torpedo and there goes your confined environment.

M.E. Lasbury, *The Realization of Star Trek Technologies*,
DOI 10.1007/978-3-319-40914-6_4

Boxers use petroleum jelly to reduce friction and minimize the force of landed punches by helping them slide off to the side. Coating a starship in Vaseline isn't really an option, so even a glancing strike on an unprotected spaceship or space station is a very big deal. Can a submarine, another type of ship with a protected environment, survive a direct torpedo impact? It's unlikely if the pressure hull is breached to any great degree. The force of the water entering the hole will rip the hull to shreds. The crew might be able to close a hatch in time, but structural damage and the loss of some ship functions will probably doom them. The same is true for a spacecraft; the astronauts must often patch holes in the exterior wall caused by a micrometeoroid with great energy. Thankfully, these are very small holes. A large impact would cause a catastrophic failure of the hull and the International Space Station (ISS) protocols call for the astronauts to immediately retreat to the Soyuz capsule as an escape vessel—if there is time to get there.

Submarines have several different systems to avoid being hit by torpedoes. The hull is usually several layers thick to absorb the energy from an impact without exposing the inside environment. Their passive countermeasures include stealth by running silent or having a surface coating that poorly reflects sonar. There are also active countermeasures such as noise makers to act as a target decoy and anti-torpedo torpedoes that were first deployed on surface ships in 2013 (LaGrone 2013). This is a hard–kill interceptor designed to destroy incoming torpedoes, not deflect them.

For a starship that is unable to outrun, outmaneuver, or destroy a photon torpedo or phaser bank pulse, deflecting the energy away becomes much more important than absorbing the force and energy, and for that you need a deflector shield like those in *Star Trek*. The Enterprise's tactical defense screens are sometimes written in as major elements of the plot, although most of the time they are a device that allows the franchise to run for many episodes while adding action and tension to battle scenes. Without deflector shields, space battles would last as long as it would take one vessel to hit the other with a torpedo or energy weapon. First contact wins, but that makes for lousy television. With deflector shields, battles are intense and drawn out, vessels can retreat and live to fight another day. Most importantly, the writers don't need to introduce new ships and crews every other episode because the last ones were blasted into space dust.

However, there is also a less obvious reason that deflectors, and more primitive technologies like armor, are important for the *USS Enterprise*. They protect the ship from the dangers of space itself. Current human space exploration is reaching the point where manned voyages will be out of Earth's atmosphere long enough to be exposed to many of the perils inherent in space flight, primarily impacts and radiation. Until we are foolish enough to start conducting wars in the heavens, it will be the natural killing aspects of space against which we will need a defense. Harmful energy waves and interstellar dust and rocks need to be deflected from ships and bases, and if they do make impact their damage must be minimized. Current research is developing real life technologies to mimic the protections afforded to ships and people in *Star Trek*. While some aspects of the fictional technology are speculative at the moment, *Star Trek*-inspired deflector shields and armor to protect astronauts will be important additions to interplanetary missions and the establishment of Moon bases.

4.2 Star Trek Defensive Technologies

An invisible shield that could deflect matter and absorb energy is not a *Star Trek* invention, they have been around for decades in science fiction stories. Yet Gene Roddenberry and his writers did add a degree of scientific validity to the concept, explaining how they worked within the *Star Trek* universe and showing the viewer how they were practical for every day use, not just in battle. It is in this spirit that real world science is developing space armor and deflector shields.

4.2.1 Deflector Shield Function and Capability

Deflector shields appear from the very beginning of the *Star Trek* timeline (*ENT: Broken Bow*), yet it is the original series that gives the best early glimpses of their functions and failings. The first three appearances with Kirk's crew give the audience a good look at the different ways the shields are used. People most often think of deflector shields as a mechanism to avoid damage in battle, yet Kirk's first use of the screens is to protect a ship without power as it wanders helplessly into an asteroid field (*TOS: Mudd's Women*). Despite the strain on his own ship's power systems, the captain orders the deflectors projected out to envelop the drifting vessel.

A few episodes later, the defense field is seen as a static screen, projected as a dome over Outpost 4 on an asteroid near the Neutral Zone (*TOS: Balance of Terror*). This episode also shows the viewer that deflectors are not impenetrable, as the Romulan primary weapon takes out the shield with a single blast. Commander Hansen gives us a glimpse of how important shields are when he states that a single shot will destroy the entire base if they are without their screens. It isn't for another three episodes that we see the more traditional use of the deflector. Kirk orders the screens to be put up when the Enterprise is under attack from an unrecognized vessel (*TOS: Arena*). Here the audience learns that the screens can be activated, deactivated, and modulated, but that nothing can be transported to or from the ship when the deflectors are engaged.

4.2.1.1 Navigational Deflector Shields

Two sets of shields are emitted by two different systems on Federation starships. The navigational shields are used whenever the ship is in motion. This system is tied into the long-range sensors and can anticipate when an asteroid or some other sizable object lies in their path. Three "graviton polarity source generators" produce the force field for the navigation shields. The graviton beam is moved forward to the main deflector dish/emitter array, located on the bow of the keel directly beneath and behind the saucer (TNG Technical Manual, pg. 87). The three beams undergo phased array interference (Sect. 1.3.2.1) to alter the direction of the shield; the movement of the beam is controlled by the long-range sensors to deflect incoming asteroids.

If this were the only function of the navigational shields, it would be fine if they were activated only when an object of sufficient size to do damage was detected by the sensors. Yet they are used whenever the impulse or warp drive engines are engaged, so there must be another reason. The key phrase is "of sufficient size to do damage." Even small dust particles can do injury to the hull when both it and the ship are moving at high speeds. If a 1 cm (0.4 in.) diameter piece of space rock with a density of 3 g/cm^3 (0.036 lb/in.^{3}) hit the Enterprise traveling at impulse speeds, maybe 10 % the speed of light (still hundreds of times faster than we can travel yet), it would convey about the same force as an SUV hitting a concrete wall at 60 mph (roughly 48,000 N). All of that force striking a 0.75 cm^2 (0.12 in.^{2}) area of the hull could do some serious damage; it would certainly breach the hull of the ISS, travel through whatever it hit inside, and breach the opposite hull as well.

The ISS is struck by every few years by a speck of dust of 1–2 mm (0.06 in.) diameter, and this doesn't cause catastrophic damage because their mass is low and the speeds are only in the 10s of km/s ($\sim$24,000 mph). However, NASA does take pains to steer around any of the 500,000 objects in orbit that are 10 cm (4 in.) diameter or larger because they would have the force to breach the hull. The fictional Enterprise travels so much faster than the ISS that even a 500 μm (0.02 in.) speck of dust could breach an unarmored hull section. Having the navigational shields run full time is just being prudent.

4.2.1.2 Tactical Deflector Shields

Since the navigational screens project in the same general direction as the Enterprise's course, the defensive deflector screens are used both for battle and when a large object approaches the ship from a direction oblique to its path. Several generators are located throughout the ship so that all areas may be covered. Each generator has twelve graviton polarity sources that tie into a series of "superconducting molybdenum-jacketed waveguide conduits" in a grid formation that disperse the field across the hull. This is how the shield can form over the entire ship hull all at once. At least one generator must be operational in each section of the ship with at least one other in reserve in case of damage or failure. To save power or direct stronger shields in a particular direction, the crew can limit the field to some of the waveguides and project a screen in an arbitrary direction (TNG Technical Manual, pg. 23).

Different ships can project different geometric shapes under full shields. A contour-following field is most likely produced by using the grid emitters in the hull section waveguides, while 24th century ships have an ellipsoid screen which projects farther from the ship. Regardless, even early shield generators can be adjusted to project out into space and protect a selected volume. Regardless of the shield geometry, defensive screens can protect the ship against both matter and energy. Matter is harmlessly deflected aside, while energy beams are absorbed after making contact with the field. This is possible because the shields are not just graviton fields; the beams are fed through a pair of "subspace field distortion amplifiers" that use the graviton field to create a layer, or several layers, of energy distortion. The distortion is capable of absorbing energy and distributing it over the entire shield. It is not explained in the canon how the subspace distortion amplifiers work, but as is discussed in the next section, perhaps gravity waves alone would be enough to deflect both energy and matter.

The graviton field is not a static entity; the crew can alter the frequency of the shields to match the frequency of their weapons so that outgoing fire is not be absorbed (*Star Trek Generations*). In a similar fashion, if they can match the frequencies of the opponent's shields, the Enterprise can fire right through them to the ship itself. This is always a problem when battling the Borg, as they quickly determine and alter the frequencies of their weapons and shields. Federation ships counter by randomly altering their shield harmonics to stay ahead of the Borg (*TNG: The Best of Both Worlds, parts I and II*). Altering frequencies suggests that the deflector shields use electromagnetic radiation of some sort, but the canon only describes energy distortions generated by graviton fields. The "graviton" sounds made up, an invention of science fiction to power an imaginary technology. Twenty-first century shields will most likely have make use of EM fields, although one has to wonder, what is this graviton they keep talking about?

4.2.2 Coherent Graviton Fields

Gravity is the force that draws any two objects with mass toward one another. A ball tossed in the air will be pulled back down by the gravitational force of the Earth. Even though the ball pulls on the Earth just as hard, the ball does the majority of the moving because its mass is so much smaller, this is where the Law of Gravity meets Newton's third law of motion. We call it the "Law" of gravity because we can quantify and predict how it will manifest itself, but that doesn't mean we can explain *why* it acts the way it does. Science has not yet formulated a *theory* of gravity that holds for both the Newtonian world and the quantum world, though several possibilities that have been and are being tested. Einstein spent a good portion of the second half of his life looking for a unified theory of fields, joining gravity to electromagnetism.

It is believed that all the fundamental forces have their basic unit; light comes in photons, and their interactions with matter and space make up the electromagnetic force. The strong nuclear force, that which holds protons and neutrons together in an atom's nucleus, is mediated by the oh-so-well named gluon. The quanta of the weak force are the W and Z bosons, if the Higgs field is confirmed, then its basic unit would be the Higgs boson. Despite these discoveries, no particle or wave, or particle/wave duality has been found for the gravitational field. Nevertheless, scientists have been able to make predictions about the existence of a particle that mediates gravity. They call it the graviton.

The term "graviton" was coined in the 1930s by those first defining the field of quantum physics; it may have come from a Russian paper in 1934 (Blokhintsev and Gal'perin 1934), but no one has nailed it down for sure. Very few people were even aware of the possibility of such a hypothetical particle when Roddenberry hired Harvey P. Lynn of the Rand Company to act as scientific advisor, and it doesn't appear that Lynn set Roddenberry on the graviton track. The term never appears in the original series or in any of the films before the third season of *Next Generation*, so it was not until the 1990s that the technical consultants and writers assigned a mechanism of action to the deflector shields or the tractor beam (Chap. 5), the technologies that work via the force of gravity.

Gravity itself is pretty weak, it decreases as the distance between two bodies increases, proportional to $1/d^2$. Consider the fact that a baby can throw her spoon up in the air, in direct opposition to the gravitational force mediated by an entire planet. Is it a super baby or a weak gravitational force? Nevertheless, despite its relative weakness, the gravitational force between any two objects is basically infinite; it decreases with distance, but it never goes away completely. The infinite reach of gravity means that the graviton, if it exists, must be a massless particle like the photon. They are without mass, yet the graviton and photon still exert force. The momentum of the photon pushes electrons around to create the electromagnetic force and the equivalent gravitational force is thought to reside in the graviton. Any one graviton would have a very small force and will therefore be hard to detect; however, what it lacks in might it makes up for in numbers.

According to some of the current theories of gravity, the interaction of matter and energy causes a graviton to be emitted. This graviton starts a chain reaction in which one graviton emits another graviton, confining a large amount of energy and momentum in a very small volume. This confinement causes a new graviton to be emitted, which again gives two gravitons confined in a tiny space. On and on it goes. Something as large as the Earth is a constant source of huge numbers of gravitons.

This idea of gravity mediated by the exchange of gravitons between matter goes against what most people have learned about the nature of gravity. Einstein's theory of general relativity says that gravity is the result of large masses warping space-time. But leading theorists now believe that both theories could be valid, that graviton exists as a field based on curved space-time *and* as a particle, just as light exits as a wave and a photon particle. Those large funnels at the mall that fling your coin around in smaller and smaller circles until it drops into the "black hole" are a good representation of space-time warped by a large mass. Warped space-time gives NASA's vessels a gravity assist to sling them around the Sun or a planet to gain speed. Despite this visual model, Einstein's general relativity theory does allow for the existence of gravitational waves; with gravitons representing gravitational waves of specific wavelengths. Gravitons would represent the small packets of the space-time gravitational field just as the photons represent small packets of the electromagnetic field. Finding the graviton would help join the Newtonian world with the quantum world.

Box 4.1 The Graviton is Weird

Subatomic Particles—The CERN Large Hadron ColliderLarge Hadron Collider is one of the best places to look for subatomic particles. It smashes atoms together at incredible energies and tries to track what they break into. A recent refit of the LHC Traces of the particles can be followed and their characteristics used to identify them. If something with mass went off in one direction, then Newton's third law says something must have gone off in the opposite direction. This how the gluon and the W and Z bosons were discovered, and it is how the Higgs boson was found as well, if it was the Higgs Boson.

Into another dimension—The Graviton is smaller than a gluon or boson, and believe it or not, it is probably able to move into other dimensions. String theory and classical theory come to a head when considering the graviton. Classical general relativity says that gravity is a curvature of space-time by mass. String theory says that gravitons are points smaller than the vibrating energy strings that make up matter and energy. But the two theories agree that gravitational forces have a wave particle duality. If it is a wave, then the graviton has a wavelength (distance from crest to crest).

More than One Graviton—If the graviton is to fit into the other dimensions, it must complete some whole number of wavelengths as it travels around the cylinder and back to its original position; one wavelength works, so would two or three, any whole number is possible. This means there may be more than one type of graviton, one type is the one wavelength version, another is the two-wavelength type, and so on. The theory predicts that the higher number gravitons might even have mass in those other dimensions, just not in ours. As a consequence, scientists speculate that gravity may not really be such a weak force, the graviton is just spreading its force through dimensions that humans cannot yet detect.

4.2.3 The Search for Gravitational Waves

One of the brain-bending things about the graviton is that scientists believe it can jump back and forth between dimensions (see Box 4.1). If this is true, then how do you look for a graviton in the remnants of particle collisions, as has been done for other particles using the Large Hadron Collider at CERN? Can you confirm a particle by its absence? Even if researchers could develop methods to detect a graviton, the physicist Freeman Dyson recently estimated that to see a graviton directly it would take a detector larger than Jupiter and the search would take decades with the signal still likely drowned out by background noise (Dyson 2012).

Despite this issue, some astrophysicists were convinced they found the graviton in 2014. The BICEP2 telescope, located at the South Pole, made a series of observations in March of 2014 that perhaps detected gravitational waves left over from the Big Bang. The mission of BICEP2 (now BICEP3) is to detect changes in the cosmic microwave background (CMB) radiation that will give indications of inflation in the very early universe. Detection of a certain polarization type in the CMB would suggest that gravitational waves/particles exist.

The news of the discovery of gravitational waves, if confirmed, took the science world by storm. The importance of the phrase, *if confirmed*, became evident later that year when analysis of the data by teams from both the ESA and BICEP indicated that some, if not all, of the BICEP2 signal came from dust within our own galaxy (the dust would polarize the CMB in nearly the same way). But it wasn't a complete loss, the data helped them establish an upper limit for the contribution of the gravitational wave as a result of inflation and they concluded that *is* possible to detect gravitons with this technique. So—they keep looking.

The possibility of finding the graviton with one method doesn't keep people from trying other methods, including an effort called LIGO. The original **L**aser **I**nterferometer **G**ravitational-Wave **O**bservatory began looking in 2002 for gravitational wave vibrations in the CMB caused by cataclysmic events such as supernovae, neutron stars or even the Big Bang. The waves are strong and large at the location of the event, becoming smaller and less distinct as they travel across the universe. Accordingly, the original LIGO efforts were unsuccessful. The interferometers for the *Advanced* LIGO project have just finished a five-year redesign to increase their sensitivity and started collecting data on September 15, 2015 (Fig. 4.1). They can pick up CMB ripples 225 million light years that are as small as a billionth of the diameter of an atom (LIGO Scientific Collaboration 2015).

Over 900 scientists at eight institutions worldwide are part of the Advanced LIGO project. They have agreed to drop their individual work at a moment's notice and provide confirming observations for Advanced LIGO. All 900 should stay close to their computers, as LIGO estimates that only one or two relatively strong gravitational wave pulses will pass by the Earth every twelve months (LIGO Scientific Collaboration 2015). If detected by Advanced LIGO, the gravitational wave observation would confirm one of the basic tenets of Einstein's general relativity theory. Amazingly, that seems to be just what happened a mere four and a half months and after Advanced LIGO started making observations. On February 11, 2016, the LIGO team announced that a gravitational wave deformation of the interferometer was detected by the instruments in Louisiana and Washington (LIGO Press Release 2016). They measured the residual effect of two black holes crashing into one another or some other such cosmic event. It should be

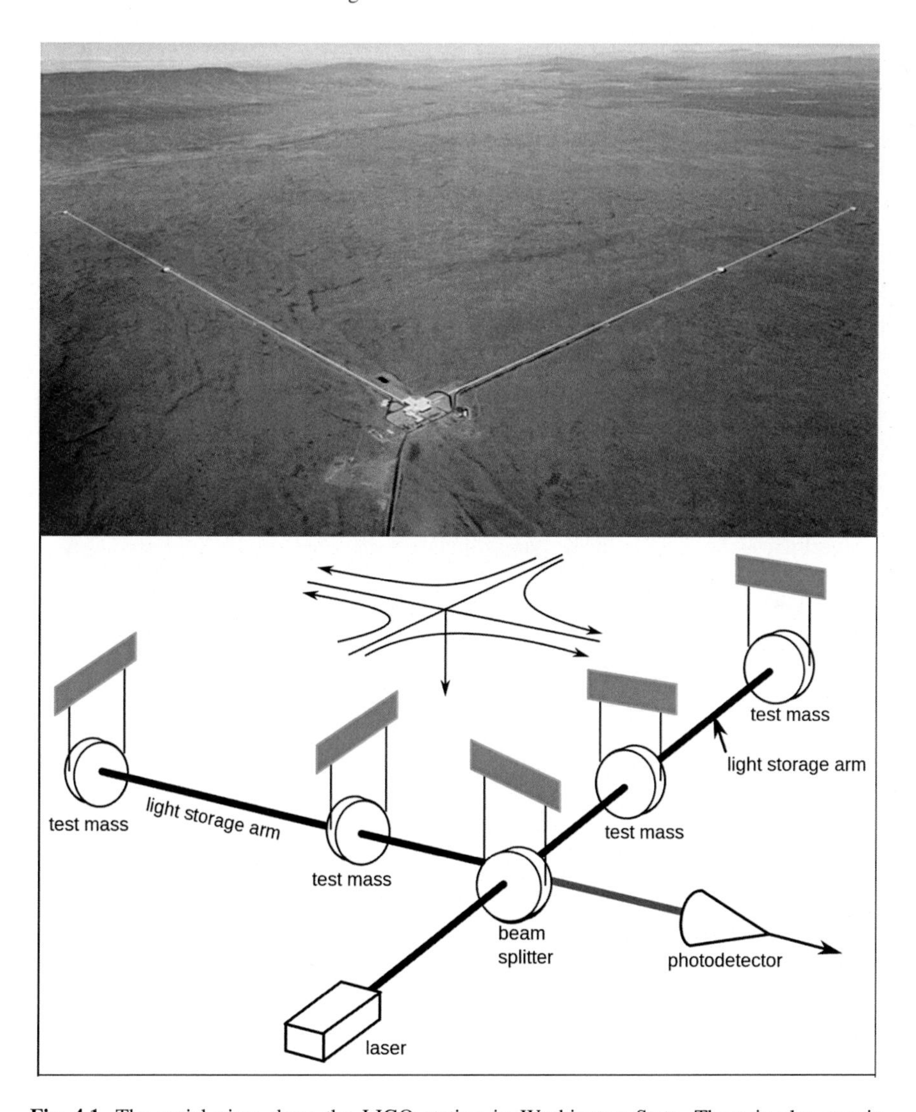

Fig. 4.1 The aerial view show the LIGO station in Washington State. There is also one in Louisiana. The lower cartoon shows the set of the interferometer. It measure the time it takes a split beam of light to travel down and back each arm. Since a gravitational wave will stretch space in one direction and compress it in the other, if a gravitational passes, the time light takes to travel each arm will be slightly different. They look for a change in the interference patterns of the waves as they rejoin each other on the return trip. The total distance stretched is less than the width of an atom. (*Image Credit* Courtesy CalTech/MIT/LIGO Laboratory)

noted that while this confirmed the existence of gravitational waves, it says nothing about whether those waves express a wave/particle duality—the search for the graviton particle will continue.

4.2.4 The Graviton and Deflector Shield Plausibility

If gravitons exist, and *if* they could be generated on command, and *if* they could be harnessed to form an arbitrary field, the question would still remain as to whether they could be used as a deflector shield. *Star Trek* says that the shields redirect matter away from the ship via graviton fields and energy distortion. The energy distortion might have an effect on charged particles and radiation, but asteroids are probably not going to be swept aside by electromagnetic waves even if they are being used in some 21st century tractor beams (Sects. 5.5.3 and 5.5.4). On the other hand, the gravity field itself might be able to save a ship. As Lawrence Krauss stated in his 1996 book, *The Physics of Star Trek*, a coherent gravitational field would indeed warp space-time around the star ship (pg. 72). Krauss concedes that the canon is at least "talking the right language" when describing a mechanism for a deflector shield that would bend an objects path and cause it to miss the ship. He is less enthusiastic about whether gravitons could be generated and used in such a way, nevertheless the point is made—an artificially generated gravity field used as a deflector is not an absurd notion.

Energy deflection is another matter. In *Star Trek*, color spreads over the shield when a phaser hits it because energy is absorbed not deflected by the screen (*DS9: Civil Defense*), even though Einstein showed that light energy is deflected by gravity, not absorbed. His theory of general relativity predicted that light traveling past our Sun would be bent by its gravitational field. It took five years, astronomers being arrested as WWI spies, and expeditions to Brazil and Africa for solar eclipses to get the data, but his prediction was proven correct—light is bent by gravity. With even stronger gravitation fields, light can be deflected even more. This is why black holes are black; light is bent into the singularity. So does this present a problem for *Star Trek*? If one really wants to believe in the deflector's mechanism of action assigned by the technical manuals, then this is where the subspace energy field distortion might play a role. A distorted energy field might be unequal across its area and could then absorb and dissipate energy. Maybe.

4.3 Current Military Energy Deflector Shields

Deflector shield capabilities are dependent on adequate power supply, few direct hits to the screens, and no technical faults in the graviton generators or emitters. Several episodes across the different series indicate that quite often one or more of these problems arise. This vulnerability of the shields to power drainage or damage requires that starships also have at least some armoring on the hull and crucial interior compartments. The *USS Enterprise* has traditional armor surrounding the matter/antimatter container (TNG Technical Manual, pg. 59) and the impulse engine reactor chamber (TNG Technical Manual, pg. 76). Traditional armor is based on wrapping yourself in something stronger than whatever is coming at you.

The harder and/or thicker the armor is, the better protection it will provide, at the expense of added weight and fuel consumption.

Several ships from *Star Trek* have ablative armor (*Star Trek: The Motion Picture*; *VOY: Endgame*) meant to be chipped away or boiled off as they protect the primary hull, and Klingon Birds of Prey (*ENT: The Augments*) have dispersive armor to spread out the force of a weapon to prevent penetration of the pressurized hull. Real world dispersive and ablative armor have been around for decades; in fact, they started out as mechanisms to protect space-going vehicles. Heat shields are a form of ablative armor first used to keep intercontinental ballistic missiles and returning Apollo spacecraft from burning upon re-entry. Dispersive armor, like the multilayered Whipple shield, is used on manned and unmanned craft to dissipate the energy of micrometeoroids that strike the ISS or other protected targets.

Advanced armoring systems are a matter of much research by the military, including explosive reactive armor that senses an incoming missile and sets off a charge to counter the energy of the incoming round. There are also non-explosive reactive armor systems that use other forms of energy to deflect, absorb, or repel missiles or artillery rounds. More to our purposes here, several armors under development or already in the field use electromagnetic radiation and energy to overcome projectiles. Hmmm… energy to defeat incoming fire, that sounds very much like a *Star Trek* tactical deflector shield.

4.3.1 Electromagnetic Shielding

Active protection systems (APS) use radar and optical sensors to track incoming fire and strike it down before it contacts the hull. Several systems, like the Israeli Trophy APS and the American Iron Curtain, use a screen or ring of small projectiles to destroy incoming missiles. Both systems use radar to identify incoming targets and optical systems to identify the type of round that has been fired at them. Even for missiles fired from very short distances, both systems can react, identify, and target the incoming round in time to destroy it.

The Trophy system of the Israeli Defense Force is the only version that has been used successfully in battle (Goure 2015), though the American Iron Curtain system has passed several tests and is looking to be deployed (Shachtman 2011). The systems are similar, yet there is one dramatic difference. The Trophy APS is mounted on the roof edges of the vehicle and faces out. It fires its projectiles away from the protected vehicle. The Iron Curtain, on the other hand, is mounted as a shelf sticking out from the roof of the vehicle. While the sensors point away to identify the incoming weapon, the projectiles are fired straight down, killing the incoming round just inches before it contacts the hull. A German made system, the AMAP-ADS, has a more *Star Trek*-like feature. It employs a directed energy beam to destroy the incoming weapon. Rheinmetall, the company that produces the AMAP-ADS has shown that it can respond in just 1/20th of a second and is effective against a greater range of ordinates than either the Trophy or Iron Curtain systems (Dodson 2015).

Yet another emerging reactive armor system uses electricity instead of an energy beam. First described by the British in 2010, this electromagnetic supercapacitor system uses two plates of armor separated by a short gap. The plates are charged with thousands of volts from an onboard battery, and an electrical circuit is completed when an incoming weapon deforms the outer plate causing it to touch the inner plate. A huge electrical discharge within a few thousandths of a second incapacitates or even vaporizes the incoming missile before it can penetrate the primary hull (The Economist 2011).

4.3.2 Military Plasma Shielding

The above reactive armor systems are impressive, just not as impressive as two plasma shields currently being developed that might just be stepping stones to true deflector shields. When energy is put into a gaseous system, electrons are ripped from their parent nuclei, creating a sea of positively charged nuclei and free floating electrons—a plasma. Plasma is a common topic on *Star Trek*; the energy extracted from the warp core is stored as plasma, conducted throughout the ship by a grid of plasma conduits, and reconverted to electricity when needed (*VOY: Good Shepherd*). This idea is firmly based in reality via devices called plasma converters (Sect. 8.4.2). Neon signs are great examples of plasma as an electrical conduit. The glass tube is filled with neon or similar gas and there is an electrode on each end. When electricity is supplied, the gas is ionized and completes the circuit. A small bit of the energy is released as light of a certain wavelength (color). This is confined plasma, but free plasma can be used in other ways, perhaps even as a shield.

Box 4.2 Plasma

Discovery—William Crookes described the luminous gas in Crooke's tube as radiant matter in 1879. American chemist Irving Langmuir coined the term "plasma" in 1927 while working on a way to make tungsten light bulb filaments last longer. The ionosphere of the Earth was identified as a plasma after studying the way it messed with radio signals in the 1920s.

Abundance—Plasma is known as the fourth state of matter (solid, liquid, gas). It is a gas that has had all or some electrons stripped from the atom. Plasma consists of free electrons and positively charged atomic nuclei. It is the most abundant form of matter, making up 99 % of the visible universe. Stars, flame, lightning are natural examples of plasmas, while neon bulbs, and some arc welders use artificially generated plasma.

Generation—Stripping electrons from atoms to create and maintain artificial plasma takes a lot of energy. Energy must continually be added to the system since cooling of the atoms will lead to re-association with the electrons. If the electrons are in thermal equilibrium with the protons and other nuclei, they are considered thermal plasmas. Nonthermal plasma can still be at very high temperatures, they are just not in thermal equilibrium (more energy in the electrons than in the ions and nuclei). Whether a generated plasma is thermal or non thermal depends on the amount of energy induced, the gas pressure and the nature of the atoms in the gas.

Hot and Cold Plasma—In general, the more energy plasma contains (the hotter it is) the greater the number of electrons will be stripped away and the plasma will become more ionized. Highly ionized plasmas are considered "hot," while plasmas, some with as little as 0.1 % ionization, even though still several thousand degrees Kelvin, are considered "cold." Ionized gases must include an appropriate density of particles, freely oscillating electrons, and a distance between particles that can ensure interactions in order to be considered plasma. Both cold and hot plasmas have properties in common even though they can be made of different percentages of electrons, have different densities, and include different nuclei.

Behavior—Charge separations in plasma permit it to respond as a single entity when there are changes to its environment. Plasma is a good conductor and has its own electrical field. Electrical fields generate magnetic fields, and these two fields tie the plasma together as a single unit. A plasma globe will send discharges to your fingers because the entire globe acts as one. Your hand drains the build up of electrons on the outside of the glass, so the inside is out of balance electrically. The heat from your hand makes the gas inside less dense in that area, which provides a discharge path of least resistance. The discharge then travels from source to your hand.

The United States Army Armament Research, Development and Engineering Center has developed a system called **PASS (plasma acoustic shield system)**. Originally designed in 2007 to be a deterrent against attack by creating a disorienting flash bang, the technology has come far in the past couple of years. PASS uses two high power lasers; the first creates an intense energy pulse that strips the air molecules of their electrons, creating a plasma cloud. Plasma generation (very hot at the point of plasma, and dissipating rapidly as you move away) creates a loud bang at the point where the gas is ionized. A second laser pulse then hits the plasma cloud just milliseconds later with additional energy. The plasma absorbs the energy, expands rapidly and creates a shock wave and a bigger bang. Many hundreds of these pulses can be delivered in short order in a two dimensional array, producing a wall of plasma and sound that can serve either as a warning to adversaries, similar to the laser dazzlers discussed in Sect. 1.5.1, or as a wall shield for soldiers. They couldn't be approached through the wall, and an intense plasma wall would be hot and also block both visible and infrared light. Eventually, PASS will be able to fire repeatedly in arbitrary patterns, creating a wall or other shapes. Depending on the energy levels of the lasers, the shapes could appear at various distances from the source (Hambling 2013).

Current PASS equipment can deliver 10 bangs per second, with 40/s an attainable goal in the next couple of years. Unfortunately, this level of energy violates the Protocol on Blinding Laser Weapons (Sect. 1.5.1), so modifications will be needed to increase the power of the system. If the problem can be overcome, PASS could go from purely disorienting to stunning or even lethal. Or perhaps it could disrupt incoming fire. This was the aim of the US Navy Plasma Point Defense System that was abandoned in the 2000s. However, advances in plasma generation, including PASS, have made the possibility of deflecting energy or projectile weapons an appealing subject again. Important to remember is that projectiles are usually uncharged; a charged energy field won't do much to slow them down *unless* it is strong enough to destroy them outright.

4.3.3 *Shockwave Shielding*

A second plasma shield under development is not meant to dissuade people or destroy missiles, it deflects the shock wave created by nearby explosions. The initial blast wave of pressurized air is dissipated by traditional armor; however, the subsequent high-energy shock waves (supersonic and carrying more energy than sound waves) are not. They will travel through the body and transfer their energy to organs and tissues, disrupting their structure and function. The lungs and GI tract are particularly vulnerable, as is the brain. The reflected waves are catastrophic in the case of a missile that penetrates the armor, but damage can occur even if the barrier remains intact. Medicine is recognizing the subtle effects of shock waves on the brain; blast exposure without overt injury can still produce mild to moderate concussion. Repeated concussive damage is now known to mediate chronic traumatic cases of Parkinson's disease and Alzheimer's disease (Gavett 2010) as well as the chronic traumatic encephalitis now recognized in so many NFL football players. Accordingly, the US military instituted a policy in 2010 to remove from active duty any soldier receiving blast exposure until all symptoms of concussion are gone.

A plasma shield to dampen this shock waves is described in a 2014 patent from the Boeing Corporation and inventor Brian Tillotson (US Patent #8806945) wherein a sensor and countermeasure system will deflect or absorb an incoming blast shock wave. Optical sensors will see the visible light of an explosion, and then calculate a direction, range, and time to impact for the shock wave. At the appropriate place and time, a blast from the vehicle will produce an expanding change in air to counteract the shock wave. The change could be from the introduction of a gas cloud or more interestingly, superheating the air itself. An arc generator blast can create a pressure wave of superheated air that will expand in front of the incoming shock wave. It could absorb some of the energy, exchange momentum with the shock wave to dampen it, or it could deflect part of the wave. Hot air has lower density and acts as a diverging lens. When the shock wave encounters the superheated air pressure wave, it will be deflected out and around the protected vehicle, similar to what happens to light traveling through a concave lens (Emspak 2015).

A laser is yet another way to produce an expanding cloud. A powerful laser will superheat the air and strip electrons from the gases, producing a plasma cloud. Into this cloud an electrical pulse is released, heating more air and producing an electromagnetic field. The interference between the incoming shock wave and the developing EM field will absorb and deflect energy, while at the same time the superheated plasma will deflect the wave away from the protected vehicle (Whitwam 2015). This technique would represent a defensive use of the laser-induced plasma channel (LIPC) that was discussed in relation to directed energy weapons (Sect. 1.7.2).

4.4 Shields as Protection from Space

The tactical deflector shields get all the credit and air time on *Star Trek*, but the navigational shields are the workhorses in everyday situations. When Janeway loses the navigational shields, a micrometeoroid shower does extensive damage to the *USS Voyager* (*VOY: Year of Hell, part II*); in fact, the navigational shield was referred to as the "meteorite beam" in the original series (*TOS: The Menagerie; The Cage*). Even if we forgive them for referring to meteors as meteorites, this function of the shields is given far more attention than perhaps the most important job of the deflectors—protecting the crew from cosmic radiation. There are a few episodes where the tactical shields are used to shelter the crew from intense radiation fields (*TNG: Booby Trap; VOY: Inside Man*), yet they don't ever discuss the background cosmic radiation that is always passing through the ship. Kirk orders the Enterprise to a safe distance when a solar flare threatens the crew (*TOS: The Empath*), so apparently Federation starships don't have cosmic ray shielding built into the hull. The problem is that cosmic rays are a constant in space, not just an occasional inconvenience. Science is currently struggling with finding ways to protect astronauts on the long missions that are planned to begin in the next decades; maybe a deflector shield could help. With the Earth as a model, researchers are working on exactly that.

4.4.1 Cosmic Rays

Unlike navigational shields, cosmic rays shields have to be functional all the time, not just when the ship is under power. It is true that just as the navigation shields point forward because the leading edge of the ship receives most of the damage from micrometeoroids, the leading face of deep space craft will receive a larger portion of the radiation. However, some cosmic radiation will strike the ship from every direction. Without at least some protection on all sides, cosmic rays can and will harm and eventually kill a human in space.

Cosmic radiation can refer to EM waves of differing energies or to subatomic particulate (electrons, protons, neutrons, nuclei) radiation. Our star gives off both types; the EM radiation includes visible light and the UV rays that tan your skin, while the particulate radiation helps to create the Northern and Southern lights. The small particles from the Sun are grouped under the term *solar energetic particles* (SEPs). Our star produces SEPs all the time, with solar flares greatly increasing the amount thrown off for a short period. SEPs consist of many different energetic nuclei and the composition of SEPs will vary from one flare to another. Luckily, most of the higher energy SEPs are deflected around the Earth (see below). The EM radiation is not deflected, which is a good thing unless you are allergic to sunshine. Even though other stars emit EM and SEPs just like our Sun, little of this gets to us because the energy is relatively low and only a small percentage of their radiation is pointed directly at us.

Some low energy particulate radiation does reach the surface of the Earth (neutrinos and muons); however, these are of such low mass and energy that they pass right through our bodies and the Earth itself without doing any damage. On the other hand, very high-energy galactic cosmic rays (GCRs) are generated by cataclysmic cosmological events in our Galaxy. The magnetic fields of the Milky Way scatter GCRs so completely that they come at us from all directions, making their sources hard to pinpoint. GCRs are somewhat of a mystery because they travel so fast that not even supernovae can account for their high energy. Every element in the periodic table is represented in the GCRs, but 99 % of the nuclei are hydrogen (protons) or helium. GCRs can strike random atoms in space or in planetary atmospheres to produce secondary high-energy particles and gamma rays. When scientists speak of cosmic radiation that can do damage to people and electronics, it's the SEPs, GCRs, and their secondary particles to which they are referring. Most gamma rays are absorbed by the atmospheric gases and gamma rays are fewer in number. They present much less of a danger for astronauts than do high energy nuclei of GCRs and SEPs.

Box 4.3 How Cosmic Radiation Damages Astronauts

Types of Rays—Galactic cosmic rays (GCR) contain high energy (E) and high charge (Z) particles and are collectively known as high-energy particles (HZEs). There is also solar radiation of the UV and other types. The HZEs of GCRs present the primary danger; the high energy means that they will penetrate deeper into the body before losing their energy and they can bounce off many cellular components, doing damage all along the way. The damage is cumulative over time.

Studying GCRs—HZE of GCRs present different problems than radiation on Earth. To learn more about them, NASA started the Mars Radiation

Environment Experiment (MARIE) in 2001. An instrument aboard the Odyssey spacecraft collected information on the radiation particles that struck its sensor as it orbited Mars. Unfortunately, data collection stopped in 2003 after a computer malfunction probably caused by cosmic ray-induced damage to a microchip!

Effects on DNA—The direct effects of HZE include mutation of DNA bases and single strand or double strand breaks in DNA. When the base sequence changes, the protein it codes for will be altered. Most mutations have no effect, but some will result in proteins that function poorly, not at all, or aren't even made. Even worse, some mutations make entire biochemical pathways dysfunctional by altering their control elements. Broken strands lead to altered DNA replication and more mutation.

Cellular Damage—HZE can transfer energy directly to cells with which they interact, and often induces enough molecular changes to kill the cell (Rabin et al. 2012). It is estimated that 5 % of an astronaut's cells would die during a prolonged space flight.

Cataract Formation—A Mars mission will lead to accelerated cataract formation on the lens of the eye. NASA studies confirm that smaller doses of radiation than previously believed lead to cataract formation (Chylack et al. 2009) and this will only be magnified by the length of a Mars mission.

Central Nervous System Damage—Neurons that die from cellular damage are usually not replaced. With increased neuronal loss, physical and cognitive functions will degrade (Cucinotta et al. 2014). Mouse models of radiation damage indicate the few neurons that are replaced (neurogenesis) seem to be altered and do not function maximally (Acharya et al. 2011). Finally, exposure to levels of GCR lower than will be encountered on a Mars mission caused Alzheimer plaque formation within six months in a mouse model (Cherry et al. 2012).

4.4.2 The Magnetosphere and the Ionosphere

As outlined in Box 4.3, GCRs and gamma rays do nasty things to the human body, from DNA mutation and strand breaks to blinding cataracts and neuropathologies. So why aren't we all dead? The plasma of cosmic radiation particles (not the gamma rays) is composed of charged nuclei, and ions can be deflected by a magnetic field. What generates the largest magnetic field close to us—our planet. The core of the Earth is composed of two parts. The solid inner core is about 2/3 the size of the moon, while the outer core is about 2300 km (1400 mi.) thick layer surrounding the inner core. They are both made mostly of iron-nickel alloy, but the outer core is at lower pressure that allows it to remain molten. The outer core swirls around the inner

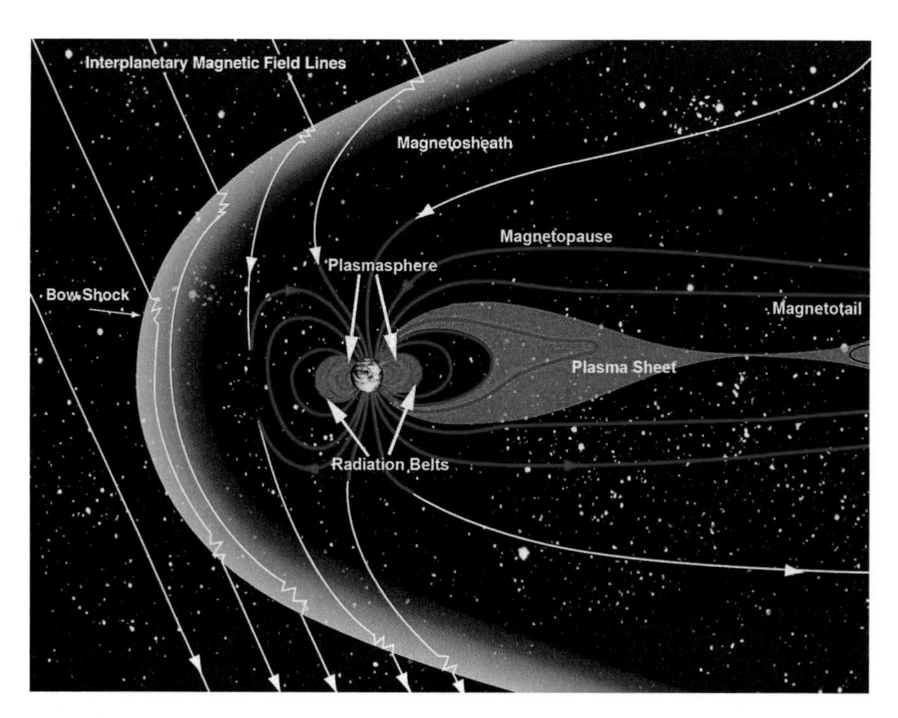

Fig. 4.2 The magnetic field of the Earth (*red lines*) forms a shield around the Earth. Radiation from space interacts with the bow shock and is shifted around the planet. The magnetosphere stretches out 1/6th the way to the moon. The ISS is well within the protected area. (*Image Credit* NASA)

core (both actually spin, with the outer core moving much faster and in complex currents) and this creates a fluxing electrical current and field. It is this electrical field that then produces the large magnetic field that projects out into space. Radiation from the Sun (the solar wind) compresses the magnetic field on the dayside of the Earth, while it stretches out 10× further on the night side, so it looks like a long teardrop on its side (Fig. 4.2).

It is this magnetic field that interacts with the GCRs and SEPs and deflects them around the planet. The particles form a turbulent layer of plasma and radiation called the magnetosheath as they are deflected around the Earth. The gamma rays are neutrally charged and pass through the magnetosphere. Luckily, they are few in number and represent only a 1/10th of the average yearly radiation dose for each human. It is the GCRs and SEPs that astronauts that leave the magnetosphere must fear. Only the Apollo lunar voyages have carried people beyond the magnetosphere, and they were outside its protective layer for only a matter of days. The ISS orbits 400 km (249 mi.) above the Earth, while the magnetic field reaches out to 90,100 km (56,000 mi.). During Scott Kelly's and Mikhail Kornienko's year on the space station (March, 2015–March, 2016) they received only a small fraction of radiation that astronauts on a Mars mission would encounter.

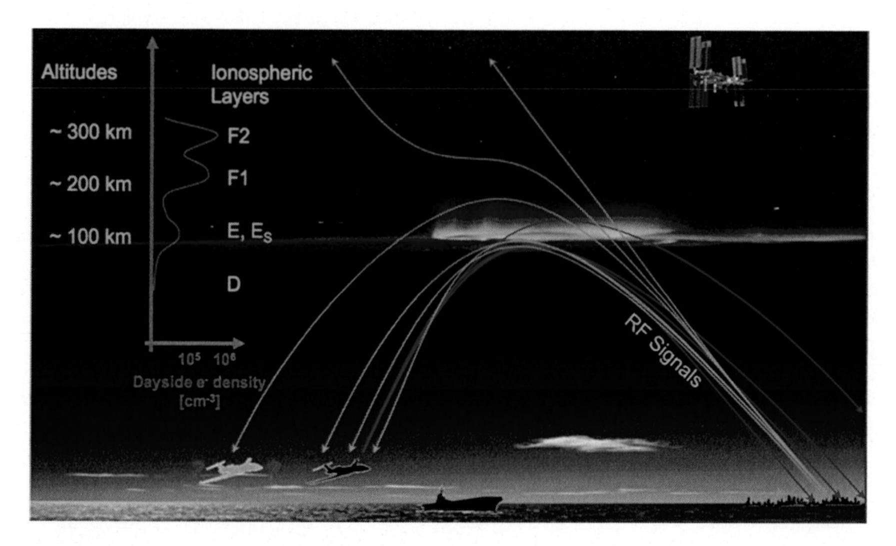

Fig. 4.3 When radiation gets through the magnetosphere and interacts with the upper layers of the atmosphere, it produces ions—the ionosphere. This layer is protective because it contains charged particles that can divert other charged particles around the Earth. The outermost ions form a cold plasma—the plasmasphere, which scientists are trying to replicate in order to protect astronauts from radiation. (*Image Credit* Navy Research Laboratory)

The magnetosphere is not the only part of the atmosphere that is capable of deflecting cosmic radiation. Some of the SEPs and GCRs traveling toward Earth penetrate the magnetosphere or are weakly deflected. When particles get through the magnetosphere and reach the upper layers of the atmosphere (75–1000 km, 46–620 mi.), they can interact with atoms of gas (nitrogen, oxygen, hydrogen). The energy of the radiation is transferred to the gas, stripping electrons away and creating positive ions (Fig. 4.3). The layer of the atmosphere where these ions reside is called the ionosphere. The dayside of the ionosphere is more charged than the night time ionosphere because the day side receives both SEPs and GCRs, but the night side only sees GCRs. AM radio signals can be picked up farther away on Earth because they are deflected off the ionosphere and come back down. The night side ionosphere is thinner and is at a higher altitude so AM radio waves are reflected further away from their source at night, so HAM radio operators like to broadcast at night. Shorter FM waves are used for space communications because they higher energy and can travel through the ionosphere out into space.

4.4.3 *Plasma Deflector Shields*

The ionosphere is plasma; in fact, the outermost portion of the ionosphere (innermost region of magnetosphere) where the plasma is the coldest and most dense is called the plasmasphere. Some physics students at the University of Leicester seem to think that the plasmasphere might make a good model for deflecting energy beams away from a starship or other asset. The physics department publishes a student magazine called *Journal of Physics Special Topics*, which includes recent articles as diverse as "Can Airheads Blow Minds," "Pyramid of Geezers," and "The Tide is High But I Am Still Short." A 2013 article proposed model where a plasma shield could deflect many forms of EM radiation, including laser beams. This would represent a big step up from the ionosphere, it can't stop lasers because the plasma is not dense enough. In fact, bouncing a laser off a mirror left on the moon's surface is how scientists measure the changing distance between the Earth and Moon—and yes, the moon is moving away from us at a rate of 4 cm/year (1.5 in.).

According to the fourth year student authors, the denser the plasma cloud is made, the higher the frequency of energy that could be deflected. A sufficiently dense cloud could turn away any energy beam we can produce at this time. The plasma would be held in place around the protected target using strong electromagnets, because plasma is made of charged particles and charged particles move along magnetic field lines, not across them. However, this presents a practical problem for the construction of a plasma deflector shield. In order to hold a dense plasma cloud in place, the magnets would be so large and heavy as to make them unfeasible for space travel; all the available power would be needed just to power the magnets (McGuire et al. 2013). A second issue to overcome is that any plasma cloud dense enough to turn away strong lasers or X-ray beams would also deflect incoherent visible light. The ship would be flying blind. This is the same problem that plagues several of the real world cloaking devices currently being developed (Sect. 2.4.2) and it points out that deflecting and cloaking are similar processes. This somehow not a problem for *Star Trek*, Kirk entered into his log that, "We are now in extended orbit around the Earth, using our ship's deflector shields to remain unobserved;" (*TOS: Assignment Earth*) they could see out yet no one could see in. This must be a property of graviton shields.

Just as the physics students proposed the ionosphere to model a plasma shield, scientists are now interested in looking at whether an artificial magnetosphere might also deflect energy and matter. In fact, the best screen may be a combination of magnetosphere and ionosphere. This might protect astronauts from cosmic radiation on a trip to and from the Mars or make a Moon landing base habitable over a long period. But before discussing these potential deflector shield designs, there are some more conventional methods to protect humans from radiation that are being considered.

4.5 Shields Against Cosmic Rays

A spacecraft hull made from solid aluminum a foot thick or two feet of concrete would stop GCRs pretty well—as long as the space program has a limitless budget. Lighter radiation shields are preferred for SWaP reasons (size, weight, and power consumption, see Sect. 1.6.2), like the boron and lithium impregnated polyethylene that lines the ISS sleeping quarters. Boron and lithium are good radiation absorbers because their large atomic cross sections can absorb many rays, and polyethylene contains hydrogen, which is another good absorber. Hydrogen gas (H_2) and water (H_2O) would be decent choices for shielding between hull layers, except that ice is less dense than water (space is cold), water is heavy, and hydrogen gas is explosive. Unbelievably, the best solution may be human waste. Billionaire engineer Dennis Tito's Inspiration Mars mission plans to use the astronauts' excrement as radiation shielding by packing it between the walls of the spacecraft. Human feces is basically water and hydrocarbons, it stops radiation very well and the shield would get more robust every day. However, it's a bit unsettling to think that a Mars mission could be jeopardized by a bout of constipation. Perhaps it is a good thing that science is investigating other possibilities.

4.5.1 Superconducting Magnetic Fields as Shields

Since the graviton particle remains elusive, the most practical ideas for the production of defense shields involve electromagnetic fields and plasma, as modeled by the magnetosphere and ionosphere. A viable example uses superconductors (SCs). Electricity passed through the all conductors will produce both a electrical field and a magnetic field. Since the cosmic radiation is really a plasma beam (free electrons and positively charged nuclei), the electric field of the superconductor works to repel the free electrons, while the perpendicular magnetic field shunts the positively charged nuclei off to the side and around the field. An SC produces a strong magnetic field because there is zero resistance to the current when in its superconductive state. The current can be raised to high levels without heating up and the magnetic field produced will be stronger than with an equivalent copper wire coil.

SCs have been used to create the strongest manmade magnetic fields, although there are conditions attached. Our current mastery of zero resistance materials requires temperatures below −254 °C (−425 °F), despite the fact that Wesley Crusher's handheld superconducting magnet operated just fine under room temperature conditions (*TNG: The Dauphin*). To use confining magnetic fields of this strength on Earth, the major issue is to increase the working temperature to at least *relative* warmth. This issue is being considered and new materials are being produced all the time. Luckily, a space-based deflector shield based on superconducting magnets will operate just fine in the 2.7 K of deep space, although near a star this can get tricky. Heat can be transferred by radiation, so a ship will be much hotter for as long as sunlight strikes the ship or if it radiates through the hull from the inside. Nearer Mars, this will be less of a problem, since it only receives less than half the sunlight that reaches Earth's upper atmosphere. When Mars is farthest from the Sun, it only gets about 36 % as much sunlight as Earth.

The idea of using a magnetic field to protect astronauts from GCRs and SEPs is not new. It's been a topic of research for more than 40 years, although Shayne Westover of the Johnson Space Center was quoted as saying that, "it remains an intractable engineering problem." The advances in superconducting materials have spurred NASA to investigate using this method to generate a magnetic field via its Innovative Advanced Concepts Program. One of NASA's designs is a cylindrical spacecraft surrounded by several coils of superconducting wire backed by a Kevlar coil, each generating a magnetic field. The encircled ship would be protected from all sides except for small areas of the front and back, which would be shielded by large mass of the propulsion and docking systems that would be located there (Westover et al. 2014).

It isn't just the US that is pursuing magnetic shields. The CERN laboratories joined the European Space Radiation Superconducting Shield (SR2S) project in 2014 and have made significant progress on a potential radiation shield. To create a large magnetic field, CERN proposes using an electromagnet based on coils of magnesium diboride (MgB_2) superconducting tape. MgB_2, the same material used in the Large Hadron Collider is an inexpensive binary compound that achieves the highest superconducting critical temperature of any of the conventional SC materials (39 K/−234 °C/−389 °F) (Del Rosso 2015). A sunshield of V-shaped design will keep the MgB_2 below its critical temperature by reducing the influence of the Sun or Mars (Bruce and Baudouy 2015).

CERN has modeled various configurations of the MgB_2 coils to see which produces the most uniform and protective fields. The stronger the field and the more appropriately shaped it can be made will translate to more cosmic radiation being deflected away from the hull of the ship. Recently, the SR2S group decided on a final structure for the shield. Four coil assemblies, each three SC coil panels, are positioned so that the entire structure looks like a ceiling fan with the blades replaced by kitchen mixer paddles. Each assembly will produce a toroid shaped magnetic field, with the habitat part of the ship protected in the middle (where the ceiling fan motor would be). The combined magnetic field is spheroid with depression in the top and bottom, and has been dubbed the "pumpkin configuration" (SR2S website, Fig. 4.4). The current design reduces the mass of the entire structure as compared to previous designs, helping with fuel consumption and by eliminating many structural elements that would interrupt the magnetic field lines and lower the shielding efficiency.

4.5.2 *Magnetic Shields Build Plasma Shields*

Like the magnetosphere has the ionosphere, a magnetic shield might also produce a plasma shield. Magnetic fields generate electric fields by moving ionized atoms and free electrons along the field lines. The vacuum of space does have a few molecules per cubic meter, and most of these are ionized by cosmic radiation. Theoretically, a magnetic shield will generate a plasma shield on its leading edge as the ship travels through space. The stronger the magnetic field, the tighter the field lines become and the more densely the plasma will be held. A superconducting magnet well below its critical temperature would generate a strong field and produce a tightly

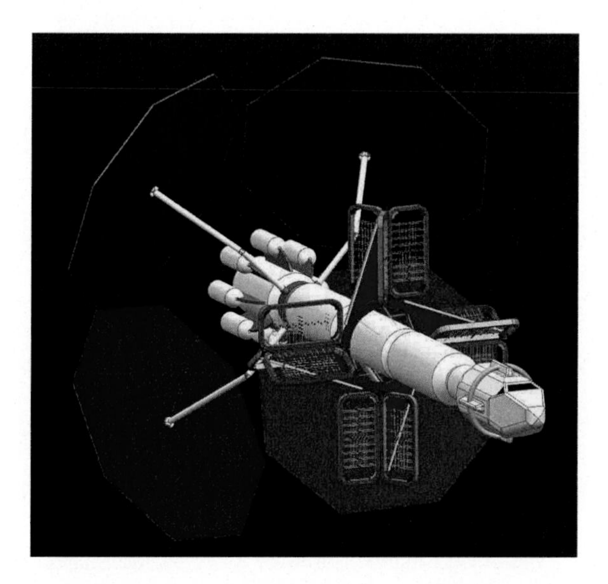

Fig. 4.4 The Superconducting Space Radiation Shielding (SR2S) project recently released a design for their radiation shield. Circling the ship like a girdle, it has four sets of tri-paddles of superconducting cables that will generate a large magnetic field bubble in front and behind the ship. The *dark red* and *yellow* structures are the supports that anchor each tri-paddle to the girdle. The large *blue* shields on the rear are more traditional solid solar shields. (*Image courtesy of the proprietors* Italian national institute of nuclear physics (INFN) and CGS S.p.A. Compagnia Generale per lo Spazio, with EU funding: SR2S research project-EU FP7 grant agreement no 313224)

held plasma shield that will increase in strength as more ions are encountered by the spacecraft.

Again, the Earth serves as a good model to show the importance of this interaction. When solar storms greatly increase the amount and energy of the radiation approaching Earth, the plasmasphere plumes out due to the increased number of ionic interactions between the SEPs, the plasma, and the magnetic field. More SEPs from a solar flare actually make the plasma/magnetic shield stronger (Walsh et al. 2014). A plasma field acts collectively as a single uncharged entity (Box 4.2), so it will act to absorb or deflect the incoming radiation. The GCR free electrons are small and will be stopped quickly by the large nuclei of the plasma shield, while the larger GCR and SEP nuclei will stopped by charge separation of the more numerous electrons (the acquired electrical field) or will be shunted around the Earth by overshooting the field. It only takes a small movement of a few particles of the plasma to incorporate, neutralize, or deflect any incoming radiation. The ship's shield will start out as only a magnetic field, and then acquire a plasma shield as they travel through space. The fact that the magnetic field will also work as a barrier to confine the acquired plasma shield to a defined space outside the hull will be additional bonus.

Box 4.4 Manned Missions to Mars

1880—*Across the Zodiac*, a book by Percy Greg from 1880, described a trip to Mars on a ship called the "Astronaut." *Astro*—meaning star and replaced the aero- of the 1780s word "aeronaut", meaning those who flew in balloons.

1952—Wernher von Braun proposed a manned mission to the red planet in his book *Das Marsprojekt* (published in 1952; 1962 in English as *The Mars Project*). In the book, von Braun proposed using rockets to ferry parts to space to build a Mars ship at a space station orbiting our planet.

1960s—Dr. von Braun proposed to NASA a two ship Mars mission, after the successes of the Apollo program and when similar programs were being proposed by Soviets. As part of the critical review of this project, the first analytic studies were conducted on just what it might take to put a man on Mars. It was scrapped in favor of developing the space shuttles.

2010—A space policy speech by U.S. President Barack Obama in 2010 reflected NASA's vision of a manned orbit of Mars by the 2030s. The prediction revealed how advanced the planning had become.

2011—European Space Agency and Russian Roscosmos agency enter into agreement to send manned mission to Mars in 2030s. Two robotic or orbiter missions are planned for 2016 and 2018. The 2018 flight will likely be delayed by financial problems until 2020, and perhaps further if tensions in the Ukraine lead to the US denying access to some American made components needed for the flight.

2021—Billionaire Dennis Tito's Inspiration Mars Foundation scheduled launch date. The program, as endorsed by NASA, will be a manned flyby of Mars, using a gravity assist to return to Earth. The mission would also include a flyby of Venus.

2027—Proposed arrival on Mars of the Dutch Mars One mission. Intended to be a one way trip, the astronauts will live out the rest of their lives on the red planet. Funding for the project will come from selling broadcast rights of the people lives and work on Mars. Thousands of people have submitted applications, but the timetable is seen as unrealistic.

2030s—NASA's official goal for manned Mars landings, with transient exploration and colonization sometime later. The plan calls for reliance on the planet to provide construction materials as well as water. This idea was bolstered by the detection of running water on Mars in 2015.

4.5.3 Mini-Magnetospheres

No researcher has thought more about producing a protective plasma shield than Dr. Ruth Bamford of the Rutherford Appleton Laboratory in England. Since 2008 she and her group have been working on producing mini-magnetospheres that buffer a small amount of plasma in space, using the magnetic field to hold it in place and build up its density. Nearly decade into this work, her group has produced a model shield but there is much more work to do (Bamford et al. 2008). As Dr. Bamford once said, "*Star Trek* has great ideas, they just don't have to build them."

She obviously agrees with NASA's Shayne Westover as to the engineering problems that will keep a ship-sized magnetic/plasma shield from being practical. For example, there is the issue of getting the superconductors to produce a magnetic field in space. On Earth, superconducting materials only produce a zero resistance electric current if below their transition temperature, and even for the best of materials (YBCO and BSCCO) this is somewhere in the range of 110 K (−163 °C/ −262 °F). When the superconducting coils are exposed to the Sun in space, they will rise several hundred degrees at least. The development of MgB_2 has increased the strength of the magnetic field that can be generated and they can be partially shaded from the Sun, but that adds weight and mass shielding that will interrupt some of the magnetic field lines. Rather, it is likely that some method of cooling the superconductors will be needed, especially on a return flight from Mars (towards the Sun). Cooling equipment adds weight, and more weight means more fuel is needed.

The weight problem also plagues the magnetic field generation as a whole. Early estimates of the magnetic field needed to protect a spacecraft from cosmic radiation were in the 10,000 nanotesla range, a scaled version of our best model, the Earth. To generate this type of field over an entire ship's hull would require a huge coil and generator, adding hundreds of tons of weight to the space craft. In 2008 this was considered untenable economically, and put the idea of producing a magnetic field around the ship on the backburner.

4.5.4 Magnetic and Plasma Shields on the Moon

As they so often do, a discovery in 2010 in a different area of research brought the spacecraft designers back to older ideas. Though they did not understand the mechanism at work, astronomers had noted for decades that small patches of the Moon's surface are lighter than the rest and that these patches had static yet whorled shapes. Additional research finally showed that the Moon exhibits small magnetic fields (200–400 nT) that extend over the "lunar swirls" like a bubble (Fig. 4.5). What is more, those fields hold a thin layer of plasma in place above them. The field concentrates the plasma, and together they produce a protective electric field to deflect particles and keep the surface of the moon at those spots from being irradiated and darkened (Wieser et al. 2010).

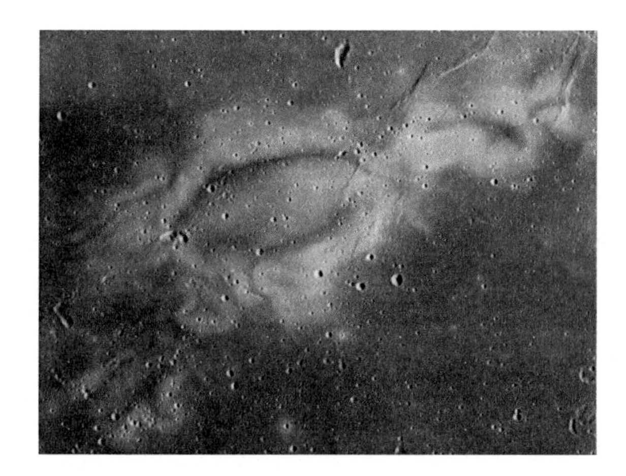

Fig. 4.5 Small electrical fields (km across) on the moon create field bubbles that can trap ions and act as a protective radiation shield. Areas of the moon's surface under a plasma shield are not struck by cosmic radiation that tends to darken the surface. The protected areas remain lighter in color. (*Image Credit* NASA)

The protective magnetic forces on the Moon are much weaker than previous estimates had thought necessary, yet they still cover areas effectively, some as large as 300 km (186 mi.) in diameter. The lower intensity, yet effective magnet field made Bamford's idea of a mini-magnetosphere plasma shield attractive again. A smaller (lower SWaP) superconducting coil than previously assumed could be used to create a magnetic field and hold a thin layer of plasma in a bubble around a spacecraft. Bamford estimates that a coil of 16 kW would generate the necessary field, while it would take another 5 kW to cool the unit. The whole apparatus would come in at less than a ton and a half, feasible for a manned mission to Mars and back (Bamford et al. 2014).

As proof of concept, Bamford's group built a model mini-magnetosphere in the lab and bombarded it with a plasma beam in their "solar wind tunnel." They showed conclusively that a mini-magnetosphere of just over 100 nT could shield an area behind the field (Fig. 4.6). What is more, the self-captured plasma field, made up of a buffer of the solar wind particles confined within the magnetic field added to the deflector shield efficiency and would further reduce the needed size and weight of the superconducting coil and its cooling unit (Bamford et al. 2014). Her paper predicted standoff distances and widths of the plasma layer without stating an efficiency of target protection; however, aluminum shields are only 25 % efficient and impregnated polyethylene is 36 % effective. The clues point to a high efficiency of the Bamford model since the magnetosphere it is designed to mimic is very effective and because Bamford's model continues to be investigated.

What is even more amazing is that this plasma/magnetic shield goes *Star Trek* one better. The coherent graviton field of the Enterprise-D shield is held in place by a grid of molybdenum-jacketed superconductors on the hull (Star Trek: The Next Generation Technical Manual, page 23), so it requires a graviton generator *and* a superconductor magnetic field. Bamford's plasma/magnetic shield needs only the magnets, it plays the roles of both shield and confinement grid. Unfortunately, we can't become too impressed with the plasma/magnetic shield; it will be useless against even a small meteoroid in the ship's path since the deflection and absorption

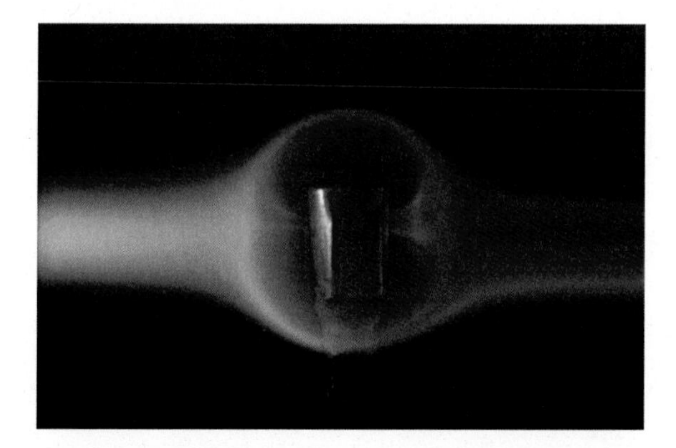

Fig. 4.6 The plasma stream (*purple*) created in a laboratory solar wind tunnel is diverted around the center object by a minimagnetospheric bubble created by a strong magnetic field. A thin sheet of the plasma trapped by the magnetic field diverts the rest of the plasma around the magnetic field and protects the center volume from cosmic radiation. (*Image credit* STFC)

will work only for charged particles and most EM waves—just not the highest energy waves.

The highest energy EM waves encountered in space, gamma rays, are neutral and will not be deflected by a magnetic field. Even though they can be absorbed by gases and plasma to produce more ions and free electrons, the thickness of artificially generated shields will be insufficient to stop all of them. While not a danger at all times, galactic events like supernovae, coronal mass ejections, solar flares and gamma ray bursts are large gamma ray sources. Astronauts can seek shelter in a radiation safe room to ride out solar flares, but nothing we know now would save a spacecraft, or Earth, from a gamma ray burst. Though no one knows for sure the source of all gamma ray bursts, they are definitely the brightest, most energetic events in universe. In just a few milliseconds, a gamma ray burst can impart as much energy as our Sun will emit over its entire lifetime (Cain 2015). If Earth were in the path of one, it would be an extinction level event, and this may explain some of the extinctions in Earth's past. Perhaps we should be thinking about a planet wide deflector shield to augment our magnetosphere—just in case.

4.6 Metamaterial Shields

Metamaterials (Sects. 2.4–2.6, 4.6.1) are engineered solids made from component materials with characteristic refractive or absorptive properties. Designed with repeating patterns of construction where the individual unit cell is much smaller than some wavelengths of EM radiation, the 3D arrangement of the material can interact with EM waves in unnatural and unique ways. We have discussed the possibility of using these characteristics to bend light around an object as cloaks, and science is learning that they may also be able to deflect away or absorb radiation as well. Could a cloak be used as deflector, just as Kirk used his deflector as a cloak (*TOS: Assignment Earth*)?

4.6.1 Metamaterial Black Holes

Qiang Cheng and Tie Jun Cui at Southeast University in Nanjing, China produced the first electromagnetic black hole in 2009. Using sixty concentric rings of metamaterials, low energy microwaves striking the outer ring were bent to the inner rings and eventually absorbed (Cheng et al. 2010) (Fig. 4.7). A Polish group published a study in 2014 that used S-shaped metamaterial cells to absorb even higher frequency energy. The terahertz energy waves (just a bit longer than infrared) entered the double bend chambers and bounced around until their energy is dissipated (Grzeskiewicz et al. 2014). If either system could be expanded to work with higher frequency energy, the applications could be extensive. A metamaterial structure that traps all the light that enters it (a black hole) could make a very efficient photovoltaic cell. Or perhaps small shields could be layered over electronic components to protect them from EM pulses. An object cloaked with an absorbing metamaterial would look black; in space that might be better than appearing silver with blue light nacelles, but it isn't a great cloak. However, a total absorber could make a good EM shield—if you can find a way to keep the absorbed energy from heating your shield up and burning it to a crisp. If the heat and wavelength issues (present metamaterials work only for longer wavelength radiation) can be resolved,

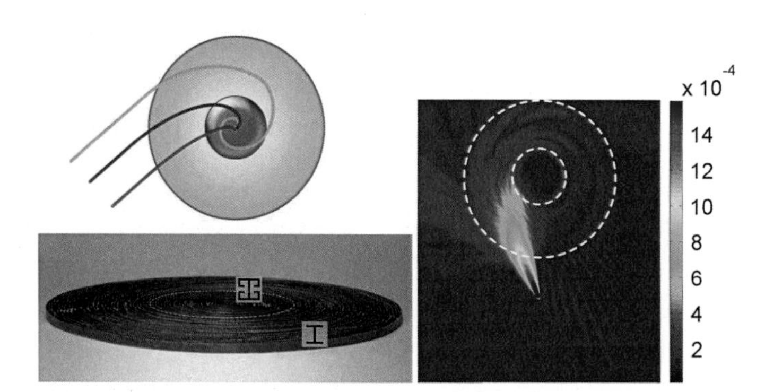

Fig. 4.7 To protect an object from EM radiation, absorption might be as good as deflection. Some metamaterials can be designed to funnel EM energy and trap it. The model for this is shown at the *top left*. The physical device (*bottom*) is made of 60 concentric rings, with different structural patterns from the outside to the inside. The forms keep bending the light further toward the center. The electrical fields generated by such a device is shown at *right*. All the energy is funneled toward the center and none escapes—a *black hole*. (*Image credit* (Cheng et al. 2010), © IOP Publishing & Deutsche Physikalische Gesellschaft. CC BY-NC-SA (https://creativecommons.org/licenses/by-sa/4.0/legalcode), http://iopscience.iop.org/article/10.1088/1367-2630/12/6/063006)

then perhaps a metamaterial shield to absorb laser fire or cosmic radiation would be possible.

A different approach would use certain metamaterials that are termed hyperbolic, based on the math used to describe their behavior. Igor Smolyaninov proposed an electromagnetic black hole using hyperbolic metamaterials in a 2011 paper (Smolyaninov 2011), and made an interesting connection to magnetic field shields in a later report. He knew from previous research that extremely energetic collisions in the Large Hadron Collider (LHC) can induce very strong magnetic fields (Deng and Huang 2012), strong enough to vacuum in charged mesons (a subatomic boson made of a quark and an antiquark). He was also aware that this meson cloud could be dense enough to form a real condensate (Cherndub 2011). Smolyaninov hypothesized that the meson condensate could act as a hyperbolic metamaterial and create an electromagnetic black hole within the vacuum area (Smolyaninov 2013). Not only do we see that magnetic fields play a role in achieving energy shields, by absorbing not deflecting, we also note that those "crazy" people who worried that the LHC might open a black hole were right … sort of (Sects. 1.2.1.3 and 8.4.4.1).

The story of electromagnetic black holes created by metamaterials brings us right back to the wonders of the graviton. A graviton-based shield system would protect a ship or base from energy weapons like lasers of phasers, beam weapons like proton or neutron beams (Sects. 1.2.1.1 and 1.2.1.2) and from micrometeoroids or other debris that could crash through the hull. Electromagnetic black holes using metamaterials could only work as tactical defense shields against lasers or phasers, they would have no effect on GCR or SEPs. Particles with mass with would not be subject to the unique permittivity and permeability characteristics of the metamaterials.

4.7 Conclusion

For all the attention given to tactical defense shields in *Star Trek*, it would seem that the real heroes of the story are the navigational shields that keep the crew safe on a day-to-day basis. By deflecting micrometeoroids and GCR from the leading edge of the Enterprise, some of the dangers of traveling in space are reduced to manageable levels. It is also a good lesson to remember that for all the energy beams and fancy technologies, *Star Trek* still realizes the need for traditional armoring. Ablative and dispersive armors may seem like science fiction when they are made by the replicator and transported outside the hull (Johnson 1987, pg. 14), yet they were born in the American and Russian Space and military programs and will still be the main protection for astronauts in the years to come.

Since heat shields and Whipple shields predate the *Star Trek* franchise, we can't say that Roddenberry and his writers were seeing the future when they wrote them into the later series. Even more, the advanced armors and energy shields being developed at the present time far surpass any of the armor technologies to which Kirk, Picard, or Janeway have access. The evolution of LIPC shockwave shields, electromagnetic armor or plasma acoustic shielding were probably not inspired by

Star Trek, if they had been the military would have given them Trek-like names or initialisms—they love to do that. Yet the media picked up on the similarities when they asked if Boeing had patented a *Star Trek* deflector shield (Hill 2015).

Science fiction isn't our best model for developing shields to protect astronauts in space. It turns out that our planet has its own shield and we can steal ideas from it. Stealing from Nature is a time-honored tradition in the physical and life sciences; the ways of the universe or a mystery yet we can still exploit them for fun and profit. Is it just lucky that Earth developed a magnetic and plasma shield that protects life on the surface? If it hadn't, we wouldn't be here to consider the question. The laws of physics and the nature and components of the universe are just as big a part of evolution of life on this rock as Darwin's theory ever was.

On the other hand, the magnetic and plasma shield ideas provided to us by the Earth and Moon only meet part of or requirements. Science is developing many interesting variants on shields, it is unfortunate that none of them can do everything we need them to do. Magnets and plasma work against charged particles and some energy, but not against gamma rays. Traditional and advanced armors are great for micrometeoroids and other projectiles, they just don't do well with GCR. Metamaterials might absorb energy, yet would do nothing against charged particles. Maybe *Star Trek* had it right all along, gravitons are the answer.

References

MM Acharya, LA Christie, ML Lan, E Giedzinski, JR Fike, S Rosi, CL Limoli. Human neural stem cell transplantation ameliorates radiation-induced cognitive dysfunction. *Cancer Research.* 71(14); 4834–4845, 2011. doi: 10.1158/0008-5472.CAN-11-0027. http://cancerres.aacrjournals.org/content/71/14/4834.long

R Bamford, KJ Gibson, AJ Thornton, J Bradford, R Bingham, L Gargate, LO Silva, RA Fonseca, M Hapgood, C Norberg, T Todd, and R Stamper. The interaction of a flowing plasma with a dipole magnetic field: measurements and modelling of a diamagnetic cavity relevant to spacecraft protection. *Plasma Physics and Controlled Fusion*, 50 124025 (11 pp) November 4, 2008. Online. DOI: 10.1088/0741-3335/50/12/124025. http://iopscience.iop.org/article/10.1088/0741-3335/50/12/124025/meta

R Bamford, B Kellett, J Bradford, TN Todd, R Stafford-Allen, EP Alves, L Silva, C Collingwood, IA Crawford, and R Bingham. An exploration of the effectiveness of artificial mini-magnetospheres as a potential solar storm shelter for long term human space missions. *Acta Astronautica* 105(2); 385–394, 2014. doi:10.1016/j.actaastro.2014.10.012. http://www.sciencedirect.com/science/article/pii/S0094576514003798

DI Blokhintsev and FM Gal'perin. Gipoteza neitrino i zakon sokhraneniya energii (Neutrino hypothesis and conservation of energy). *Pod Znamenem Marxisma* (*Under the Banner of Marxism*) 6; 147–157, 1934.

R Bruce, and B Baudouy. Cryogenic design of a large superconducting magnet for astroparticle shielding on deep space travel missions. *Physics Procedia* 67; 264-269, 2015. doi:10.1016/j.phpro.2015.06.085. http://www.sr2s.eu/images/documents/Bruce2015.pdf

F Cain. Are gamma ray bursts dangerous? UniverseToday.com Guide to Space. January 12, 2015. Accessed September 30, 2015. http://www.universetoday.com/118140/are-gamma-ray-bursts-dangerous/

Q Cheng, TJ Cui, WX Jiang, and BG Cai. An omnidirectional electromagnetic absorber made of metamaterials. *New Journal of Physics* 12; 063006, 2010. doi:10.1088/1367-26301/12/6/063006. http://iopscience.iop.org/article/10.1088/1367-2630/12/6/063006

MN Cherndub. Spontaneous electromagnetic superconductivity of vacuum in a strong magnetic field: evidence from the Nambu-Jona-Lasinio model. *Physics Review Letters*. 10; 142003, 2011. doi:10.1103/PhysRevLett.106.142003. http://journals.aps.org/prl/abstract/10.1103/PhysRevLett.106.142003

JD Cherry, B Liu, JL Frost, CA Lemere, JP Williams, JA Olschowka, and MK O'Banion. Galactic cosmic radiation leads to cognitive impairment and increased Aβ plaque accumulation in a mouse model of Alzheimer's disease. *PLoS ONE* 7(12); e53275, 2012. doi:10.1371/journal.pone.0053275. http://journals.plos.org/plosone/article?id=10.1371/journal.pone.0053275

LT Chylack, LE Peterson, AH Feiveson, ML Wear, FK Manuel, WH Tung, DS Hardy, LJ Marak, and FA Cucinotta. NASA study of cataract in astronauts (NASCA). Report 1: cross-sectional study of the relationship of exposure to space radiation and risk of lens opacity. *Radiation Research* 172(1); 10-20, 2009. doi:10.1667/RR1580.1. http://www.bioone.org/doi/10.1667/RR1580.1?url_ver=Z39.88-2003&rfr_id=ori%3Arid%3Acrossref.org&rfr_dat=cr_pub%3Dpubmed&

FA Cucinotta, M Alp, FM Sulzman, and M Wang. Space Radiation risks to the central nervous system. *Life Sciences in Space Research* 2; 54-69, 2014. doi:10.1016/j.lssr.2014.06.003. http://www.sciencedirect.com/science/article/pii/S2214552414000339

A Del Rosso. A superconducting shield for astronauts. CERN website. August 5, 2015. Accessed January 27, 2016. http://home.cern/about/updates/2015/08/superconducting-shield-astronauts

WT Deng and XG Huang. Event-by-event generation of electromagnetic fields in heavy ion collisions. *Physical Review C* 85; 044907, 2012. doi:10.1103/PhysRevC.85.044907. http://journals.aps.org/prc/abstract/10.1103/PhysRevC.85.044907

B Dodson. Rheinmetall tests new active defense system under live fire. *Gizmag* January, 31, 2012. Accessed November 4, 2015. http://www.gizmag.com/rheinmetall-ads-live-fire-test/21278/

F Dyson. Is a graviton detectable? Poincare Prize Lecture. International Congress of Mathematical Physics. Aalborg, Denmark. August 6, 2012. http://publications.ias.edu/sites/default/files/poincare2012.pdf

J Emspak. Sci-fi cloaking device could protect soldiers from shock waves. *Live Science*. March 24, 2015. Accessed January, 26, 2016. http://www.livescience.com/50236-boeing-device-could-block-shock-waves.html

BE Gavett. Mild traumatic brain injury: a risk factor for neurodegeneration. *Alzheimer's Research & Therapy* 2; 18, 2010. doi:10.1186/alzrt42. http://alzres.biomedcentral.com/articles/10.1186/alzrt42

D Goure. Now is the time to provide US armored vehicles with active protection systems. *Lexington Institute*. August 3, 2015. Accessed January 35, 2016. http://lexingtoninstitute.org/now-is-the-time-to-provide-u-s-armored-vehicles-with-active-protection-systems/

B Grzeskiewicz, A Sierakoqaski, J Marczewski, N Palka, and E Wolarz. Polarization-insensitive metamaterial absorber of selective response in terahertz frequency range. *Journal of Optics* 16 (10); 105104, 2014. doi: 10.1088/2040-8978/16/10/105104. http://iopscience.iop.org/2040-8986/16/10/105104

D Hambling. The Pentagon's wall-of-light laser shield. *Popular Mechanics Online*. January, 22, 2013. Accessed May 15, 2015. http://www.popularmechanics.com/military/research/a8626/the-pentagons-wall-of-light-laser-shield-15008409/

K Hill. Did Boeing just patent a deflector shield? *Nerdist* March 24, 2015. Accessed on December 10, 2015. http://nerdist.com/did-boeing-just-patent-a-deflector-shield/

S Johnson. *Star Trek: Mr. Scott's Guide to the Enterprise*. Pocket Book, NY, NY. 1987. ISBN13: 9780671635763

S LaGrone. Navy develops torpedo killing torpedo. *USNI News*. June 20, 2013. Accessed January 20, 2016. http://news.usni.org/2013/06/20/navy-develops-torpedo-killing-torpedo

The LIGO Scientific Collaboration. J Aasi, BP Abbott, T Abbott, et al. Advanced LIGO. *Classical and Quantum Gravity* 32(7); 074001, 2015. doi: 10.1088/0264-9381/32/7/074001. http://www.iopscience.iop.org/article/10/1088/0264-9381/32/7/074001/meta

J McGuire, A Toohie, and A Pohl. Shields up! The Physics of Star Wars. *Journal of Physics Special Topics* 12(1): P6_11, 2013. https://physics.le.ac.uk/journals/index.php/pst/article/view/678/486

BM Rabin, JA Joseph, B Shukitt-Hale, and KL Carrihill-Knoll. Interaction between age of irradiation and age of testing in the disruption of operant performance using a ground-based model for exposure to cosmic rays. *Age (Dordrecht)* 34(1); 121–131, 2012. doi:10.1007/s11357-011-9226-4. http://link.springer.com/article/10.1007/s11357-011-9226-4

N Shachtman. Genius computer stops rockets right before impact. *Io9, we come from the future*. June 25, 2011. Accessed January 25, 2016. http://io9.gizmodo.com/5815529/genius-computer-stops-rockets-right-before-impact

I Smolyaninov. Quantum electromagnetic "black holes" in a strong magnetic field. *Journal of Physics G: Nuclear and Particle Physics*. 40; 015005, 2013. doi:10.1088/0954-3899/40/1/015005. https://www.researchgate.net/publication/258294051_Quantum_electromagnetic_'black_holes'_in_a_strong_magnetic_field

I Smolyaninov. Critical opalescence in hyperbolic metamaterials. *Journal of Optics* 13; 125101, 2011. doi:10.1088/2040/8978/13/12/125101. http://iopscience.iop.org/article/10.1088/2040-8978/13/12/125101

Technology Quarterly, Q2. The armour strikes back. *The Economist*. June 4, 2011. Accessed on October 2, 2015. http://www.economist.com/node/18750636

BM Walsh, JC Foster, PJ Erickson, and DG Sibeck. Simultaneous ground- and space-based observations of the plasmaspheric plume and reconnection. *Science* 343(6175); 1122-1125, 2014. doi:10.1126/science.1247212. http://www.sciencemag.org/content/343/6175/1122.long

SC Westover, RB Meinke, R Battiston, WJ Burger, S Van Sciver, S Washburn, SR Blattnig, K Bollweg, RC Singleterry, and DS Winter. Magnet Architectures and Active Radiation Shielding Study (MAARS). *Final Report for NASA Innovative Advanced Concepts Phase I*. May, 2014. http://ston.jsc.nasa.gov/collections/TRS

R Whitwam. Boeing patents sci-fi force field that deflects explosive shockwaves. *ExtremeTech website* March, 23, 2015. Accessed June 16, 2015. http://www.extremetech.com/extreme/201777-boeing-patents-sci-fi-force-field-that-deflects-explosive-shock-waves

M Wieser, S Barabash, Y Futaana, M Holmstrom, A Bhardwaj, R Sridharan, MB Dhanya, A Schaufelberger, P Wurz, and K Asamura. First observation of a mini-magnetosphere above a lunar magnetic anomaly using energetic neutral atoms. *Geophysical Research Letters* 37; L05103, 2010. doi:10.1029/2009GLO041721. http://onlinelibrary.wiley.com/store/10.1029/2009GL041721/asset/grl26727.pdf?v=1&t=ijnn3m5c&s=6dcb5451a6da7d69bc754592911c1a2012df2d66

Chapter 5
The Tractor Beam: Pulling Earth Out of the Fire

Captain, our tractor beam caught and crushed an Air Force plane. It'll be impossible to explain this as anything other than a genuine UFO.

—Mr. Spock
(TOS: Tomorrow is Yesterday)

5.1 Introduction

October 31, 2015 was a typical Halloween night around the world. Kids in costumes gathered candy. Teenagers toilet-papered houses and adults either dressed in more risqué costumes or followed the kids around. The ghosts of the dead supposedly rose from the grave to walk the Earth for one night. Perhaps an escaped murderer returned to take revenge on the residents of his hometown. Oh wait, that's just a scary movie. Halloween can be scary if you want it to be, and this particular Halloween could have been a lot scarier than a horror movie.

High in the sky, although not as high as you would like, a comet passed just outside the orbit of the Moon. Called 2015 TB145, this mass of compressed interplanetary dust flew past Earth just 300,000 miles above the surface, a mere 1.27× the average distance of the Earth to the Moon. A hurtling boulder 1960 ft (600 m) in diameter—that's two New York City blocks worth of rock, metal and organic compounds—traveling at 78,000 miles per hour (125,500 kph). Pictures taken by astronomers as it passed our planet showed a silhouette eerily similar to a human skull, perfect for a Halloween night visitor.

A celestial object passing so close to Earth is scary enough, yet the truly worrisome aspect of the episode was that scientists didn't know it was approaching us, or that it existed at all, until October 10th. The orbit of TB145 is very oblong and is tilted with respect to the plane of Earth's orbit. This is unusual for objects within the solar system and is one of the reasons scientists didn't notice it until so late. Instead of approaching Earth with an east/west trajectory, it came upon us from the south, and there are fewer observatories in the Southern hemisphere.

M.E. Lasbury, *The Realization of Star Trek Technologies*,
DOI 10.1007/978-3-319-40914-6_5

TB145's orbit makes scientists believe it is a dead comet, or maybe an undead comet, considering the night it made its flyby. Comets have more oblong orbits and higher encounter velocities (average 35 kps/22 miles per second), which makes them more dangerous in collision scenarios. A dead comet is one that has lost all of its volatile elements, suggesting that TB145 has flown by Earth many times. Its next visit in 2088 presents no threat, it won't get any closer than 25,000,000 miles (46,300,000 km). Don't rest too easy, TB145 will be back in 2152 and could pass *within* the Moon's orbital path on that transit.

Had TB145 collided with Earth, the damage could have been extensive. If it hit land, the crater would have been almost eight miles (14.8 km) across—a definite city killer. An ocean impact would have sent a 200–300 ft (60–90 m) high tsunami rushing toward shore in every direction. For comparison, the Chicxulub impactor that killed off the dinosaurs of the Cretaceous-Paleogene Era was probably 6.2 miles (10 km) in diameter and produced a 112 mile (180 km) wide crater. Because astronomers first viewed TB145 only three weeks before it crossed Earth's path, there is nothing we could have done if the comet had been intent on ruining Halloween—just something to think about when you stare up at the night sky. Either that big rock or another like it will very probably try to kill us all in the future. Wouldn't it be nice if we had something to move an asteroid or comet out of our way?

That's what the Federation of Planets does when a member species is threatened with extinction by a cosmic object. They put the tractor beam to work. For example, Commander LaForge and Hannah Bates manage to save Moab colony IV only by employing an ingenious tractor beam to deflect a stellar core fragment (part of a collapsed star) away from its collision course with the planet (*TNG: The Masterpiece Society*). Tractor beam functions aren't confined to dramatic rescues; some are for everyday ship operations while others are nefarious tools that evil empires like the Borg use to capture and assimilate ships, crews and technologies (*VOY: Dark Frontier*; *Collective*). But no matter the purpose, the tractor beam is a tool used to hold, push, pull, or deflect objects that either need to be moved someplace else or kept just where they are.

The idea of an "attractor beam," or tractor beam for short, is simple enough. It first appeared in the 1931 serialized science fiction story *Spacehounds of IPC*, which was later republished in book form (Smith 1947, p. 26). In that story, they use the tractor for several purposes, through the action of "inter-atomic" energy. While the ray or energy may be different in different stories, the uses of tractor beams haven't changed much since *Spacehounds*. The function is easily understood and demonstrated; a beam applies a pushing or pulling force to any object it is shone upon. Writing one into a story is easy; actually producing one is much more challenging. Engineering a non-contact, non-magnetic force that pulls things toward itself instead of pushing them away is quite a feat, and scientists better be working on it. There's an asteroid out there with Earth's name on it, and we might not see it coming until it's right on top of us.

5.2 *Star Trek* Tractor Beams at Work

In physics, there is little difference between a push force and a pull force, each acts on an object in the same direction as the force. The distinction is found when putting them into practice. A push is a single move, you push a ball and it moves away. You push harder and it moves away faster. For a pull, you first must expend energy to reach out and get the ball, and then invest more energy to pull it back. Reaching out faster or harder doesn't pull the ball back any faster, only the force in the direction of intended movement matters. Now imagine you could send energy out, like reaching for the ball, and that energy alone would pull the ball back to you —that's a tractor beam. It moves an object *toward* the source of the outgoing energy instead of pushing it away. Pushing an asteroid might be easier, and scientists are working on that, although the energy costs could negate it as a practical option in some cases. A tractor beam that pulls an object could be a neat toy, or it could be a planet saver.

A tractor beam might take the form of a traveling wave that can move a targeted object along its length back to the source. In its truest *Star Trek* form, it could also hold an object a designated distance from the source, or push it in the direction of the wave front. *Star Trek* makes more varied use of the tractor beam, even if they were not the first to introduce the concept. Capturing, holding, and towing a rogue vessel (*VOY: Scorpion*) is the way many people see a tractor beam being used, and they can also be used to change the heading of an oncoming impactor in order to rescue a world or a ship too (*TNG: The Naked Now; The Neutral Zone*).

Though towing huge celestial bodies into or out of the way is important work for a tractor beam, there are more mundane tasks to perform on routine days in space. A beam is often emitted from within the docking bay to help guide shuttle landings (*TNG: Time Squared*). This is analogous to the parking assist programs on many newer car models that help with parallel parking—something few people do well on their own. The shuttle bay beam emitters are small, rectangular boxes that sit atop short pedestals. There are also small emitters on the RCS (reactions control system) of the Enterprise (Galaxy Class) to help it dock with starbases when there is no mooring tractor beam available. The RCS emitters are located on the nacelles (TNG Technical Manual, p. 86), to bring the ship in for docking and to hold the ship in place when moored; nobody wants to have warp coils knocking into the hull of the space dock.

A starship crew must also know how to use a tractor be to hold a large body in a static position or to push it away from an object, showing that tractor beams are versatile. Tractors can capture, push, or pull, and are therefore crucial ship equipment. When called upon to keep the Enterprise away from the gravimetric distortions associated with an "energy ribbon," Kirk calls for the tractor beam to be switched on. When told they don't have one, he queries, "You left space dock without a tractor beam?" (*Star Trek Generations*) The question implies that a tractor beam is standard equipment and is a first line tool for keeping the ship safe.

The important role tractor beams play on starships means that the engineers don't begrudge them their need for extraordinary amounts of energy. Federation starships tractor beams are powered directly from the warp propulsion system (TNG Technical Manual, p. 56). Both the shields and the tractor draw energy from the same matrix (*VOY: Collective*), so they can't both operate at the same time. At times this can create problems for the crew. Shuttles returning to the Constitution class Enterprise while the shields are up must fly in under manual power rather than being tractored (*Star Trek V: The Final Frontier*), and the Galaxy Class Enterprise has an automatic system for disengaging the tractor beam any time the deflector shields are activated (*TNG: The Hunted*). Of course, the shields and tractor beams have something else in common that might preclude them from being used at the same time—they both emit graviton particle fields. Refer to Sects. 4.2.2–4.2.4 and Box 4.1 for a discussion of the graviton, a gravity producing subatomic particle.

5.2.1 Tractors Generate Attenuated and Collimated Gravity Beams

In *Star Trek*, the tractor is a linear graviton beam, produced by a graviton particle generator, not so different from the deflector shield arrays. The two uses are similar, yet have several individual characteristics. A deflector shield surrounds an object on all sides or perhaps is limited to specific sides of an object, while a tractor is projected from one object to another in a narrow cross section, like the metallic grapplers of the NX-01 Enterprise. The grapplers' rope-like qualities are a solid representation of the later tractor beams. They are fired from the starship, travel to the target, and latch on with magnetic hooks to tow or hold the captured object (*ENT: Broken Bow*).

Tractor beams are preferred to grapplers (*ENT: Breaking The Ice*), but there are still issues that must be dealt with when using tractors. A gravity beam that gets wider as it propagates and propagates forever could be very dangerous. This is why the graviton beam of the tractor is *attenuated* and *collimated.* All sorts of space debris, battle debris, or small vessels could be pulled in if they encounter a tractor beam wider than the target. It is better if only the target is exposed to the beam, like how a grappler reaches out to latch onto a single object. In physics, attenuated describes how a certain property is lessened as the distance from the source increases. Kirk wouldn't want to attract, hold, or repel every object in wide beam's path from the emitter to the edge of the galaxy, so the tractor beam's energy is attenuated; it decreases with distance, just as gravity does. This sets up an inverse power and distance relationship for capturing objects. The *Next Generation Technical Guide* states that a small mass has to be within 20,000 km of the tractor beam, while even with power augmentation a large mass must be less than 1000 m or so away to be held (TNG Technical guide, p. 90).

Your dentist collimates the X-ray beam when he makes radiographs of your teeth. The circular or rectangular pipe on the end of the X-ray generator is the collimator that confines the X-ray energy to a single path through the correct teeth and the small piece of film. Allowing the beam to diverge would reduce the energy focused on the film and would expose more of your mouth to the X-rays for no reason. Laser light is also collimated as it bounces between two parallel mirrors in the resonant cavity (Box 1.2), or by an optical collimating lens that refracts the diverging rays and redirects them in a parallel fashion. A laser pointer is a good example. It produces a spot of light on the wall across the room not much larger than the opening on the pointer. In similar fashion, the graviton beam of the Enterprise's tractor is collimated to reduce its divergence through space. This narrow column of energy is much more suited to pulling in just the shuttlecraft that is trying to land and not any space debris around it. Perhaps that is why there are no scenes of tractors beams being used during battle, too much floating debris could be drawn in and destroy the emitters.

5.2.2 *Gravity Beams Exert Force*

Graviton beams create a gravitational force and this force creates spatial stressors on the target in specific areas, allowing an object to be held in a fixed location relative to the emitter, or to alter its trajectory or position (TNG Technical Manual, p. 89). The direction and magnitude of the stressors determines how the target object moves. Sometimes the stress can be too much. Scotty and Kirk accidentally crush a US jet fighter that makes visual contact with the Enterprise even though Spock warns that the primitive craft might not be able to handle the stresses. Before the plane breaks up and crashes in Nebraska, they save the pilot by transporting him aboard the Enterprise (*TOS: Tomorrow is Yesterday*).

In truth, could any tractor beam really work in space? Lawrence Krauss wrote in his book, *The Physics of Star Trek*, a lack of gravity in space would make it difficult to engineer a useful pulling beam. When an astronaut on the International Space Station turns a bolt, she had better be anchored to the floor or else she'll be the one turning. On Earth, the gravitational force creates a different scenario. When a young boy flying a kite pulls it back down, the force applied to the kite and to the boy through the string are equal and opposite according to Newton's third law of motion. The reason the kite moves down instead of the boy rising up is that boy has more mass and his mass is amplified by the gravity of Earth. The smaller mass of the kite is affected less by gravity. By anchoring him down, gravity increases his apparent mass and lets him rein in a kite even though the force of the wind is trying to blow it away from him.

Box 5.1 Newton's Third Law and Pulling an Asteroid

Consider a Federation starship (4.0×10^6 kg) using its tractor beam to pull an asteroid (4.0×10^4 kg) next to it from a distance of 10 km away. Force is equal to mass times acceleration (F = ma). By Newton's third law, the force on the asteroid and starship are the same, just in the opposite direction, so $F = M_{ship} \times a_{ship} = M_{asteroid} \times a_{asteroid}$. Assume the tractor beam exerts a constant force of 4.0×10^4 N on the asteroid and that both objects are initially at rest. Newton's law says they will both move, but how far?

$F_{starship}$ on asteroid $= 4.0 \times 10^4$ N and $F_{asteroid}$ on starship $= -4.0 \times 10^4$ N

$F_{asteroidonstarship} = M_{starship} \times a_{starship}$

$-4.0 \times 10^4 \text{ N} = M_{starship} \times a_{starship}$ where $M_{starship} = 4.0 \times 10^6$ kg and $M_{asteroid} = 4 \times 10^4$ kg

substitute and rearrange

$$a_{starship} = F_{asteroid}/M_{starship} = -4.0 \times 10^4/4.0 \times 10^6 = -1.0 \times 10^{-2}\,\text{m/s}^2$$

and

$$a_{asteroid} = F_{starship}/M_{asteroid} = 4.0 \times 10^4/4.0 \times 10^4 = 1\,\text{m/s}^2$$

$$a_{net} = a_{asteroid} - a_{starship}$$

because both are moving toward each other, the magnitudes are additive

$$a_{net} = 1.0\,\text{m/s}^2 + 1.0 \times 10^{-2}\,\text{m/s}^2 = 1.01\,\text{m/s}^2$$

distance $= 0.5 \times a_{net} \times \text{time}^2$ rearrange and solve for time (t)

$t^2 = 2 \times \text{distance}/a_{net}$

$t = \text{sqrt}(2 \times \text{distance in meters}/a_{net}) = 140.7$ s to impact

use the time to impact and the acceleration of the ship to solve for distance it travels

$$d_{starship} = 0.5 \times a_{starship} \times t^2 = 0.5(1.0 \times 10^{-2}) \times 19802 = 9.901\text{ m}$$

so the starship moves 9.901 m and the asteroid moves 9990.1 m

If the asteroid were as massive as the starship, they would each be pulled an equal distance. Who would be towing whom?

Without an anchor like Earth's gravity, a ship pulling in an asteroid will be pulled in with the same force that it exerts on the rock. This was Krauss' problem with a graviton tractor beam and will have to be overcome if an off-earth tractor is to be accomplished (Krauss, p. 12). See Box 5.1 for an explanation of how far the starship might be pulled. The power from the emitters can be increased for capturing larger objects or for those moving away from the vessel (*TNG; The Masterpiece Society*), but no matter how much force is used, it will act on the ship to same degree as the object they are capturing. The forces and the stresses are greater as the power is increased, and this is why the main emitters are anchored to the main structural elements of the Enterprise. By distributing the force over the entire skeleton, the ship can withstand the stresses of the graviton beam (TNG Technical Guide, pp. 89–90). How can the effects of the third law be managed? Perhaps the engines are simultaneously fired in the opposite direction to balance the force pulling starship towards the asteroid, and as Dr. Krauss points out, this is not discussed in the canon (Krauss, p. 13).

There are three main tractor beam emitters of 16 MW each; two are on ventral aft, close to the nacelle attachment. They each have a precise pattern control system to control the beam pattern. A third emitter is more forward and higher—still on the ventral aspect, just aft of the saucer. All are attached along the keel. The reason for multiple emitters is two fold. Multiple positions allows for capturing and towing objects that are located in various locations relative to the ship, and more importantly, using two wave emitters that partially overlap will create an interference pattern. Where the target object is placed in the interference pattern determines whether the object is held. By controlling the focal point and the interference pattern, the target can be fixed in space, brought closer or moved away (TNG Technical Guide, p. 89). The interference patterns are an integral part of the tractor beam function, and once again Star Trek leads the way. Many of the tractor beams on 21st century Earth also make use of wave interference patterns.

5.3 Current Tractor Beams on Earth

Watching an object move toward a force that should be pushing it away is fascinating. The mental image doesn't square with our experience; we might attribute it to trickery or magic. We may not perceive the forces at work, but just as a fixed wing rises against gravity due to a difference in air pressure on its surfaces, current tractor beams are using physics and forces to produce interesting results—even without gravitons. Many different approaches are being taken, some using mechanical waves and others using electromagnetic radiation; however, only a couple of the current techniques will translate to tractor beams in space. With the limitations of space in mind, dedicated scientists are pursuing many additional methods of tractoring to be put to use in a long line of potential applications. Some of the first efforts even have practical uses right here on Earth.

5.3.1 The Aquatic Tractor Beam Pulls with Waves

One thing Earth has a lot of is water, and that resource that has been exploited to create a tractor beam. If you throw a pebble in a pond, the ripples move from the impact spot and propagate out in all directions. If a leaf is resting on the water, it will be pushed by the traveling wave front. Specifically, it will be pushed away from the source of the wave front. This is a mechanical repulsing system; it is no big deal, we see it at work everyday. On the other hand, a propagating wave that draws objects back toward the source would be a big deal—a tractoring system.

The National University of Australia recently demonstrated a tractor beam that can draw a table tennis ball back towards a wave generator. This idea goes back to the 1800s and George Stokes' idea of "Stokes drift," where the motion of a particle on a wave will travel with the propagating wave. The idea of manipulating the Stokes drift came about when the way dust particles moved in certain laser beams was observed. By generating a wave front with many small waves, three-dimensional vortices will be created and will have a force that meets the ball on the sides instead of in the front. It pushes the ball with an additive effect *toward* the wave generator. By manipulating the wave pattern and size, the ball can be pushed in any X/Y direction, or could generate vortices that hold it in place (Punzmann et al. 2014).

Small particles placed in the water were mapped with computerized tracking software developed by the group in Australia headed by Dr. Michael Shats. The particle movement showed that wave generators induce surface currents along the X and Y axes when the waves exceed a certain height. The height needs to exceed a certain minimum to induce the proper interference pattern that ejects fluid molecules sideways and creates the surface vortices (Fig. 5.1). The mathematics of small

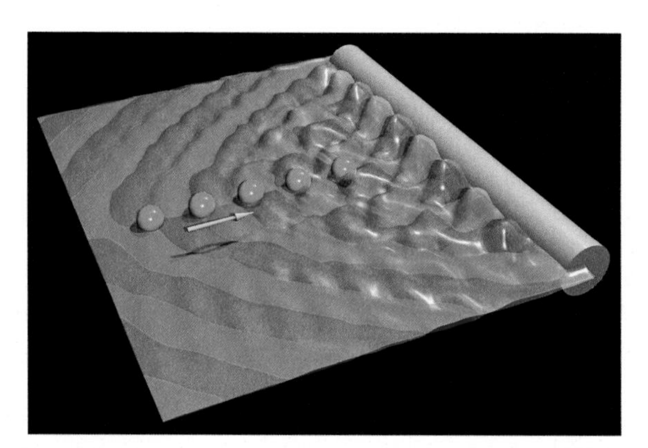

Fig. 5.1 Dr. Michael Shats explained that this image represents a computer reconstruction of the experimentally measured wave fields in their aquatic tractor beam. The *pink* and *blue colors* correspond to two consecutive periods of the oscillations of the cylindrical wave maker. The direction of the surface flow driven by this wave field is shown by the *arrow*. (*Image credit* Michael Shats and Australian National University)

waves is simple; most of the water particles simply move up and down as the peaks and troughs pass under them. But the mechanics of the larger waves and the manner in which they generate surface flows is much more complex. Dr. Shats told me that describing this wave driven flow is a very challenging mathematical problem, luckily it doesn't keep science from using the phenomenon to our advantage.

Think of the usefulness of such a tractor after an oil spill. By tuning the wave generation to counter that of the natural waves, the oil could be confined in an area for easier cleanup. It could be kept away from shore, thus sparing the shore birds, plants, and other wildlife from lethal sticky coats of petroleum. This same tractor may well save human lives at some point too. Over one hundred Americans are drowned each year after being carried out to sea by riptides that move strongly toward the wave generator (open ocean). More than 90 % of rescue runs performed by United States Lifesaving Association agencies from 2000 to 2010 were due to riptides (USLA 2013). The force of the rip current is a result of the energy of the mechanical waves. By investigating and understanding how the aquatic tractor beam moves water, the riptides can be predicted and perhaps measures can be developed to prevent them.

5.3.2 Pulling with Sound, an Acoustic Tractor Beam

Sound waves are mechanical waves, similar to waves in water. They move air as they propagate; air has mass and imparts force as it strikes an object. The force generated by the energy of sound waves in air can be used to do work on objects, such as when it turns turbines and generates electricity. In medicine, ultrasound (frequencies beyond the sensitivity of human hearing) conveys enough force to image tissues and organs, or at higher energies, break up kidney stones or cut through tissue in surgery. High intensity focused ultrasound (HIFU) has recently become popular and effective for noninvasive tumor ablation (reviewed in Hoogenboom et al. 2015). Collaborators from universities in Southampton and Dundee, as well as Illinois Wesleyan University used the force of HIFU to create a tractor beam. The apparatus used in their study of pulling force by sound waves was a commercial medical ultrasound ablation system because the researchers anticipate varied medical uses for the acoustic tractor.

To produce an acoustic tractor beam, two sound waves (550 kHz) are propagated from horizontally placed speakers, but as the wave fronts move out in all directions. Some waves strike the angled sides of the triangular target (about 1 cm long) placed 51° between the speakers, at exactly a 51° angle from each. The deflected waves travel vertically after bouncing off the sides of the target and this transmits momentum to the target in the opposite direction, straight down toward the source of the sound waves (Démoré et al. 2014). The concept works with objects of other shapes, but the phenomenon is best demonstrated with flat angled surfaces because the direction of the deflected waves and their forces were easiest to visualize, straight up and straight down.

Doing the test in water served three purposes. Sound travels better in water because it is denser than air and therefore transmits more force. This made the test results more robust. But just as important, performing the acoustic tractor beam experiment in water let the investigators measure the force imparted to the buoyant triangular target. Any downward movement meant that a force was being applied in the direction of the wave source, with gravity but against buoyancy. By measuring the degree to which the target was pulled down, the tractor force on the target is easily calculated (~ 1 mN, Démoré et al. 2014).

The third advantage to doing the demonstration in water relates to the fact that some real world applications for this type of tractor involve the human body, a watery environment. Instead of ingesting a drug, having it spread throughout the body and hoping an effective amount reaches the target tissue, wouldn't it be better to hold the drug in the target tissue? This technology could do that. Effective doses could be higher, ingested doses lower, and non-specific effects of drugs could be minimized.

A different type of acoustic tractor beam behaves much like the water tractor beam of Puntzmann and Shats above. Interfering patterns of sound waves from multiple miniature speakers create traps that can hold a small particle in place, constructs that the collaborators from Great Britain and Spain call acoustic holograms. A hologram is a three dimensional representation created by the interference of energetic waves. We think of holograms in terms of light, but they occur with many types of waves. In this case, an array of 64 small speakers produces three-dimensional acoustic fields that form specific interference holograms that can trap or cage a particle (Marzo et al. 2015). If the target object begins to move sideways, it comes under the influence of a higher intensity sound front, and is nudged back into the center. The greater pressure surrounds the particle to push it anywhere the user wants it to go. Each speaker gives a single wave front, but based on the pitch and intensity, they can interact differently with adjacent speaker outputs to produce patterns that are additive (peak meets peak) in some cases, cancel each other out (peak meets trough) in others, or produce intermediate forces.

Different patterns of sound intensities and interaction from the 64 source speakers can produce rotations, or even a cage that can trap the target in all six directions (Fig. 5.2). Lead author Azier Marzo described them with colorful names like twin traps, bottle traps, vortex traps and acoustic cages. By adjusting the sound intensities from each speaker, the different holograms can be moved in space so that the trapped target is pushed or pulled in any direction. In many demonstrations, the polystyrene target object is levitated against the force of gravity and is pushed in the direction of the propagating waves like a repulsor beam. Yet the speaker array can be inverted without the particle falling to the floor. In this inverted state, the acoustic holograms can be manipulated so that the target object rises against gravity yet moves toward the speakers—a true tractor beam. Azier told me that the pulling force is harder to manipulate and less stable with the device upside down because the pulling forces are weak compared to the side-to-side and pushing forces. Only a small portion of the waves produce the interference that strikes the particle on its sides and imparts a pulling force.

Fig. 5.2 A polystyrene bead (*white*) is held in place by the acoustic tractor beam. Acoustic holograms formed by many small, focused speakers can be manipulated to push the bead away or draw it closer, even against gravity. (Image courtesy of Asier Marzo, Matt Sutton, Bruce Drinkwater and Sriram Subramanian © 2015)

Azier explains how the sound waves can *pull* a target up in the air, "If the particle is not too large, the traps can recreate their structure on the opposite side of the encountered particle." This can hold the particle in place against gravity, with the vertical twin traps working better than the vortex of bottle traps for this because they have less structure to recreate on the side further from the source. Being able to recreate their structure after interacting with an interrupting particle is much like how metamaterial cloaks can redirect and light waves after passing around an object (Sect. 2.4.2) and how Airy beams permit lasers to turn corners (Sect. 1.7.2). It will also be important for the development of true tractor beams in space using lasers, especially those that rely on theoretical laser outputs called Bessel beams.

5.3.3 Laser-Based Tractor Beams

The aquatic and acoustic tractor beams will have important functions on Earth. Unfortunately, their reliance on mechanical wave systems means that they won't work in space since there is no medium through which mechanical waves can propagate. Even in a thinner medium like air it would is harder to generate enough force to have controlled movement of macroscale objects, which is why the acoustic tractors above are so amazing. Lasers have the *potential* to go even farther; a little force can go a long way in the microgravity of space.

Though photons of light are without mass, they still produce force because they have momentum. The proof is seen in solar sails powering spacecraft and comet tails pointing away from the Sun. The formula for momentum (p) is given by p = ma (mass x acceleration), but how can light have momentum if it doesn't have mass? Think of light as a form of mass that has been converted to energy, as with Einstein's $E = mc^2$. That means that light has mass on a relativistic level, sometimes called inertial mass. Remember that the formula for force is F = ma, so light should impart a force too, right? Not really, since F = ma is a Newtonian physics definition. The mass in this case is called a rest mass, because it does not change whether the matter is at rest or moving with respect to an observer. Since light is never at rest, it has no rest mass. However, when the momentum of light is transferred to an object with mass, it becomes a force acting on that rest mass.

Photons strike a surface and, if the energy is absorbed, the momentum is transferred to the surface. Interestingly, it imparts 2p to the surface if the light is reflected perfectly because momentum is always conserved. The photon approaches with a momentum of p and is reflected with $-p$ (momentum is a vector quantity), so the surface must gain $2p$ to balance the Eq. ($2p_{surface} - p_{re\text{-}emitted\ photon} = p_{absorbed\ photon}$). However, an object might change velocity when it acquires momentum so the system must take in more than just momentum, it must take in energy as well. By the law of conservation of energy the photon must then lose some energy, so the re-emitted photon will be of slightly lower energy (will have a longer wavelength: a red shift). In all cases, there is a pressure transferred to the object, somewhere between p and $2p$ because there are no perfect absorbers or perfect reflectors. The momentum acts as a force because the object has rest mass and accelerates. The end result is that light can push matter around.

Science has learned how to use the momentum of light to capture objects and move them around on a surface with light-based tractor beams, also called photonic tractors or optical conveyors. One technique, euphemistically named optical tweezers, is used in biology quite often, and will also be applied to molecular manufacturing for the development of replicators (Sect. 3.5.2). The tweezers use the differences in force over the radius of a laser beam to move small particles or cells in the X and Y axes, and sometimes even up and down.

5.3.3.1 Optical Tweezers

The momentum exerted on a surface of an object by electromagnetic radiation is termed its radiation pressure. If the radiation imparts an acceleration to a bit of mass, then it is a force that moves the matter, and that force for perfectly reflected photons is twice as large as it is for absorbed photons, similar to the case with momentum. The radiation pressure within a laser beam creates force in two different ways. Momentum along the propagating axis of the beam can be translated to a force that pushes objects in its path, as long as the force of the light is directed through the object's center of mass, analogous to how wind blows a leaf through the air. The second type of force via radiation pressure occurs *within* the laser beam itself.

Many laser beams are circular in cross-section and cylindrical in three dimensions. They have a normal Gaussian radial distribution of intensity; higher intensity in the center with a regular dissipation as you move to the edges. This intensity gradient in a Gaussian laser produces a gradient of pressure that can hold small spherical objects in the center of the beam. This may seem counterintuitive - if there is greater radiation pressure in the center, why doesn't it push the object off to the side and out of the beam's path? The answer is once again found in Newton's third law, but this time the light is being refracted through the target object, as with a living cell (Fig. 5.3). Cells in biology labs can be moved around on slides or in petri dishes using optical tweezers. If part of the beam is off to one or the other side of the cell, the light refracts through it at an angle. The resulting force is directed 180° to the reflection, just as described above for the aquatic and acoustic tractors and this can move the cell. So why does it move back to the center?

As a cell moves out of the center of the beam, higher intensity photons strike it on the more central side of its center of mass and lower intensity beams hit it on the opposite side. The refraction of the beam at each spot imparts an equal and opposite momentum to the target object. Since the radiation pressure is greater toward the center of the beam and that portion is being refracted toward the edge, the equal and opposite force vector actually pushes back *toward* the center. This force is larger than the force vector directed toward the edge of the beam via the less intensity photons being refracted toward the center. The difference in force vectors pushes

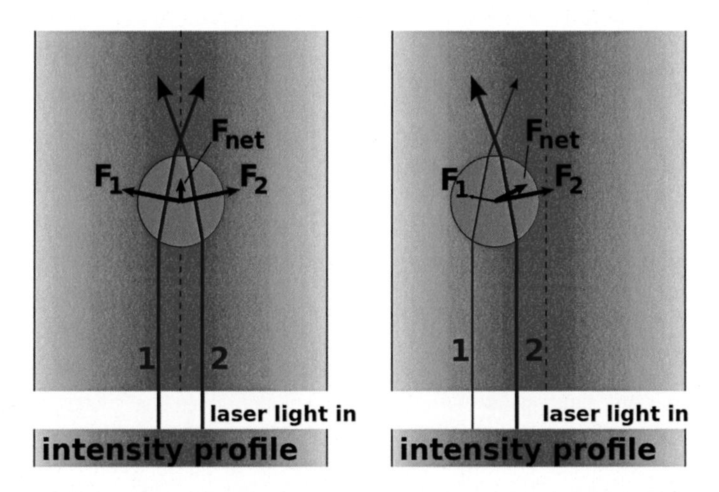

Fig. 5.3 In optical tweezer technology, the Gaussian nature of the beam results in more pressure in the center and less toward the edges, As the target moves out of the center of the beam, the refracted waves create a pressure in the opposite direction, resulting in more pressure pushing the bead back to the center. By slowly repositioning the beam, a target can be moved in specific directions. (*Image credit* Roland Koebler [CC BY 3.0 [http://creativecommons.org/licenses/by/3.0], via Wikimedia Commons)

the object back toward the center of the beam. If you slowly move the beam to the left, the target will also move to the left since momentum changes keep pushing it back to the center of the beam.

There are many variants of optical tweezers using multiple beams, focused beams, and even those that use electrical or magnetic forces in addition to, or instead of light, but all use some version of radiation pressure. If the laser beam is focused through a set of microscope optics, then it becomes focused to a narrow volume called the beam waist. Prior to reaching the beam waist, the radiation forces described above are present and keep a target object in the center of the beam (gradient force) and move it downstream toward the beam waist (scattering force). Beyond the beam waist, the *scattering gradient* still pushes the particle along the propagation direction, while the *force gradient* now draws the particle back toward the beam waist. By tuning the power of the laser and the plane of focus, the forces can keep the particle trapped near the beam waist. The user can move the particle in the X, Y, and Z axes (reviewed in Ramser and Hanstorp 2010). Focusing the beam through the lens also produces an electrical force with its strongest point at center of the beam waist. If the object(s) to be moved or held are dielectric in nature, the generated electric field will also work to keep them in the center of the beam, making the tractor even stronger.

Optical tweezers have many functions in the biology lab. The tweezers can be used to segregate cells on a microscope slide according to cell type so that pure cultures can be grown. Specific cell types can be placed with other cell types to manipulate or observe their chemical or physical interactions. The force a single molecule or cellular structure places on another can be measured by tuning the laser to an intensity that barely breaks or allows the interaction. This method has been used to determine the force that molecular motor proteins (kinesins) generate when they walk on scaffolds of microtubules (Hesse et al. 2013) or measure the binding force between DNA and certain enzymes (Bai et al. 2006).

Box 5.2 Light Can Attract And Repel Itself

In 2005, a group from Yale University led by Dr. Hong Tang proposed the hypothesis that light could both attract and repel *itself*. In short order the group was able to show that infrared light split into two beams and passed through a silica nanowire waveguide produced an attractive force between the two parallel guides. However, it took another four years to demonstrate that light can repel light (Li et al. 2009).

Light travels at the same speed with the same wavelength in each waveguide and will stay in phase, *unless* one part of the split beam is forced to travel a slightly longer distance. Then the beams will be out of phase when brought back together. *In phase* light beams have crests and troughs that match up; *out of phase* light waves do not line up. When the researchers forced the light in one waveguide to travel further, the two beams repelled one another when laid side by side.

This is an unusual situation; usually opposites attract. Positive ions in chemistry will repel other positive ions and be drawn to negative ions. The north pole of a magnet is drawn only to the south pole of another. In contrast, light beams with opposite phases are repulsive, with beams that are 180° out of phase being the most repulsive. The phase can be manipulated by changing the length of the one of the waveguides or by changing the wavelength of the light slightly in one guide. Using a tuning knob, the system can be changed from repulsive to attractive.

The force generated is perpendicular to the direction of propagation. The waveguides, lain side by side, are drawn to or pushed away from one another and can be used to do work (Fig. 5.4). A mechanical switch for a microchip can be connected to the waveguides and activated or deactivated by the force —microchips could be powered by light instead of electricity. In proof of this concept, the Yale group demonstrated workable, nanoscale "optomechanical" controls for electronic couplers in an all-optical system (Fong et al. 2011). The attractive and repulsive forces are very small, work over very micrometer distances, and only work in the silica guides, not in air. I doubt Commander Spock will ever order Sulu to activate the tractor chip.

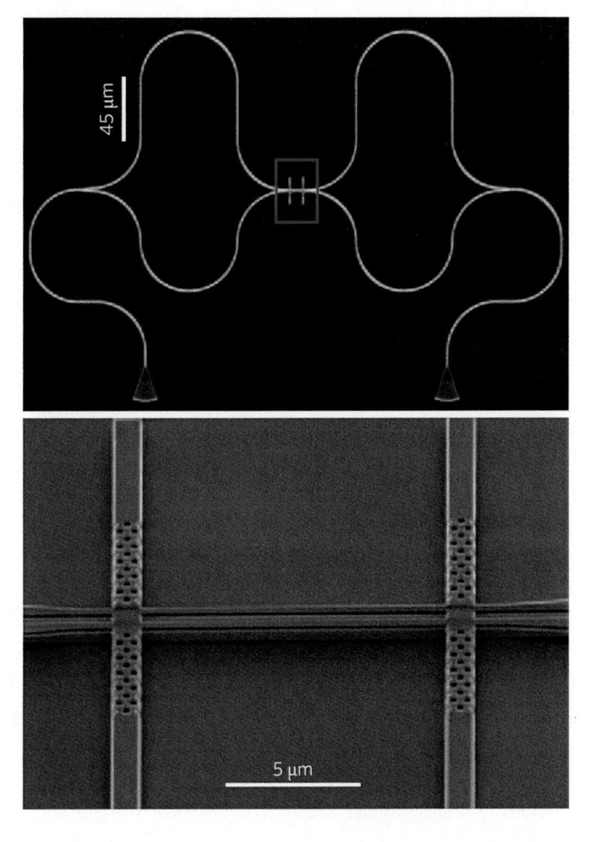

Fig. 5.4 Light passed through a waveguide can be split into two beams. If the two portions of the split beam travel different distances as seen by the different loop lengths up and down in the top image, the waves will be out of phase when brought close together, as shown in the lower image. The difference in phase generate a force that pushes the two waveguides apart. (*Image credit* HX Tang, Yale School of Engineering and Applied Science)

Most optical tweezers techniques use radiation pressure to push (scattering pressure) or move particles laterally (gradient pressure). In order to pull a particle toward the source without changing the focal plane, multiple beams must be used to create interference patterns and a gradient pressure that imparts a force vector toward the beam source. These techniques may be important on a cell level, but the force generated is not large enough to make them useful in space. Perhaps the power of the laser could be increased greatly, although the increased energy absorbance creates its own problems—you don't want to ignite what you are trying to pull in.

5.3.3.2 Bessel Beam Approximations

All the tractors described above; tractors in water, those using sound, optical tweezers with single or multiple beams; they are all just early attempts at the holy grail of laser tractors, a Bessel beam. Named for Friedrich Bessel, the German astronomer and physicist who first developed differential equations to deal with "many-body" gravitational systems, Bessel beams are theorized to do some amazing things. The central core will never diffract, it will stay focused and the same size from here to the end of the universe. Your laser pointer *seems* to stay focused because the distance to your projected powerpoint slide is so short. If you could see the beam when it reached the Moon (unlikely given the dust and atmosphere, and low power of the laser), the dot would cover the entire face of our only natural satellite.

Even more amazing, a Bessel beam's central core can reconstruct itself *beyond* an object that lies in its path. Like a metamaterial cloak that bends light around an object and then lets it continue on its way, a Bessel beam could scatter off an object placed in its path, while parts of the outer beam (if not also blocked) assemble another central core directly behind the object. How Bessel beams would do this is related to the interactions of the parts of the beam, not by bending light around the target as some metamaterial cloaks do.

Bessel beams have a unique distribution of light that produces a series of concentric rings. Portions of the beam converge on one another through space to produce an interference pattern. When two peaks interfere, the resulting point is twice as intense as the surrounding area, so the central core is the brightest part of the beam. When a peak and a trough interfere, they cancel each other and not beam is propagated. The canceled waves correspond to the dark portions between the concentric rings. The outer rings represent interference where non-crests and non-troughs become additive to greater or lesser degrees, depending on the phase of each when they interact.

The reconstruction of the central core and the additive and subtractive interferences that produce the rings also mean that if one precisely controls frequency of light, a target object in the beam can be struck with photons that scatter forward instead of backward. The target would experience negative scattering radiation pressure and move toward the source, since the force is again equal and opposite.

This is a laser version of the interference patterns seen in the aquatic and acoustic tractors described above, but uses the radiation pressure of electromagnetic waves instead of mechanical waves. Bessel beam negative radiation pressure was proposed in studies a few years ago at the University of Central Florida (Sukhov and Dogariu 2010) and from the Netherlands and Singapore (Novitsky et al. 2011), with both studies concluding that the target object to be pushed toward the laser source would have to be smaller than the wavelength of the light—not a good sign for trying to move asteroids.

A true Bessel beam, one that does not diffract as it propagates, would take an infinite amount of energy. Thankfully, approximations of the interference patterns can be made with lenses that are slightly conical, called axicon lenses. Using axicons, the beam will refract toward the center (Fig. 5.5). In the area where the converging waves interfere, the central core and concentric rings will be formed. This is different from a true Bessel beam in which the rings and central core would be maintained over the entire length of the beam. A single beam approach did not achieve the fine control for movement that was needed by a group at NYU in 2011, so they chose to use two separate converging beams, which they termed "coherently superimposing coaxial Bessel beams" (Ruffner and Grier 2012).

By crossing the two beams (not as catastrophic as crossing the streams in Ghostbusters), a strobe effect was possible in the central core on the back end of the target object when the two beams manifested periodic intensity changes (concentric rings). The rings act as the traps. Changing the phase of the beam(s) then creates the negative radiation pressure on the back surface of the target and pushes it toward

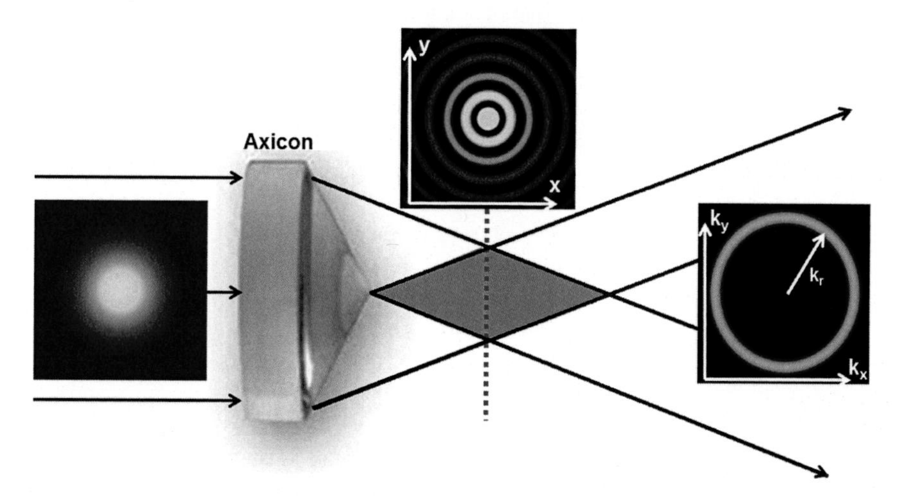

Fig. 5.5 An axicon lens can turn a laser beam into a series of converging interference patterns, some constructive, some destructive. Where the rays interact most, a zero order Bessel beam is approximated, with a bright center because that is where the interference is most constructive. At this point, if an object is place in the path, the Bessel pattern can actually reform behind the object. (*Image credit* Andrew Forbes/Structured Light Laboratory/University of the Witwatersrand, Johannesburg, South Africa/from *Optics and Photonics News*, June 2013)

the source using the strobe effect. Their apparatus was able to move 30 μm silica spheres in water toward the laser source. Water was used with buoyant targets to counter the force of gravity, suggesting the negative radiation pressure is weak. However, with the lack of gravity in space, it might work just fine. The system could increase the mass moved by increasing the energy of the beams, although a limit would be approached where the energy of the beam would negatively affect the target (melt it).

Increasing the power of a visible light or infrared laser would impart more force, but more energy would not always mean a bigger pull. One way to increase the energy of the laser beam is to use a higher frequency wave, an X-ray laser for example. Now that these lasers are more practical (Sect. 1.7.1) wouldn't they impart more momentum to the target? Not really. With a much shorter wavelength, X-rays are absorbed or pass through more materials than do visible light or ultraviolet lasers. Since the greatest force is generated when the beam is perfectly reflected (2 times the momentum) having a significant portion of the beam absorbed will actually decrease the movement of the target. Apparently, "Scotty, I need more power!" is not always the best idea, so perhaps it is fortunate that Kirk never actually says that line.

5.4 Saving the Earth with Tractor Beams

The current tractor beams are well designed for their intended purposes. For applications like medicine, biology, nanocomputing, or molecular manufacturing, small forces are adequate and may be preferred for precise control and fine-tuning. On a slightly larger scale, NASA has gotten into the business of small particle tractor beams. In the past, NASA has used several means to acquire small particles from space for analysis. The 1999 Stardust mission flew through the tail of a comet and trapped dust in an aerogel matrix, while the 2006 Genesis mission used ultrapure discs of several elements to collect solar wind particles. In 2011 NASA contracted for a new mechanism for capturing particles (Keesy 2011), with three methods coming under review; optical tweezers with two beams, a solenoid beam based on electromagnetic principles that were demonstrated several years ago (Lee et al. 2010), and a more theoretical version of a Bessel beam. While these options are needed and practical, they don't represent what the public thinks about when tractor beams in space are mentioned. People want science to move ships and asteroids.

5.4.1 Magnetic Field Architecture for Repulsor Forces

When someone asks whether science can produce a tractor beam, what they're really asking is, "Can science produce a tractor beam like in *Star Trek*?" Can current engineering and physics build a device that can pull a large object back to the source of the force? The answer is a solid yes, although it isn't the kind tractor that many people conjure up in their mind's eye. People imagine Spock or LaForge remodulating a graviton beam to move huge chunks of space debris out of the way or catching a renegade starship and towing it back to a starbase so the rogue captain can receive his just punishment. Presently, we don't have a need for a tractor beam to capture starships, but we could use one for docking miniature satellites together or to the International Space Station. To make this happen, NASA has partnered with Arx Pax (Los Gatos, CA), the maker of the Hendo Hoverboard (Arx Pax press release, Sept. 02, 2015).

Developer Greg Henderson is an engineer and an architect who started studying magnetic levitation as a way to protect buildings from earthquakes, and this is just one of many practical applications of his hover engines. In fact, the hoverboard was just a way to bring attention to the engine so that it could be exploited for the public good. The hoverboards were first made available to the public on "Back To The Future Day," October 21, 2015 for $10,000 each. Marty McFly traveled forward in time from 1985 and was amazed by many things, with the hoverboards being one of the most memorable items for Marty and the audience. Several companies have tried to make hoverboards, with most "hovering" in name only since they have wheels. Several also have lithium batteries that spontaneously catch fire and are appropriately banned on US airlines. The Hendo hoverboard is a bit different.

Each Hendo board has two hover engines, the first commercial use of a technology Henderson calls magnetic field architecture (MFA). Lenz's Law states that a changing magnetic field will produce a similar magnetic field in a conductor placed within the lines of the oscillating field. The changing field creates electrical eddy currents on the surface of a conductor and these currents generate a second magnetic field, again based on Lenz's Law. The two magnetic fields will repel each other so the hoverboard will float over a non-ferrous, conducting surface. The result is levitation, although you can only ride the board on a sheet of aluminum, copper, silver or similar metal surface because it requires a conducting surface. Fortuitously, satellites and space faring vessels have aluminum skins.

NASA believes that MFA technology could solve a problem for them. The space program will be putting cubesats, 4-inch cubic satellites, into space starting in 2017 (Siceloff 2015). They are excellent for relaying signals and performing experiments, but are too small to include thrusters (Fig. 5.6). What NASA requires is a non-contact way to move the cubesats around in space, to dock them to each other, and to capture them for maintenance. Over the next few years, Arx Pax and NASA will work together to use the attractive and repellent forces of magnets to position cubesats and join the modules together into multiples of three or six. The cubesat maneuvering system will finely balance the attractive forces of magnets to some metallic surfaces with the repulsive forces of MFA to capture, hold and pull and push these tools in space. This is a true tractoring method over short distances that will do crucial work in the vacuum of space. It may not be flashy, but it is important.

Fig. 5.6 This cubesat is being prepared for a thermal vacuum test to see how it will hold up against the cold and vacuum of space. The size of the cubesat is demonstrated well in this image, although they are modular and can be joined to form larger units. NASA hopes to capture and move these satellites in space using magnetic forces similar to those that move the Hendo Hoverboard. (*Image credit* NASA)

5.4.2 The Real Threats to Earth

Threats to Earth from outer space are numerous and varied. Asteroids, meteors, comets, and space debris from dead satellites could impact us and do damage. Those objects that present the largest threat to come close to Earth are being catalogued at several institutions across the world, under names like SpaceGuard, SpaceWatch, Near Earth Asteroid Tracking, and NEOShield. The number of Near Earth Objects (NEOs) is expanding all the time, including NEOs of 10 km diameter or larger that could lead to an extinction level event were they to impact Earth.

A four-meter diameter meteoroid, 2008 TC3, impacted Earth on October 7, 2008. The predictions for when and where it and Earth would attempt to occupy the same space were very accurate, based on observations from several places on Earth and using the parallax of the different observers. The entrance into the atmosphere was predicted with an uncertainty of just 15s, and it impacted in the Nubian Desert in Sudan just as predicted (Jennesikens et al. 2009). The problem with the scenario is that the first observation of the object was just twenty hours before it made impact. A larger asteroid would undoubtedly have been seen earlier, although there still not in time to do anything about it. TC3 2008 was the first, and is still the only, space rock we have discovered *before* it impacted Earth. Finding them is hard, and the problem is multiplied many times when it comes from the direction of the Sun.

Over 20 % of NEOs that come within 0.01 AU approach Earth from within 30° of the Sun. Observation during the day will not detect these due to sky-brightness

(Isobe and Yoshikawa 1996). To try and compensate for this blind spot, a private enterprise, the B612 Foundation, proposes to build and orbit a space telescope around Venus. Since its lens will always point away from the Sun, they predict that 90 % of the NEOs of more than 460 ft. in diameter will be detected. Unfortunately, the project failed to meet certain key deadlines and NASA terminated its Space Act Agreement with the Foundation, pulling its thirty million dollars of promised support in the process.

Box 5.3 Near Earth Objects

There are several different types of Near Earth Objects (NEOs), classified by their origin, composition, orbit, and potential danger to Earth.

NearEarth Asteroids (NEAs) Asteroids are defined as meteoroids that have diameters larger than 10 m, although definitions can vary. They are most often classified by their orbit. The asteroid must come with 0.3 astronomical units (AU, the mean distance from the Earth to the Sun) of Earth's orbit to be considered and NEA. There are more than 10,000 known NEAs; 1 % are 1 km in diameter or larger. The types of NEAs are:

Atiras—asteroids with orbits completely confined within the orbit of Earth, also referred to as Interior-Earth Objects (IEOs). The first was discovered in 2003 at the Lincoln Observatory at MIT. Seven more have been discovered since then.
Apollos and *Atens*—asteroids with orbits that cross that of Earth, spending some time outside the orbit of Earth and some time within it. Apollos have more circular orbits than do Atens. Apollos make up just under 2/3 of all known NEAs.
Amors—asteroids with orbits that are completely outside that of Earth's, but can approach our planet at some point of their path.

Near Earth Comets (NEC) usually have velocities two to four times that of asteroids. They have a tendency to orbit out of the plane of Earth, which means they will be less likely to be spotted on routine scans and have a higher angle of inclination when the crash into us, thereby transferring more energy to the planet. Their periods are usually longer, so if detected early they could be deflected away over a longer time. Currently, there are 171 known NECs being tracked by NASA.

Potentially hazardous asteroids (PHAs) are any NEO with a predicted orbit that passes within 0.05 AU of Earth. As of 2014, 1642 PHAs have need identified, with 150 or more considered planet killers.

In the time period between December 14, 2015 and January 14, 2016, NASA estimated that 80 NEOs passed within 0.19 AU of Earth. The closest object was a 10–20 m diameter meteoroid and passed within the orbit of the Moon. For comparison, an iron containing meteorite just 30–50 m in diameter caused the 1.86 km diameter crater at Winslow, AZ. Whether a meteor burns up completely in the atmosphere or reaches the planet's surface depends on its composition, size, speed, and the angle at which it strikes the atmosphere.

(Sources: International Astronomical Union and NASA)

Once the orbit of a NEO is determined observationally, the danger it poses to Earth on future orbits can be determined with a bit more accuracy. For each unique orbit, based on speed and path, there is a small window of three-dimensional space called a gravitational keyhole. If the NEO passes through that keyhole, the likelihood of a collision course with Earth is greatly increased. An asteroid called 99942 Apophis passed through a keyhole in 2004, and was apparently headed for a collision with Earth on its 2029 or 2036 orbit. Fortunately, the scenario was re-evaluated on its 2013 passage by Earth and found to be inaccurate. Based on the new observations of the orbit, the chances of a collision between 99942 Apophis and Earth in 2029 or 2036 were reduced to near zero. Therefore, NEOs that pass through a gravitational keyhole may not present certain danger, while NEOs that miss a keyhole may still pose a threat. This is why collision avoidance strategies are a subject of strong research.

5.5 Near Earth Object Collision Avoidance Strategies Using Beams

Just as there is more than one way to command a starship, there are many ways to try and avoid global catastrophe. Kirk would blast an asteroid, so there are kinetic impactors to reduce NEOs to small pieces that will burn up in the atmosphere. Picard would motivate the asteroid to follow a different path, so deflecting techniques are being researched just as vigorously. Kathryn Janeway is tough yet caring; she might prefer methods that combine impacts and deflections. The kinetic impactors try to reduce NEOs to rubble; run a large rocket into its surface, explode a nuclear weapon on or near it, combine an impactor with a subsurface nuclear weapon. With more great minds being put to work on this problem, the methods for deflecting asteroids are growing more diverse, including several that meet the criteria for tractor and/or repulsor beams. Since this chapter is about tractor beams, we'll stick to deflecting strategies.

Both the Earth and the NEO are hurtling through space. If either one speeds up or slows down, its trajectory will change so that the two will no longer be on a collision course. Since the Earth is big and the asteroid or comet is hopefully much smaller, it would be easier to alter the course of the NEO than our planet. The asteroid needs to be moved at least two Earth radii; just how long that takes depends on the angle at which the two bodies approach one another. The closer to perpendicular the trajectories are, the less an asteroid will need to be sped up or slowed down to avoid impact. The time required to effect a one Earth diameter change also depends on how early the deflection is initiated. The earlier a push is begun, the greater the trajectory change will be. If impact is a short time away, the push will need to be much stronger. These are all things that science will have to manage in each and every case of an NEO; however, it is always better to start deflection earlier rather than later.

Several of the current tractor beam methodologies use mechanical waves (aquatic, acoustic, tunable hollow lasers) and these techniques just won't work in a space without a consistent medium. This leaves particle beams, light beams and gravity beams as the viable methods for deflecting NEOs. Radiation pressure from current lasers doesn't produce enough thrust to deflect an NEO in any practical time frame, although that doesn't mean that lasers are useless when it comes to NEO avoidance strategies.

5.5.1 Ablation by Repulsive Laser Beams

Strong laser light, or enough focused starlight, will produce sufficient heat on the surface of an NEO to burn away its surface. The surface particles that absorb the thermal energy turn directly to gas (sublimation) and shoot away from the surface of the object, producing a force that pushes the object in the opposite direction. Two major efforts are underway to study this as an NEO deflection technique, at the University of Strathclyde in Scotland and the University of California at Santa Barbara. The UCSB DE-STAR project (Directed Energy System for Targeting of Asteroids and exploRation) proposes using a 1 km^2 array of 1 kW near infrared lasers to ablate the surface of celestial objects to produce thrust (Lubin et al. 2014). One of the main advantages of this method is that the propellant is the NEO itself, as long as the laser has power, the propulsion will continue.

The DE-STAR system is designed to fly in low Earth orbit and blast away at NEOs as they pass close to Earth—before they pass through a keyhole confirming their collision course with Earth. If the keyhole has been navigated and the NEO needs to be deflected before it comes back to Earth, the DE-STARLITE is a more appropriate system. This is a short distance system, where the laser array system is sent to the NEO and enters into its orbit (Kosmo et al. 2014). Though a longer distance from Earth, the DE-STARLITE system is a smaller and lower power system, so it is more feasible right now as compared to the long distance DE-STAR system shot from Earth. Each system is designed to raise the surface temperature of the NEO to approximately 3000° K, the threshold for vaporizing all known asteroid and comet constituents.

The Strathclyde Advanced Space Concepts Laboratory is in favor of using multiple smaller (1 kW) lasers that focus on the same area of the object. Their "laser bees" would present an advantage because if one had a mechanical problem, the rest would still function. They also hypothesize that multiple lasers are better because they would be lighter and therefore cheaper to put into space, and this system would be scalable, more lasers could be added to achieve adequate thrust against a bigger target. The disadvantage is that the smaller lasers must be placed close to the NEO, so they will likely be contaminated with the ejecta, causing their function to degrade over time (Kahle et al. 2006). The usefulness of the laser bees has been investigated in the lab (Housen 2004) and high-energy light power has been modeled for results in space (Vasile and Maddock 2010). Both systems remain in development.

Laser ablation produces thrust via sublimation of the rocky material to create thrust in the same direction as the laser beam propagates. If the laser is placed ahead of a NEO, the thrust will slow it down. If it is placed behind the object, the force will speed it up. If the lasers are adjacent to the moving NEO, then the orbit will be changed either against or with the gravity of the Sun. The variables are the generated force, based on the power of the laser(s), the make up of the objects surface, and the time to impact. Any sufficient change in acceleration (a vector quantity, meaning that it is a specific direction) will prevent a collision; however, laser ablation methods aren't true tractors because the force is always away from the source of the laser. Both the UCSB and the laser bee system will use solar electrical propulsion. As long as the Sun shines, the spacecraft and the lasers will remain powered and pushing the NEO. This is important, especially for the DE-STARLITE and laser bee systems since it may take decades for them to deflect the NEO away from a collision course with Earth.

5.5.2 *Ion Beam Shepherds*

Instead of a laser pushing the target by ejecting surface material away from the surface, an ion propulsion system on a spacecraft could pummel the surface with xenon ions. These may not be hot enough or strong enough to ablate the surface, so they rely on the force they transmit to the NEO in the opposite direction when they bounce off its surface. Individual ions may not carry much momentum, but billions of them have a detectable force over time. A similar method has been proposed using a conventional rocket engine firing against the NEO surface, although the ion thruster would be more energy efficient and could act over a period of years or decades. Using two known NEOs of 130 and 140 m diameters to model the system, the ion thrust could achieve a two Earth radii change in position over a two-year thrust followed by an eight-year coast (Bombardelli et al. 2013).

As is always the case, other factors must be considered when choosing a method for NEO deflection. The best collimated ion engines produced now have a divergence of ten degrees, so some of the ions will miss the target if the craft hovers more than a couple of NEO diameters away from the surface. Also, two engines are required for the process, one to produce thrust against the target, and a second in the opposite direction to maintain the spacecraft's position relative to the NEO. The need for twice the propellant makes another technique more appealing in this regard. Instead of creating a pushing thrust with chemical propellant, pulling the NEO off target using gravity is a technique under serious consideration.

5.5.3 Gravity Tractors

Dr. Dan Mazanek leads NASA's efforts to deflect NEOs away from their collision course with Earth. A true child of *Star Trek*, Dr. Mazanek told me that the series definitely influenced his feeling that humankind needs to venture out into the galaxy. The original series was what he grew up on, and he watched later iterations as they came along. One can only imagine that the tractor beams of *Star Trek* appear to him as he works on ways to prevent collisions with NEOs. However, instead of using an exotic graviton particle beam to pull on NEOs, Dr. Mazanek's team is investigating a strategy that will tug on asteroids with plain, old-fashioned gravity.

The "gravity tractor" that NASA is designing makes use of the natural gravitational attraction of two bodies for one another. First proposed by Ed Lu and Stanley Love a decade ago (Lu and Love 2005), NASA is now planning the Asteroid Redirection Mission (ARM) for launch in 2020 to demonstrate the power of a gravity tractor on a Near Earth Asteroid (NEA). Gravity isn't a particularly strong force, so this technique requires a long time to work. Any targeted NEO will have to be engaged more than one orbit before its predicted collision with Earth. This highlights the importance of programs for identifying NEOs that might impact Earth, and eliminates the gravity tractor as a method of deflecting an NEO like TB 145 that came out of nowhere this past Halloween.

Any spacecraft that orbits an NEO around its leading or lagging end will gently tug on it via gravity. Depending on the position, the plan is to have the orbiter slow down or speed up the target enough to change its trajectory by one Earth diameter. Gravitational force between two objects is increased if the spacecraft is large and if the distance between the two objects is small; therefore, the distance between the gravity tractor and the target must be small enough to maximize the gravitational force yet large enough to remain safe.

Two possible techniques are being considered for positioning the craft in front of the target NEA. An in-line position would maximize the gravitational force, but the plume of the thruster would push against the target and negate the attractive force. To avoid this, the thrusters could be positioned to fire at angles from the gravity tractor craft, although this would require more propellant. The other option is to put the craft into a "halo" orbit, positioning it in front of and just outside the diameter of the asteroid. A rocket with a backward thrust would not work against itself in this configuration. In addition, the halo option increases the gravitational force by decreasing the distance between the craft and the asteroid (Fig. 5.7).

So is it better to position the tractor orbiter in front of the asteroid and speed it up, or behind the target and slow it down? Dr. Mazanek says it does make a difference. He stated, "Given sufficient warning time, the best direction to provide a deflection is along the orbital path of the NEO. This gives the greatest change in the NEO's orbit and the best chance for it to miss the Earth." However, whether speeding it up or slowing it down is a trickier question. Are you altering the speed as the asteroid moves away from Earth on another orbit, or as it approaches Earth on a collision course? Speeding it up as it moves out into the solar system will

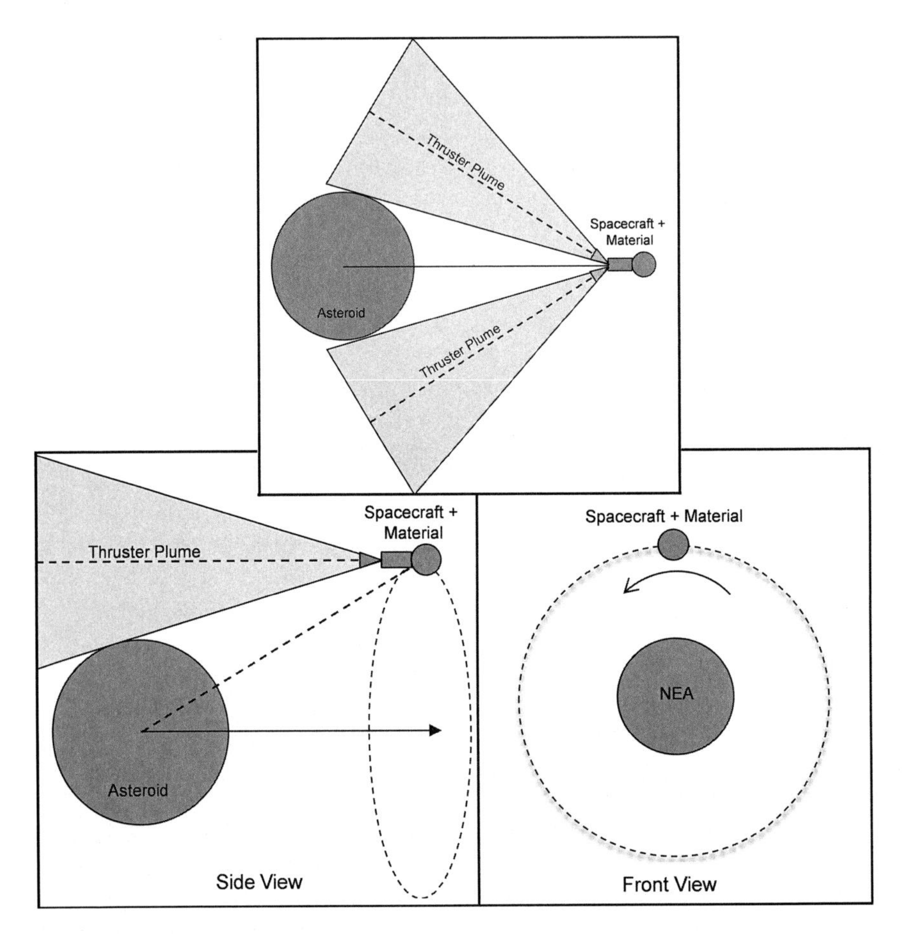

Fig. 5.7 The drawings represent two possible configurations for positioning a gravity tractor near a NEO. The inline method (*top*) maximizes gravity yet would reduce its effectiveness by spewing exhaust against the object. Therefore, NASA would cant the exhaust away from the surface even though more propellant would be needed. The halo orbit (*bottom*) allows the orbiter to get closer, increasing the gravitation pull. (Images courtesy of NASA)

lengthen its orbit, so it will arrive at Earth later. On the other hand, speeding up as it approaches Earth means it will arrive sooner. There are many questions that have to be answered before deciding from which direction to tug on an asteroid.

Mazanek is quick to point out that no matter what direction you choose, there are going to be unhappy people. The *risk corridor*, the possible point of impact and its probability cone, changes once you start moving the asteroid. Tractoring may move the most probable impact site away from Poland for example, but if the final deflection is less than one Earth diameter, you will have increased the chance of it landing on Russia or France, depending on which direction you pushed it. It's politics on an interstellar stage.

5.5.4 Enhanced Gravity Tractors

A force generated by a gravity tractor is dependent on the mass of the orbiter and the asteroid, with only the mass of the orbiter is under NASA's control. Even this control is only partial, the orbiter design must balance added mass against the fuel costs of getting a larger craft to the target. A more massive tractor craft would reduce the needed tractoring time, which may be crucial for success if the NEO is discovered late, so other ideas have been investigated to increase the mass of the vessel. One of these ideas combines a boulder capture mission with the tractoring mission, resulting in an Enhanced Gravity Tractor (EGT). This is the technique that Dr. Mazanek's team will test during the ARM. The EGT method adds a landing mission on the NEA prior to establishing a halo orbit to alter trajectory. The spacecraft will land on the asteroid and pick up a five to ten ton boulder. It will then lift off again and establish a halo orbit to alter the NEA's speed and orbit via gravitational force. The craft + boulder will be more massive than the craft alone, so the time for tractoring and the propellent needed will be significantly reduced (Fig. 5.8).

The limit on the mass of the boulder to be captured is a function of the propellant system used to put the EGT craft back in orbit after capture. The capture system as currently designed can accommodate up to 70 tons (1–4 m in diameter), and could be modified for larger masses as the propellant problem is managed. Several possible methods for capturing the boulder are being considered, including nets and grappling arms. If an appropriate boulder cannot be identified or targeted, the craft will have a backup strategy permitting it to collect a large amount of surface rubble (regolith) for the same purpose, or it may collect several smaller boulders. Depending on the size of the NEA to be deflected, the power of the EGT craft and the mass of the added boulder, the time needed for deflection could be reduced by months as compared to the unaided gravity tractor (Mazanek et al. 2015).

The difficulty in landing and taking off from the NEA is balanced by the reduced time needed for tractoring and the reduced power required for getting the smaller version of the orbiter to the target. When launched, the ARM project will demonstrate many of the techniques needed for EGT deflection of an NEA: the craft will enter an orbit around the target; identify an appropriate boulder to procure and

Fig. 5.8 The cartoon shows one possible way the lander of the enhanced gravity tractor could pick up a boulder to increase its mass and the resulting gravitational force on the NEO once it lifts off with the boulder and orbits the object. (Image courtesy of NASA/Analytical Mechanics Associates, Inc.)

do several dry runs before landing on the NEA capturing it. The craft and boulder will tractor the NEO for two months, a long enough period to measure and study its effect; and then return to low Earth orbit and release the boulder into a stable orbit around the Moon.

Successful completion of the ARM will demonstrate a remarkable facsimile of a *Star Trek* tractor beam. The use of gravity is the same as the *Star Trek* use of gravitons. The target will be moved toward the source of the beam, a true tractor. The ARM will modify the trajectory of a space object, just like *Star Trek* does in several episodes, and the ARM will deliver a piece of space rock into a stable orbit around the moon, allowing science to learn from it for years to come.

5.6 Conclusion

In the most literal sense, tractor beams aren't a technology of science fiction any longer. Science has shown us how to move objects in the opposite direction as they would be pushed by regular forces, and they can do work for us, even if it is only on the level of a cell. Powerful tractor beams will require the ramping up of power in lasers or generation of stronger mechanical waves, while current and future uses in medicine, biology and even for herding oil spills, will rely more on precision than raw power.

It is when considering tractor beams in space that we see the need for power. Space both harms and helps efforts to make more powerful tractors. The vacuum of space means that mechanical waves are useless, although the presence of micro-gravity means that small forces could have dramatic effects, especially if delivered over a long period of time. While laser and magnetic forces offer strong possibilities as useful tools for moving objects in space, perhaps the most applicable force is the one we have had all along, gravity. If not through the discovery of the graviton, then with massive orbiters strategically positioned, gravity may be the first true tractor beam that could save the Earth from destruction.

Near Earth objects and a possible collision with Earth represent the most dramatic need for tractor beams, but as space exploration becomes more commonplace in the decades ahead, more practical uses for tractors will present themselves. Construction of space stations, positioning of telescopes and satellites, retrieval of potentially dangerous space debris—these are just some of the jobs that tractor beams could do safely and efficiently. However, that saving the Earth thing still has to rank right up there with reasons to continue tractor beam research. *Star Trek* and other science fiction stories may have introduced us to the tractor beam, and that idea might very well end up saving us all. And isn't the pursuit of knowledge for the good of mankind the very essence of *Star Trek*?

References

L Bai, TJ Santangelo, and MD Wang. Single-molecule analysis of RNA polymerase transcription. *Annu Rev Biophys Biomol Struct* 35; 343-60, 2006. doi: 10.1146/annurev.biophys.35.010406.150153. http://www.annualreviews.org/doi/abs/10.1146/annurev.biophys.35.010406.150153

C Bombardelli, H Urrutuxa, M Merino, J Palaez, and E Ahedo. The ion beam shepherd: a new concept for asteroid deflection. *Acta Astronautica* 90(1); 98-102, 2013. doi: 10.1016/j.actaastro.2012.10.019. http://aero.uc3m.es/ep2/docs/publicaciones/bomb13.pdf

CEM Démoré, PM Dahl, Z Yang, P Glynne-Jones, A Melzer, S Cochran, MP MacDonald, and GC Spalding. Acoustic Tractor Beam. *Phys. Rev. Lett.* 112; 174302, 2014. doi: 10.1103/PhysRevLett.112.174302. http://journals.aps.org/prl/abstract/10.1103/PhysRevLett.112.174302

KY Fong, WHP Pernice, M Li, and HX Tang. Tunable optical coupler controlled by optical gradient forces. *Optics Express* 19(16); 15098-15108, 2011. doi: 10.1364/OE.19.015098. https://www.osapublishing.org/oe/abstract.cfm?uri=oe-19-16-15098

WR Hesse, M Steiner, ML Wohlever, RD Kamm, W Hwang, and MJ Lang. Modular aspects of kinesin force generation machinery. *Biophysical Journal* 104(9); 1969-1978, 2013. doi: 10.1016/j.bpj.2013.03.051. http://www.cell.com/biophysj/abstract/S0006-3495(13)00390-1

M Hoogenboom, D Eikelenboom, MH den Brok, A Heerschap, JJ Futterer, and GJ Adema. Mechanical high-intensity focused ultrasound destruction of soft tissue: working mechanisms and physiological effects. *Ultrasound Med Biol* 42(6); 1500-1517, 2015. doi: 10.1016/j.ultrasoundmedbio.2015.02.006. http://www.umbjournal.org/article/S0301-5629(15)00198-2/pdf

KR Housen. Collisional Fragmentation of Rotating Bodies. 35th Lunar and Planetary Science Conference. March 15-19, 2004. League City, TX. Abstract 1826.

S Isobe, and M Yoshikawa. Asteroids approaching the earth from directions around the sun. *Earth, Moon, and Planets* 72; 263-266, 1996. http://articles.adsabs.harvard.edu//full/1996EM%26P...72..263I/0000264.000.html

P Jennesikens, MH Shaddad, D Numan, S Elsir, AM Kudoda, ME Zolensky, L Le, GA Robinson, JM Friedrich, D Rumble, A Steele, SR Chelsey, A Fitzsimmons, S Duddy, HH Hsieh, G

Ramsay, PG Brown, WN Edwards, E Tagliaferri, MB Boslough, RE Spalding, R Dantowitz, M Kozubal, P Pravec, J Borovicka, Z Charvat, J Vaubaillon, J Kulper, J Albers, JL Bishop, RL Mancinelli, SA Sanford, SN Milam, M Nuevo, and SP Worden. The impact and recovery of asteroid 2008 TC(3). *Nature* 458(7237); 485-488, 2009. doi: 10.1038/nature07920. http://www.nature.com/nature/journal/v458/n7237/full/nature07920.html

R Kahle, E Kuhrt, G Hahn, and J Knolenberg. Physical Limits of Solar Collectors in Deflecting Earth-threatening Asteroids. *Advanced Science and Technology* 10; 256-263, 2006. doi:10.1016/j.ast.2005.12.004. http://www.sciencedirect.com/science/article/pii/S1270963806000125

L Keesy. NASA studying ways to make "tractor beams" a reality. NASA New Topics. October 31, 2011. Accessed December 11, 2015. http://www.nasa.gov/topics/technology/features/tractor-beam.html

K Kosmo, M Pryor, P Lubin, GB Hughes, H O'Neill, P Meinhold, J Suen, J Riley, J Griswold, BV Cook, IE Johansson, Q Zhang, K Walsh, C Melis, M Kangas, J Bible, C Motta, T Brashears, S Mathew, and J Bollag. DE-STARLITE - a practical planetary defense mission. *Nanophotonics and Macrophotonics for Space Environments VIII*, edited by Edward W. Taylor, David A. Cardimona, Proc. of SPIE 9226, Aug, 2014. http://digitalcommons.calpoly.edu/cgi/viewcontent.cgi?article=1045&context=stat_fac

SH Lee, Y Roichmann, and DG Grier. Optical solenoid beams. *Optics Express* 18(7); 6988-6993, 2010. doi: 10.1364/OE.18.006988. https://www.osapublishing.org/oe/abstract.cfm?uri=oe-18-7-6988

M Li, WHP Pernice, and HX Tang. Tunable bipolar optical interactions between guided lightwaves. *Nature Photonics* 3; 464-468, 2009. http://www.nature.com/nphoton/journal/v3/n8/abs/nphoton.2009.116.html

ET Lu, and SG Love. Gravitational tractor for towing asteroids. *Nature* 438: 177-178, 2005. doi: 10.1038/438177a. http://www.nature.com/nature/journal/v438/n7065/full/438177a.html

P Lubin, GB Hughes, J Bible, J Bublitz, J Arriola, C Motta, J Suen, I Johansson, J Riley, N Sarvian, D Clayton-Warwick, J Wu, A Milich, M Oleson, M Pryor, P Krogen, M Kangas, and H O'Neill. Toward Directed Energy Planetary Defense. *Optical Engineering* 53(2): 025103-1 - 025103-18, 2014. doi: 10.1117/1.OE.53.2.025103. http://www.deepspace.ucsb.edu/wp-content/uploads/2013/09/SPIE-Optical-Engineering-Towards-Directed-Energy-Planetary-Defense-Lubin-at-al-2014.pdf

A Marzo, SA Seah, BW Drinkwater, DR Sahoo, B Long, and S Subramanian. H. Holographic acoustic elements for manipulation of levitated objects. *Nature Communications*, 6: 8661, 2015. doi: 10.1038/ncomms9661. http://www.nature.com/ncomms/2015/151027/ncomms9661/full/ncomms9661.html

DD Mazanek, DM Reeves, JB Hopkins, DW Wade, M Tantardini, and H Shen. Enhanced gravity tractor technique for planetary defense. *4th IAA Planetary Defense Conference*, Rome, Italy, April 13-17, 2015.

A Novitsky, CW Qiu, and W Wang. Single gradientless light beam drags particles as tractor beams. *Physical Review Letters*, 107; 203601, 2011. doi: 10.1103/PhysRevLett.107.203601. http://journals.aps.org/prl/abstract/10.1103/PhysRevLett.107.203601

H Punzmann, N Francois, H Xia, G Falkovich, and M Shats. Generation and reversal of surface flows by propagating waves. *Nature Physics* 10; 658-663, 2014. doi: 10.1038/nphys3041. http://www.nature.com/nphys/journal/v10/n9/full/nphys3041.html

K Ramser, and D Hanstorp. Optical manipulation for single-cell studies. *J Biophotonics* 3(4); 187-206, 2010. doi: 10.1002/jbio200910050. http://onlinelibrary.wiley.com/doi/10.1002/jbio.200910050/abstract

DB Ruffner, and DG Grier. Optical Conveyors: A Class of Active Tractor Beams, *Phys. Rev. Lett.* 109, 163903, 2012. doi:10.1103/PhysRevLett.109.163903. http://journals.aps.org/prl/abstract/10.1103/PhysRevLett.109.163903

S Siceloff. CubeSat launchers expected to open research opportunities for all. NASA Press release. October 14, 2015. Accessed December 15, 2015. http://www.nasa.gov/feature/cubesat-launchers-expected-to-open-research-opportunities-for-all

EE Smith. *Spacehounds or IPC, a Tale of the Interplanetary Corporation.* Fantasy Press, Reading, PA. 1947.

S Sukhov, and A Dogariu. On the concept of tractor beams. *Opt Lett* 35(22); 3847-3849, 2010. doi: 10.1364/OL.35.003847. https://www.osapublishing.org/ol/abstract.cfm?uri=ol-35-22-3847

United States Lifesaving Association. National lifesaving statistics report, 2012. Published online, 2013. Accessed December 08, 2015. http://arc.usla.org/Statistics/USLA_National_Statistics_Report_2012.pdf

M Vasile, C Maddock. On the Deflection of Asteroids with Mirrors. *Journal Celestial Mechanical Dynamics* 107; 265-284, 2010. doi: 10.1007/s10569-010-9277-3. http://link.springer.com/article/10.1007%2Fs10569-010-9277-3#/page-1

Chapter 6
Universal Translators: Now You're Speaking My Language

Either the universal translator's offline, or I hit my head harder than I thought.

—Tom Paris
(*VOY: Gravity*)

6.1 Introduction

The bridge of the Starship Enterprise is a place where very different individuals participate in reasoned discussion and take definitive actions. Clearly understanding each other is a matter of life and death. They each speak passionately for their preferred course of action or shout out commands that must be followed to the last syllable, yet Kirk is from Iowa and Picard from France. Mr. Scott calls Scotland home, La Forge hails from Somalia, and Ensign Sato is Japanese. It gets even broader —Spock and T'Pol are Vulcans and Dr. Beverly Crusher was born on the Moon (*TNG: Conundrum*)! Perhaps the entire *Star Trek* franchise uses some tongue that has not been invented yet or the characters are all speaking the same language and it is converted to English for the sake of the audience. Could they all have taken the same online language course? More likely there is a device working in the background to translate all their various languages—some kind of universal translator.

An online company called Duolingo (Pittsburgh, PA) is just one of many firms whose business is to teach people to speak a second language. Being an online company is important for Duolingo because it allows them to reach people all over the world. Indeed, it is the internet itself that affords people the opportunity to interact with others around the world of whom they would not have been previously aware. Unfortunately, with this added reach comes problems; language can become a barrier to worldwide communication. Human translators (for written language) and interpreters (for spoken language) are quickly being replaced with pieces of software that move ideas from one language to another, sometimes with hilarious results that are passed around the internet. Duolingo is trying to bridge the language-learning chasm by offering courses in many languages, including one man-made, universal language.

M.E. Lasbury, *The Realization of Star Trek Technologies*,
DOI 10.1007/978-3-319-40914-6_6

The designed language called Internacia Lingvo appeared in Eastern Europe in 1887 with the express intent of becoming a universal second language. It's creator, Dr. Ludwik L. Zamenhof (1859–1917), was a Russian Jew raised in Poland whose life experience demonstrated the need for a common language. His designed language wasn't meant to replace traditional languages. Instead, it was intended to be a link so that people of different tongues would always share some common ground and a way to communicate. Internacia Lingvo is simple; all verbs are conjugated the same way (i.e. no irregular verbs), all nouns are pluralized in the same way, root words have specific prefixes and suffixes that always mean the same thing, and complex words are compounded from simple words. It can be learned in a short time and is not allowed to evolve regional dialects; everyone speaks exactly the same version of the language.

Zamenhof published his language manual under a pseudonym, Doktoro Esperanto; his devotees adopted his pen name as the common name of the language, Esperanto (meaning "one who hopes"). Shortly after WWI, the League of Nations considered adopting Esperanto as one of their official languages, but the motion was killed by the French delegation (Kontra et al. 1999, p. 32). This set back kept Esperanto from becoming well known, yet the language has continued to be spoken in pockets of many populations and is slowing growing in popularity. At the present time, an estimated 500,000–1,000,000 speak Esperanto, with perhaps 10,000–50,000 being fluent or native speakers.

Esperanto will have to serve only as a good model for a middle language since it almost certainly isn't going to become a common planetary language. In the absence of a common language for all organisms, there is a need to translate newly encountered languages—a universal translator (UT). Such a device is not imperative on Earth—the number of languages is declining (about 6900) and almost all newly discovered languages belong to a language family that can be used as a model for translation. But we will certainly need a UT for the day we find intelligent life off of Earth. Scientists have found water on Mars, and the number of known exoplanets in habitable zones is exploding, so the aliens are most probably out there somewhere. We can't count on them studying us long enough to know our languages, so a Star Trek-like UT is going to be necessary—and a number of linguists are trying to resolve the issues that would make that possible.

6.2 *Star Trek* Universal Translator Usage

The UT is the ultimate storytelling device for *Star Trek*; without it we wouldn't be celebrating the 50th anniversary of the franchise in 2016. The crew of the Enterprise could never complete a five-year mission to seek out new life and new civilizations if they couldn't talk to anyone they met. Jerry Stohl, a former *Star Trek* writer, stated early on that the writers knew they needed a device and were going to equip every character with a wrist-worn translator. After some discussion, they decided that the viewer would just assume that a translator was working in the background (The Star Trek Interview Book, pp. 127–128). So the audience sees the UT when it is an integral part of the story, and accepts that the UT is there and working properly when it isn't pertinent to the storyline. The UT ensures that the plot progresses seamlessly so that we can get on to more important things, like which alien female Kirk is going to kiss.

Many of the early episodes of the original series do not even deal with aliens. The stories feature humans that have been stranded in space or who are controlled by alien entities who all speak English. Even the Talosians encountered in the pilot communicate by telepathy, a translator is not necessary (*TOS: The Cage*). When the need for translator does arrive in the first season, it is not even the Starfleet UT. A Metron model is given to Kirk and the Gorn captain so they can communicate with one another. Kirk ends up using his as a journal while the more devious Gorn eavesdrops on Kirk's oral diary to gain an advantage (*TOS: The Arena*). When it does make its first appearance (*TOS: Metamorphosis*), the Starfleet UT is a crucial part of the story. It must be modified, which indicates how alien the alien must be, and it is the only way to understand what drives the alien's motives. This episode, and many others, underscore the need for clear and deep communication amongst beings, and speaks to the social relevance of the franchise.

The original series introduces the UT, but it gives only limited information about its development and only some insight into its workings. *Star Trek: Enterprise* does a more thorough job of showing the audience how amazing the UT is because it is an emerging technology in that series. Communications expert and linguist Hoshi Sato is an important character because she bridges the gap between standard human interpreters and early UTs; she controls the beta test UT unit onboard (*ENT: Broken Bow*). The *Enterprise* UT is a handheld device which can be attached to a communicator. It has a keypad for translation of written text (if they use compatible symbology) and a display (*ENT: Precious Cargo*) whose functions are undefined. Perhaps it's for entering cultural or speech characteristic data that will speed up the algorithm's deciphering of a language.

Hand entering the data on a keypad or carrying around a tubular device (*TOS: Metamorphosis*) looks clumsy and primitive in an era where most communication technologies are hands free, so they solve this problem by the time of the *Voyager, The Next Generation,* and *Deep Space Nine* series. The UT becomes incorporated into the now common voice-activated communicator (combadge) which is attached to each crewperson's uniform (*VOY: The 37's; DS9: For the Uniform*). But even in the experimental UT days of *Enterprise*, communicators have software to translate all *known* languages on the fly (*ENT: Civilization*). These incorporations of technology into everyday items makes two points, (1) the UT is so crucial that it is included in something the crew has with them at all times, and (2) there is a big difference between translating a known language and the process of decoding and translating a newly encountered language.

It is common in science fiction to ignore inconsistencies for full viewing or reading pleasure. This is less common in *Star Trek* than in other franchises, yet the

UT does have its issues. For instance, Chekov and Uhura must scramble to find and thumb through a Klingon dictionary in order to fool the Klingon border guards and gain access to the Rura Penthe penal colony (*Star Trek: The Undiscovered Country*) because the Klingons would recognize that a UT was being used. Yet Riker and Troi have UT-communicator implants as part of their disguise when they pass among alien natives on Mintaka III. They go undetected among the inhabitants while using the UT (*TNG: Who Watches the Watchers*) despite the fact that their lip movements don't match their speech and their voices emanate from their clavicle area. Fortunately, *Star Trek* does a good job at explaining how the technology is doing its job, so the suspended disbelief is maintained.

6.2.1 Star Trek *UT Mechanism and the EEG*

Even though the UT is more computer code than computer, the hardware is interesting and contains features that one might not realize. There are many aspects to language that must be recognized and analyzed in order to receive all the meaning. Some of these aspects are outwardly seen, while others reside in the brain of the speaker. For example, spoken words start in the language processing and speech producing areas of the brain and are then transduced via muscle-dependent modulation of sound waves, so accessing the brain could help to translate language. But communication also relies on the body. The UT must track every aspect of speech, speech formation and non-verbal communication. It takes sensors to gather this information, and the UT has several different types of sensors.

One of these sensors is for brainwaves. Kirk explains to Zefram Cochrane that there are many universal concepts common to all intelligent life, including language. The UT tracks brainwaves of the speaker and teases out the common language aspects inherent in their brainwaves. Then the machine's software then adds grammar to put the thoughts into a form the crew can understand (*TOS: Metamorphosis*). Kirk is suggesting that the UT includes a non-contact form of electroencephalography (EEG). Meaning can often get lost in the nuances of speech, so getting information directly from the brain might be helpful in some cases.

Box 6.1 Electroencephalography
Electroencephalography (EEG) measures the electrical activity of the brain by localizing electrical spikes across the cerebral cortex. Resting EEGs show baseline brainwave patterns. EEG can also be administered while performing certain tasks in order to visualize activity in a portion of the brain.

Twenty to seventy-five electrodes are attached to scalp at precise positions so that functional areas can be monitored across all lobes. The electrodes pick up electrical signals generated by neurons via electrochemical action potentials.

Neuroelectrical signals are transferred from one neuron to another in a linear, expanding, or collapsing form to downstream neurons. The recording device shows waveforms; squiggles of specific frequency, shape and amplitude in each monitored area based on neuroelectrical output. Different waveforms are seen in the various functional parts of the brain during different activities.

Waves can be defined by frequency, and may be considered normal or abnormal given the situation:

Alpha—bridge sleep and wakefulness; if they are aberrant it could indicate insomnia or OCD.

Beta—associated with consciousness; low amplitude may indicate anxious thinking.

Gamma—associated with cognitive function. Low amplitude may indicate depression or learning disability.

Delta—the slowest waves (3 Hz or less). Found mostly in children, but in adults may indicate brain injury or severe ADHD.

Theta—associated with relaxation or light sleep, or in deep meditation.

Waves are also named by morphology. The most common are:

V—sharp peaks and troughs and are associated with normal sleep, but their appearance when awake could indicate pathology.

Mu—sharp in one direction and rounded in the other. Mu waves should be apparent when at rest; usually will disappear with motor activity.

A typical EEG readout will have multiple waveform lines, with each not necessarily corresponding to an individual electrode (Fig. 6.1). *Montages* are groups of electrodes that are connected so that multiple outputs may have electrodes in common. A bipolar montage is useful in pinpointing abnormalities to a portion of the brain by showing reversal of phase. Adjacent readout lines will show a peak in one wave and a trough in the other, so they will come close to one another or touch. Other abnormalities in frequency, type, or amplitude can indicate other brain pathologies, like epilepsy.

Consider the sentence, “We had Grandpa for dinner last night.” This string of words might indicate a warm familial gesture or a horrific crime; if brainwaves indicated by electroencephalography EEG spikes confer both the meaning *and* intent of the words, confusion could be avoided in a translation. We will consider below whether or not this much information could ever be gathered from brainwaves, especially since alien brainwaves may not look anything like ours. Spock solves this problem by explaining that the UT is to be used with life forms that are somewhat humanoid; he calls them congruent life forms (*TOS: Metamorphosis*). Yet directly after stating this, Commander Spock modifies the UT so they can communicate with the alien that is very different from humans, the Companion that

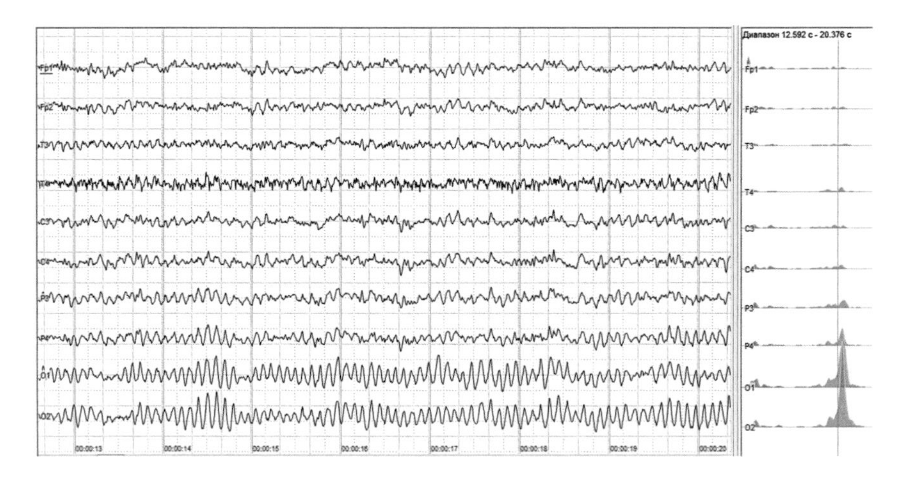

Fig. 6.1 Electroencephalogram readouts show electrical activity in many parts of the brain. This readout shows strong alpha wave (*sinusoidal waves*) activity, especially in the posterior portion of the brain (*the lower two lines*). There are likely many more electrodes than are represented by each tracing, the electrodes are joined in series to show activity over a larger area. Power readouts to the right (*red peaks*) are indicative of a restive state (*Image credit* By Andrii Cherninskyi [Own work] [CC BY-SA 4.0 (http://creativecommons.org/licenses/by-sa/4.0)], via Wikimedia Commons)

loves/imprisons Dr. Cochrane. Another example is seen when, after using binary code as a base for modification, the UT finally understands the language of the nanites created by Wesley Crusher (*TNG: Evolution*). Once again, the UT serves more as storytelling device than as a plausible piece of scientific equipment.

6.2.2 Audiovisual Sensors of the UT

Even if brainwaves can provide clues to language, they will be insufficient on their own for complete language decoding. Therefore, the blueprints of the UT shown in Franz Joseph's Starfleet Manual (p. TO:03:02:04) indicate that the UT has other sensors. One is a microphone, which is logical to include if one wants to translate verbal language. The units of a spoken language are called phonemes (consonant and vowel sounds), and these bits are put together to form morphemes. Morphemes include root words and their affixes, including prefixes, suffixes, circumfixes (a combination of prefix and suffix used together) and infixes (that are inserted inside roots). Meaning is conveyed through the ways in which phonemes and morphemes are used in a language, so it is crucial the sounds be picked up and analyzed by a language translator. In fact, spoken language is probably more amenable to decoding and translation than is written language; the difficulties in deciphering dead languages supports this (Sect. 6.4.3).

The microphone also permits recording of the language for inclusion in the permanent record and in the linguistic database of known languages with which the communicators are programmed (*ENT: Fight or Flight*). But just as with brainwaves, not all the meaning can be collected from just the recorded words. So much of language is conveyed by how we say things, not just what we say; therefore, the UT also has a "speech characteristic sensor" built into the microphone (Starfleet Technical Manual, p. TO: 03:02:04). One example of these paraverbal elements is word stress. One three-word sentence could carry three very different messages. "**Dr. Crusher** loves you," indicates one had the wrong person in mind. "Dr. Crusher **loves** you," indicates it is a more intense feeling than one might have thought, while "Dr. Crusher loves **you**," might mean, "I wish it were me."

Nonverbal communication works with paraverbal elements when deciphering meaning from language. A glance between co-conspirators can say, "Keep your mouth shut," or a wife's facial expression tells the husband, "Don't even think it," even as she says she would *love* to have dinner with his mother next week (Fig. 6.2). These meanings and nuances in communications would be lost if the UT relied on the words alone. In fact, nonverbal and paraverbal communication carries an estimated 38 % of the meaning of feelings and attitudes in any sentence or speech (Mehrabian 1971, p. 77). The Federation of Planets would know that nonverbal communication is a crucial part of all languages, so it is likely that the UT would have a camera in its package of sensors to pick up the body language and animation of the speaker. Indeed, a flow chart of UT function indicates that there are kinesthetic (body movement) and cultural pattern analysis programs included in the UT software (*Star Trek: TNG Technical Manual*, p. 101).

Fig. 6.2 Spoken language is full of paraverbal information—tone, level, body language, etc. This is actually an aid in deciphering meaning from language. In written language, there are many fewer cues as to meaning and emotion, unless specific clues are written into the text. In today's society, emojis serve a paraverbal function for written language. (*Image credit* Maksym Chechel/Shutterstock)

6.2.3 The Translation Matrix of the UT

The UT sensors gather data on what is being spoken and process this data through translation algorithm software. The algorithm then constructs a "translation matrix," a specific decoding framework designed to translate in real time. There is one translation matrix for every language ever encountered. A newly encountered species is visually instructed to speak into the UT to engage the translation algorithm, with more data leading to a better translation matrix. Just how much input is needed varies based on the complexity of the language. Although Ensign Sato states that the UT can learn new languages very quickly (*ENT: Breaking the* Ice), the complexity of the Skrreean tongue requires several hours of input (*DS9: Sanctuary*). In most cases, the learning period is implied since episodes have a limited amount of time to build and resolve conflicts. Unfortunately, it is probable that every time a UT senses a new word it will likely have to delay defining it until it hears that word enough times to assign meaning from context—this suggests very long learning curves.

To illustrate this point, linguist Dr. Christian Di Canio at the State University of New York at Buffalo points out the importance of the recursive process our brains use to build context from speech parts and identify speech parts from context. He told me, "In an over-simplified way, the whole "universal translator" idea is problematic because you can only identify things like stress by analyzing the alternations that words/phonemes undergo in context. It is a bit of a chicken and egg problem, and this is why linguists work incrementally. First you get the patterns at the word level and build from there. Then you go back to really understand what stress or other speech components are doing in that particular instance." *Star Trek* acknowledges this difficulty by including storylines where the translation matrix fails or only partially works with specific alien languages (*ENT: Vox Sola; DS9: Statistical Probabilities; TNG: Darmok*).

6.2.4 Linguacode

Highlighting the importance of communication for both exploration and storytelling, *Star Trek* has contingencies for when the UT is offline (*VOY: Hope and Fear*) or when it is less than adequate for communication with a species. NX-01 Enterprise Ensign Hoshi Sato developed a language tool called the linguacode, to serve as a Plan B. The linguacode is built from the common factors seen in many languages and acts as an interlingua to bridge the gap between two languages (*ENT: Breaking the Code*).

Though it is used for emergency situations, the linguacode can also be of help when the UT is working fine. When in a first contact situation with an alien species that has a translator device of its own, the Federation suggests that each language be put through a linguacode algorithm in the ship's computers. The output can be transferred to the UT to aid in development of the translation matrix and reducing the time to seamless translation (*ENT: In a Mirror* Darkly; ST: TNG Technical manual, p. 101). It would seem that linguacode and Esperanto are similar; they are both constructed languages and they can both serve as common ground for people who speak different languages. Additionally, just as the linguacode can shorten the time it takes the UT to learn a new language, Benny Lewis (of the website fluentinthree-months.com) claims that a few weeks of Esperanto training will increase the amount of any third language a student can learn in a given time (Dean 2015).

With few exceptions, the UT and linguacode are used for spoken language. Human translators are kept as part of the crew to translate alien control panels and written language when away from Starfleet computers. Computers can be used in some cases; Captain Janeway and Tuvok manage to use the Starfleet computer to translate written language when on the Nyrian biosphere vessel (*VOY: Displaced*). The use of computers to translate written text is something we can wrap our heads around. In fact, 21st century Earth is far ahead of *Star Trek* in translating written language.

6.3 Current Translation Mechanisms

In the 1980s, handheld translators first appeared in Sharper Image and Brookstone catalogues. One entered a word or phrase on the keypad and then selected a language for translation. The gadget would give the equivalent word and perhaps a definition. Words and definitions for a few known languages were held in an internal database, and a basic retrieval system was accessed. Many words could be stored, but the accuracy was spotty at best and decreased significantly when phrases were queried. No one was going to confuse a Sharper Image translator with artificial intelligence (AI).

Improvements in translation software and larger databases have made far more reliable electronic translators out of our smartphones and tablets. The most important progress has been made in the translation methodologies themselves and the hardware, speed, and interfaces have improved as well. Online translators like BabelFish, Bing Translator, and PROMT have gotten better, but the two programs getting the majority of the work and publicity are Google Translate and Skype Translator. Google rolled out significant added features for its Google Translate application in 2015 as proof of the continued improvement in these technologies.

Google Translate can now perform text-to-text translation in 90 languages, including Esperanto. It doesn't just translate between one of those 90 tongues and English, it can translate between any two of them. The browser-based version provides definitions and synonyms along with the translation so that the user can massage the translation for best fit. To improve the translation when nuances in meaning are possible, the software also delivers root words and versions as other parts of speech so the user can get the intended connotation just right. Finally, text to be translated can be entered phonetically or even spoken, and the output can also be presented as speech or text. This means that this app, and some of its competitors, can read text or hear speech, automatically recognize and translate it into a subset of known available languages in real time, and provide the output in text or voice.

However, before an app can begin to translate a speech, document, phrase, or word, it has to recognize the text or voice as something to be translated. For typed text the task is simple, it is automatically encoded in machine language as one type, so the translation can begin. In practice, images, handwriting, and speech present problems. The general schema for translation is:

1. acquire text, either by optical character or speech recognition
2. recognize language
3. translate from source to target language
4. produce output text or speech.

These are the same steps that the *Star Trek* UT performs, and shows just how close current hardware can approximate a UT. Discussion of the advances that are being made in each area will illustrate how current tech rivals that of fairly recent science fiction.

6.3.1 Optical Character Recognition

Knowing that some set of lines or shapes signifies a word or thought in some arbitrary language requires interpretation, a difficult task for computer. A camera can take an image of a dog, recreating all the shades and boundaries, but does the computer understand that the blob represents a dog? Show it two different dogs and the job gets harder, could a computer tell the breeds apart? Now attempt to differentiate the symbols of 90 languages, some with pictographic symbols that could be mistaken for objects. For example, is "+" a plus sign, a poorly written "t" or a symbol for first aid? The optical character recognition (OCR) algorithms of the 1990s and 2000s were earnest but inaccurate, efforts to digitize newspaper archives and scanned books often resulted in a few words followed by an unintelligible stream of ASCII characters. The outputs required massive numbers of man-hours to correct misrecognized letters and numbers. The early programs did not have the ability to differentiate and make educated guesses.

Box 6.2: Seeing Letters and Numbers
Optical character recognition (OCR) has improved greatly in the past decade. OCR algorithms convert images or input into an editable form—typed, handwritten, scanned, or image captured numbers and letters get converted to machine-encoded text that can be manipulated or analyzed (TXT, JPG, DOC files).

The file to be scanned represents more data than just the text to be recognized. The computer must figure out what is and isn't text and before it predicts what letters, numbers, or symbols the text represents. The process is complex and varies from system to system, but generally contains the following steps (Sharma et al. 2013):

1. *Preprocessing*—reduce noise, de-skew to reduce angles, convert to black/white
2. *Segmentation*—look for breaks or uneven spacing between possible characters and words. Predict proper breaks. This produces isolated bits for feature extraction.
3. *Normalize image*—reduce size and remove redundant information
4. *Extract features*—pick out the lines, loops, etc. and put them together to form characters. Can be done with several algorithms.
5. *Classify features*—predict what is represented and test predictions via first and second pass. First pass identifies the most obvious choices; second pass uses dictionary and first pass data to predict less obvious characters. This is where machine learning is most helpful.
6. *Output*—deliver the data in a selected format

The process described above is for *offline printed character recognition*, where the features are static. There is also offline handwriting recognition for characters (where there are breaks) and words (where language gives no breaks), called *intelligent character recognition* and *intelligent word recognition*, respectively. Handwriting is individualistic, which create issues beyond those of printed text.

Early OCR programs had to be trained by taking images of each character, and they worked through the alphabet one font at a time. Straight comparisons were made, with no machine learning. Neural networks now allow the outputs to be compared to known words and phrases to see if the predicted character makes sense in that position. As with other machine learning in the early 21st century, more examples are being added to datasets all the time to help the computer make better predictions.

The genesis of the first OCR failures was that not all type written text looks the same. Early versions of OCR had to be taught to recognize individual fonts; OCR programs could differentiate between fonts only after the cameras and software improved. Then the issue switched to recognizing images of handwriting or stylized typing as letters. Google Translate has incorporated a stand-alone program called Google Goggles that focuses solely on recognizing hand written or slanted text. Uneven letter height, uneven spacing, cursive or messy writing, as well as undotted I's and uncrossed T's all cause OCR programs problems because straight lines, even spacing, consistent height and so on were the very parameters used to teach the original OCR programs to recognize images as text.

6.3.1.1 Neural Networks and Deep Learning

Incorporating Google Goggles into Google Translate permits for much better writing to text conversion through software algorithms called deep learning and neural networks. These machine learning programs use increased comparison; however, the prediction and internal feedback mechanisms are most important for their success. The key to the recent leap ahead in efficiency was Microsoft engineers' return to the abandoned technique of artificial neural networks in 2009 to solve a speech recognition error problem (Hinton et al. 2012). The previously investigated and discarded system of neural networking worked wonders for reducing mistakes in speech recognition, and since that time deep learning via neural networks has made significant impact on all aspects of real time translation, from OCR to speech to text, machine translation, and voice generation.

"Artificial neural network" (or neural net) is a phrase with which every *Star Trek* fan is familiar. A neural net, being a biologically-modeled computer network for integrated action between many nodes, is the basis for Data's positronic brain (*TNG: Phantasms*). The net does not represent one specific configuration, it is more a family of computer models designed to allow for expanded comparison and predictive outputs when using very large datasets. Each connection between data points is weighted and various decision pathways can be tuned based on previous inputs and outputs, allowing the net to adapt over time as it gains experience—i.e. learning. Predictions and outputs are used to test and reinforce one another in a cyclic fashion to "learn" to make better predictions.

The input is distributed throughout the first layer of nodes (connections between pathways), while the final output is delivered after the algorithms have been accessed and all the possible outputs have been compared. In between lie hidden pathways and nodes; just which pathways are utilized for decision-making (predicting outcomes) depends on the input information, the output desired, and the experience that the net has acquired (Fig. 6.3). Deep learning models excel at

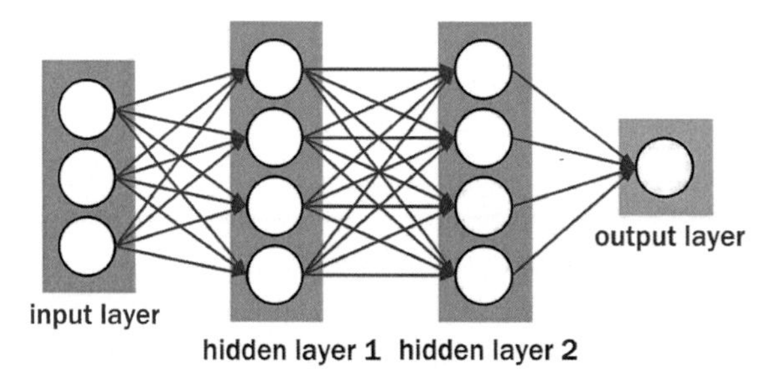

Fig. 6.3 This is a very general schematic of one possible neural network. It is a fully connected network because each possible input is connected to every possible node in the hidden layers. Neural networks can be designed to have one output as depicted here, or they have more than one possible output. Notice that each layer communicates with several nodes in the next or previous, but no communications exist within layers (*Image credit* Andrej Karpathy/Stanford University)

unsupervised learning; clustering very large sets of data by using different characteristics and drawing conclusions about those clusters—without being programmed as to how the data should be grouped. These are the exact opposite of the rules-based programing we use everyday where the exact same process is always used regardless of the input—it only knows one thing to do to the data.

Neural nets and other machine learning models allow for a reinforcement of its pathway choice decisions based previous predictions and the outcomes of those predictions, as well as newly added examples added to a dataset. The neural net outcomes are often vetted by humans, and the newly added example might also be generated by humans, so the machines may be able to learn, but they're not doing it on their own—they are definitely not artificial intelligence (AI). Current machine learning should not be mistaken for the kind of AI that would make an android like Mr. Data possible.

This is not to demean deep machine learning techniques, they are a marvel and continue to improve. They have allowed OCR to move beyond scanned books and handwriting samples where the majority of the optical data is text that needs to be captured. Recent advances in OCR allow for extracting text and symbols from images where the text is only a small part of the total information. For example, a photograph of a street scene where the name of a company is seen on the side of a city bus contains massive amounts of geometric and image data, yet a program like Word Lens (purchased by Google from Quest Visual) knows that only that small bit of the image is text and transfers it to Google Translate for translation.

The advanced Word Lens function is designed to recognize language symbols in images captured in real time and display the translation on screen using an augmented reality paradigm. The augmented reality methods mimic the font, color and style of the text and display it in the image at the proper place with the proper lighting and angles to replace the text in one language with the output from the Google Translate program. Previous versions of Word Lens could only used saved photographs, but the 2015 update allows for translation and augmented reality display on screen in real time just by pointing your camera at the target.

6.3.2 Speech Recognition

Words or sentences to be translated from source to target language can be spoken, not just written. Microsoft's Skype Translator was decades in the making and was released late in 2014 after a year of testing in the field. Skype can translate conversations among five languages using speech recognition and speech synthesis and can translate instant messages in over fifty. In most cases the results are adequate, Mandarin Chinese was still a problem as of mid-2015. For unknown reasons, translation of English into Chinese is including swear words that aren't there to begin with (Carter 2015). It probably isn't a speech recognition problem, yet somehow words are gained and decorum is lost during these translations.

Despite this minor setback, the advances in machine identification of spoken language have been amazing. The term "speech recognition" is very specific and is not the same things as speaker recognition (voice recognition). When Captain Kirk shouted to the main computer to activate the self-destruct sequence (*Star Trek: The Wrath of Khan*), it wasn't just the authorization code it recognized, it was Kirk's distinct voice and speaking style. The computer recognized his pregnant pauses, and more importantly it could differentiate his electronic voice pattern from everyone else's. Speaker recognition is for identification purposes, while speech-to-text (STT) technology, sometimes called speech recognition or automatic speech recognition, is focused on converting the content of the speech to text in machine language. Picard's constant orders to the replicator for a cup of, "Tea, Earl Grey… hot" (*TNG: Lessons*, and others)—that's STT. The replicator may or may not know it is Picard who is asking, that would be voice recognition.

Speaker recognition has been around since 1983, but speech recognition is a much more recent achievement. The American military funded efforts at STT in the 1970s and again in the 1990s; the 1990s efforts were aimed at recognizing key words in mainstream and obscure news broadcasts, with a particular eye (ear) to the Chinese (Zhan et al. 1998). The first successes came with what is called supervised, or speaker-dependent, STT. In these cases, the computer learns to recognize what one individual is saying, because that individual gives constant feedback (supervision) as to how the program is doing. This is the method that commercial software like Dragon (Nuance, Burlington, MA) STT packages use to convert your words to text (Fig. 6.4).

An STT system is limited if it can recognize only one person's speech. The goal has been a real-time, speaker-independent STT that successfully extracts sounds that correspond to words and sentences no matter who is speaking. In early speaker-independent programs, each speaker would tap the microphone button at the end of their input so the computer would know a new speaker was coming, perhaps even in a different language. Newer versions of Skype Translator and Google Translate require shorter introductory periods to identify individual

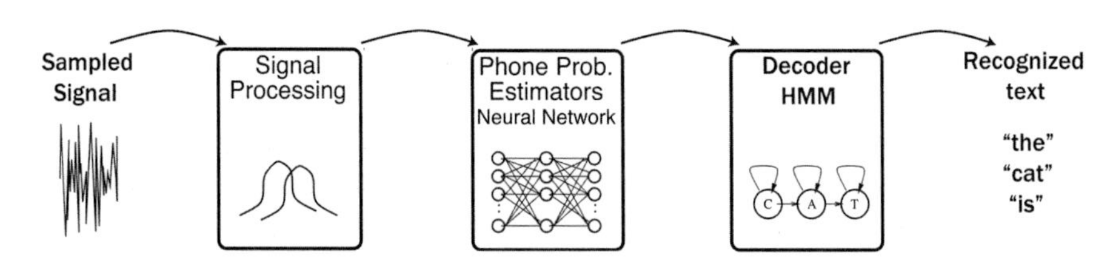

Fig. 6.4 Speech recognition begins with the conversion of speech patterns to electrical signals in computer language. Those signals represent a very large data set that can be analyzed in many ways. In the shown model, a single neural network and HMM (Hidden Markov Model) decoder are used. Many newer models use multiple neural nets and deeper learning algorithms in conjunction with speech model information to break the signals into phonemes, and then additional steps of the algorithms reassemble them as words. However, this process is only the first step in machine understanding of language (*Image credit* ICSI)

speakers and the change in speakers, rendering the microphone taps unnecessary after only a few exchanges.

One reason for these advances has been the application of the neural net and machine learning algorithms to STT just as they were OCR. The outputs have become faster and more accurate and have made conversational translators like Google Translate and Skype Translator possible.

The task is daunting; the program must (A) figure out which sounds are speech, (B) separate the sound waves for specific sounds—math algorithms convert the sound wave bits into distinct sound bits and spaces, and (C) convert each sound bit to a speech sound (phoneme). Only then can statistical software (D) combine phonemes to predict words by stringing phonemes together and guessing where the spaces between words are—this is called the word boundary identification problem (letter boundary being a parallel problem in OCR). Finally, (E) programs test the output to see if it makes a coherent phrase or sentence.

The described steps for STT would be much easier if we all spoke like Robbie the Robot from *Forbidden Planet*, but not even all English speakers "speak the same language." There are accents, stutters, verbal place keepers (um, like), dialects, and mumblers. These problems are compounded by background noise, changes in volume, and echoes. Software developers can't even use the same statistical models to predict speech as they do text because people write more formally than they speak. As Clay Dillow wrote in a *Popular Science* piece on machine learning, "Text offers data; speech and all its nuances offers nothing but problems." (Dillow 2014)

6.3.3 Language Recognition

Once the input, be it voice, text, or image containing text, is recognized as language to be translated, the software must assign that input to a specific source language. Despite the fact that some of the newer OCR and STT systems have language recognition as part of the acquisition process, Siri still requires the user to designate the language in which they will be speaking. The ability to identify speech as a particular language is one of the characteristics that make the newer programs so much better at converting speech or image to text.

Probabilistic systems making predictions as to which symbols represent speech are aided greatly by the software's ability to compare one sound or symbol prediction to neighboring sounds and symbols to see if they fit well into the first prediction, but this works best if the software knows what language is being spoken. To predict the language, the software compares the speech sounds inputted to those commonly found distinct languages and makes predictions to be tested on neighboring speech sounds. Therefore, it is a recursive system. Sets of predictions about speech and sets of predictions about language are used to test and reinforce one another to maximize probability of correctness—this is the essence of machine learning.

Language recognition algorithms are different for text and speech. For text, the letters or letter strings are often used to identify the language of origin, and these serve as good enough discriminators that these programs are adequate alone, at least for languages based on the Latin alphabet (Manning et al. 2008). Spoken samples are another matter, this issues still requires research and improvement and even then there is no guarantee that speech will be transcribed correctly. Therefore, using letters and letter strings to identify a spoken language sample may not work well. This is why you must still tell Siri what language you will be using.

Much of the research into spoken language recognition tries to mimic how humans recognize spoken language, since it has been shown that human hearing does this very well (Muthusamy et al. 1994; van Leeuwen et al. 2008). The spoken equivalent to letter strings is syllables, and research into spoken language recognition is focused on these and other acoustic-phonotactic clues (syllables, duration, pitch, intonation, spectrum of sounds) (Martin and Garofolo 2007; Zissman 1996). Interestingly, the acoustic-phonotactic clues used to identify spoken language have nothing to do with the semantics. It is not necessary for an algorithm to understand the language to identify it—people often know French when they hear it because it *sounds* French, not because they understand French.

6.3.4 Translation

Only after the translator device has detected text or speech and its particular language of origin can it take on the task of translating the source language into the target language equivalent. This is a rigorous task that requires machine learning capability and a vast amount of background data. One must dismiss the idea that translation software algorithms consist of huge databases of words in each language and the program swaps them out one for one. Instead, most use a translation model called statistical machine translation (SMT) that employs existing databases of documents on the internet that exist in the source and the target languages. Comparisons point out patterns, and the patterns can then be used to make predictions about the text to be translated. Patterns are reinforced when noted many times and when they produce correct translations. Therefore, the statistics of pattern use within the large datasets help the algorithms make better predictions. This is the "statistical" of SMT.

Box 6.3 Statistical Machine Translation

A bilingual document corpus (a large, structured set in two languages) is a powerful translation tool. Using a corpus for each language pair, models are built based on the statistical occurrence of patterns in the pairs of texts compared to the text to be translated—this is statistical machine translation (SMT). SMT is by far the most common translation paradigm.

Different probablistic models may derived from existing document databases:

1. the *source-channel method* creates a translation from a possible set of all translations by comparing each to a model of fluency in the target language. The translation model assigns probabilities to each of possible version and the fluency component works to integrate the possibilities within the target.
2. the *log–linear method* uses many aspects of the translation for evaluation at the same time. Each aspect is given a weight as to how much it impacts the final translation.

Instead of comparing words to words, most translation models use strings of words in the source sentence. Each string is a group of words that occur in order in a sentence. For example, "The car was painted red," could yield—the car, car was, was painted, etc. Strings are located in the source language corpus and paired to phrases in the target language documents. These *phrase pairs* are used to predict meaning and translation, in the following process for the *string-to-string* type of log-linear method. There are other models that take more parameters into account, but they are more complicated to describe.

1. Decompose sentence to strings.
2. Associate source strings with target strings.
3. Determine or estimate phrase distance—how far will target phrase be from source phrase based on characteristics of the phrase pair.
4. Translate each source word into zero to many target words using expectation-maximization of phrase pairs.
5. Reverse the search—each target word is translated to source word(s) and the two results are compared.
6. Use an expectation-maximization model to reach best result.
7. Sequence the phrases in the source to build the phrase pairs in the target.
8. Harmonize the translations using predicted context.
9. Use a relative frequency estimator to characterize how well each source phrase translates to each target phrase. Repeat steps 3–9, if necessary.

The SMT model searches perhaps tens of millions of documents where precise translations between the source and target languages (in both directions) have already been done. This *bilingual corpus* is often made up of United Nations or EU papers, items that are distributed in many languages. They might even be translations made during online language course work, as with Duolingo. But the documents are not compared as whole items; if that were done, the number of discovered patterns would be low and their usefulness would be minimized. Instead, the sentences are broken down into phrases and parts of phrases allowing for more possible pattern pairs. Overlapping phrases in the source text to be translated are compared to the same phrases found in dataset documents. These make up phrase pairs, and *might* represent a correct translation. Many pairs are compared and the computer predicts which translation makes the most sense based on what it has predicted before and the outcomes of those predictions. Just as for language recognition, the actual meaning of the words and sentences is not a part of the nodes and pathways of the SMT neural network. It is simply a statistical program that does not need to know the meaning of the words it is translating. See Box 6.3 for a more in depth explanation of SMT.

At times, SMT must make more predictions than in others, and it often reduces the accuracy of the translation, much like what happens in the children's "telephone operator" game. Once a phrase passes through all the game players, it may not resemble the original phrase at all. In SMT this occurs when a large bilingual corpus does not exist for a language pair, perhaps Slovak and Yiddish. To translate something from Slovak to Yiddish, SMT might go from the Slovak to Czech, Czech to English and English to Yiddish. English in this case is a "pivot language;" a middle language for which there are many bilingual corpora to make a final translation to any of several languages. This trick works—just not as well.

The predictions get better over time through machine learning, and the source of learning can come from different sources. A Skype Translator infographic produced by Microsoft states that using the application helps itself to learn and make better predictions on its own. It also states that user feedback improves the algorithms. And that addition of more documents to the dataset is a third mechanism of "learning." This happens every day and is sped up by the improved OCR programs. But it is important to point out that much of the machine learning at this point in time still requires continued human input (more documents, more feedback), so it is not an example of autonomous machine learning, i.e. true AI.

6.3.5 Synthetic Speech Generation

Translation output can take a number of forms. If it is found in a picture using Google Translate, it will be projected into the original image as an augmented reality inset using WordLens. If it is text, it will likely be delivered as text, and if speech in a Skype conversation, it will be delivered as speech. The outputs sometimes overlap, as when Skype gives a running caption at the bottom of the screen. Outputting the translation as text is child's play compared to generating speech, yet this challenge has been met, and is why Siri and Echo (Amazon) can talk to you.

Captain Kirk and Commander Spock explain to Zefram Cochrane that the UT speaks with a female voice in this instance because the brainwave patterns it is sensing from the Companion are definitely female. Furthermore, it uses warm dulcet tones because the brainwave patterns indicate that the Companion is in love with Cochrane (*TOS: Metamorphosis*). Although current devices like Siri don't choose the output voice based on input, they do give us some options as to what he/she will sound like, beyond the original female voice.

It was announced in 2013 that the voice of iPhone personal assistant Siri (Speech Interpretation and Recognition Interface), was created from words and sentences spoken by an American actress named Susan Bennett. Siri was a stand-alone application that Apple purchased in 2010 from a startup company of the same name. That first Siri personal assistant only gave text responses, but she had more attitude than the speaking version Apple re-debuted in 2011. More recent versions have male and female Siris generated by different voice-over artists from all over the world. The user can choose the sex, language, and even nationality of their synthetic conversation partner. He or she will speak with different accents and even adapt their vocabulary to match a dialect.

Generating a synthetic voice to speak on command is called a text-to-speech (TTS) algorithm, but it is not merely the reverse of STT—it is an entirely different challenge and has complex solutions all its own. The *bottom up approach* starts with a single sound wave. The waveform is modified electronically in pitch, length, intensity, boundary edges, etc. to produce vowels and consonants, each with the right stress and finish (Fig. 6.5). The forty or so phonemes of English can generate an unlimited number of meanings, so the bottom up approach modifies and melds the waveforms to string together the right phonemes in the right order, including correct

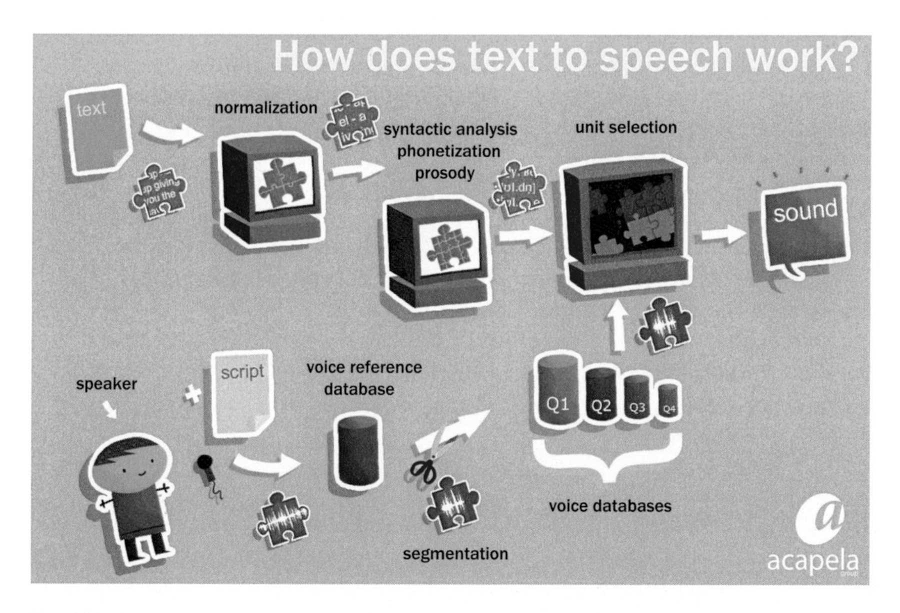

Fig. 6.5 Text-to-speech is a challenge that must be met on two fronts, the text to be spoken and the speaker or computer who provides the sounds (might also be generated from a single sound, modulated in the computer). The text is broken down into segments and analyzed for how each morpheme needs to be pronounced and stressed. The speech is built from a single sound wave through modulation, or from recorded human speech. The speaker creates a database of phonemes with different spacing, stress, level and finish. (*Image credit* Acapela Group)

situational gapping, proper spacing, stresses, starts and finishes. The bottom up approach is good for generating natural sounding speech, although another approach is getting to be almost as good, with less computing power and time needed.

Box 6.4 Giving Computers a Natural Voice

The goals of synthetic speech generation are intelligibility and naturalness, yet deducing what to say and actually saying it are very different tasks:

1. Derive meaning from the text.
2. Figure out the meaning *not* contained in the text, called prosody (things like intonation, rhythm, stresses).
3. Deduce what sounds to make (pronunciation).
4. String the sounds together with proper prosody in order to sound natural (Taylor 2009).

Prosodic prediction is harder for text than speech. Speech displays rhythm, stress, and intonation, but rarely does text have much affective content. Sometimes the writer adds an affective part in writing that he wouldn't include in speech: "She said no, *in a sad voice*." If the prosody isn't explicit, a reader makes it up based on contextual clues. A speech synthesis model must make these predictions as well.

Prosody is also found in the pauses. The computer segments the text, then looks for possible non-breaks, minor breaks, major breaks and whole sentences. Every juncture is assigned a value. Predictive models of contextual meaning then evaluate the breaks separately and together to ensure that they make sense when put together. Once a model is derived for phrasing, then it is repeated for prominence, intonation, stress, rhythm, etc.

Pronunciation is decoded next. One couldn't merely use a dataset consisting of every word recorded many different tones and stresses and gaps in rhythm; even a small vocabulary that could be 500,000 inputs. This won't lead to natural speech because one word always affects how the next is said. Most algorithms break the words down to individual sounds, taking context and emotion into account, most often using a *concatenative method*, using recorded words and sentences are broken down to sound bits and categorized.

On the other hand, a *bottom up* approach builds a single sound and then modifies it electronically to produce all the phonemes, like your mouth does with real speech. An algorithm for converting the text to sound uses an inventory of modified waveforms. The program builds words, then bigger and bigger units, all based on probabilistic results from previous predictions, still in real time.

Finally, the speech is synthesized using engineering to control sound waves and merge words with proper spacing for each situation.

The *concatenative approach* uses large datasets of recorded phonemes. This is where Susan Bennett came into the picture for Siri. Companies like Nuance (Sunnyvale, CA) build these datasets by having actors/actresses read scripts. Some sentences make sense, some don't. The same word or syllable may be pronounced slightly differently based on the words and syllables around it; a good speech generator must take these into account. The focus is on stringing different words together so that decomposition into phonemes generates different spacing, tone, stress, and finish. Mr. Data sounds natural because his probabilistic neural nets derive meaning and can shape that meaning into a proper soundscape, although nothing in *Star Trek* tells us whether he is a bottom up or a concatenative speaker.

6.4 Computational Linguistics and the Possibility of a UT

Current translators can already mimic many functions of the *Star Trek* UT. They can speak, listen, understand, and translate; unfortunately, they can only toggle between known languages. The translators we have now can learn to be better with experience, but they also must be told if what they did was correct and have new knowledge added to the system by humans. A true UT will be able to learn and translate novel language systems and be able to do it without human assistance. If our species continues to progress toward artificial intelligence devices, then science will need to answer only one question to know whether we can create a UT: are there enough commonalities between all languages that one can predict the structures within a language, and use those structures to decode?

6.4.1 What Is a Language?

The UT learns a language as the speaker speaks it, but just what is it learning? A language is nothing more than a code that uses symbols, sounds and gestures to convey information between two organisms. It can be written, oral, gesture-based or a combination of all three. To learn an unknown or second language, there must be a transmission of linguistic clues from the speaker to the learner. The clues might be explicit—one person points to the window and says, "window." The other person points to the same window and says, "*fenetre*." Each is learning something in the other language. Repetition can then reinforce the concepts. Once the learner gets "boy", "old", and "big," he/she can learn "I am" when the person says, "I am big. I am old. I am a boy." On the other hand, childhood acquisition of language needs many fewer explicit cues; simple immersion in the language is enough.

The child may not know the structure of the language it is learning, even though structure and exceptions to that structure are inherent in all languages. When learning a language as an adult, people tend to learn the grammar and other rules explicitly. However, it takes a long time to learn the exceptions, and languages are full of exceptions. If commonalities between languages could be identified and enumerated, then speaking a foreign language would be a matter of spotting the common features and understanding what they indicate. When James Kirk explains to Dr. Cochrane how the UT functions (*TOS: Metamorphosis*), he is basically saying that the UT uses the commonalities in languages to decode a newly encountered one.

The idea of looking for characteristics common to all languages is very old. The philosophers Descartes (1596–1650) and Leibniz (1646–1716) asked what were the constituents of a language and wondered if they occurred in all languages. The idea was debated for centuries, and the related languages hypothesis found a hero in American linguist, philosopher, and political activist Noam Chomsky (b. 1928). According to Chomsky, there exists a biologic faculty for innate language production and use in all organisms of high brain function. The implication is that language learning and language development are genetic and therefore will have shared characteristics. From the Chomskyan viewpoint, environment participates only in so much as it triggers language development (Owens 2011, p. 309). If Chomsky's hypothesis holds, it should be possible to write software that could break any language into its universal patterns and then decipher it.

Box 6.5 The Components of a Language

Language is complex and is made of many parts. Here is a partial list of the components of language:

Phonemes—individual sounds, vowels and consonants that make up oral language. English has about forty individual phonemes (depending on dialect), represented by only 26 graphemes (letters) because certain letters or letter combinations can have different pronunciations. This increases language complexity. The study of phonemes is called phonology.

Morphemes—individual or strings of phonemes that carry meaning. Morphemes are words, words, or affixes (prefixes, suffixes, circumfixes, and infixes). Language complexity increases here as well; "uncook" may not be in the dictionary, but there is an understanding when the root is used with the prefix. The study of morphemes is called morphology.

Syntax—the set of rules and exceptions by which morphemes are organized into phrases and sentences. Tense might be included in this; in English, future tense is part of syntax (will) while past tense is comes through morphology (*-ed*). Syntax can also be called grammar.

Semantics—the meaning component defined by syntax and morphology. This is the relationship of morphemes and syntax to provide nuance of meaning—metaphor, figurative language, and absurdity (humor). The semantic value of a sentence can determine word class for syntax, i.e. "love" used as a noun or as a verb. The *meaning* of love as object or action is a semantic distinction, yet the word class for each (noun vs. verb) is a syntactic distinction.

Semantics is the hardest thing for a translator or a human to decode; language users so often ignore rules of semantics. Some things that produce misunderstandings include: *equivocation*—phrases with two meanings; *relative language*—gaining meaning through comparison; *static evaluation*—stating things in absolutes; and *abstraction*—language short cuts providing inexact meaning.

Characteristics of speech—these also participate in semantics. These include quality of voice, pitch, intonation, rate of speech, rhythm, stress, animation (body language), and context. Speech characteristics greatly affect meaning.

Star Trek takes a strong Chomskyan viewpoint. For example, Species 116 from the Delta Quadrant could speak any language fluently after hearing only a few phrases. There are essentially organic UTs that are even better than all of Starfleet's technology. Unfortunately, the only living member of Species 116 is assimilated by the Borg, so now the collective basically has UT capability (*VOY: Scorpion and Scorpion, part II*; *VOY: Hope and Fear*). This idea impacts our space programs in real life. NASA placed a gold record with messages of welcome on Voyager I and II in hopes that a species with the ability to decode unknown languages would find and understand it. Of course, it also happens to be the plot of *Star Trek: The Motion Picture*. In the film, Station Epsilon IX broadcasts linguacode greetings as V'ger approaches, and eventually the message is received and understood. The linguacode and the NASA gold record show us that if someone is out there, we're willing to work hard to talk to them.

6.4.2 Linguistic Universals and Speech Characteristics

Members of the Association of Computational Linguistics (ACL) use computers and large datasets of language components to assess the rules and structure of different languages and determine if there are rules that are common to all languages. As Ming Xiang, Assistant Professor of Linguistics at the University of Chicago said in a 2011 interview for the magazine *Tableau*, "The fundamental similarities are what linguistics is going after. What are the basic, primitive units that can combine to produce all the languages?" (Allen 2011) The computational linguists call rules/characteristics that are common to all languages *linguistic universals*. For example, all known languages contain nouns and verbs, and all contain consonants and vowels, so these are considered universals. Linguist Joseph Greenberg (1915–2001) derived a set of 45 linguistic universals based on a study of thirty languages, almost all having to do with rules of syntax (Greenberg 1963).

If languages can be grouped based on different rules for the same aspect of language, the rules are called *tendencies*. An example of a linguistic tendency would be the groupings of language based on subject verb ordering. Some languages place the subject before the verb while other have the verb first, but all languages fit into one of the categories. Tendencies can still be useful for decoding by UT. If a language under investigation doesn't fit one tendency group, then the UT would move on to other groupings in a tree like fashion.

Many linguists believe linguistic universals are apparent in all of the components of language; phonemes, morphemes, syntax and semantics (see Box 6.5). Phonological universals include specific types of phonemes called stops that are produced by closing the vocal tract. All languages use stops, so identifying them where they occur would help a UT to pick out individual phonemes and morphemes. This would then give the UT clues as to individual morpheme structure and syntax because they are tendencies as to where stops most commonly occur. The numbers and types of vowels and consonants could also help a UT decode a language. Different languages may have different numbers of vowels and consonants, yet which versions each language has and how they are used are universal (Hyman 2008). One can see that the structural universals that linguists are aware of do provide clues for language translation, and more clues can come from the non-structural parts of language too.

In all languages, a rise in tone indicates an open statement, while a fall indicates finality. In a yes/no question the intonational feature is always a final rise. Dr. Bazarbayeva of Kazakhstan stated that, "it will become possible to build a general model of intonation types not attached to some specific language." (Bazarabayeva et al. 2015) Intonation is one example of a speech characteristic, and Dr. Bazarbayeva sounds like she endorses the possibility of a UT that translates partly based on intonation universals. There are also universals in rate of speech and in stresses—all of these universals would be part of datasets and probabilistic comparators for a UT, in the same way that bilingual language datasets are the datasets for Google Translate and Skype Translator. More tendencies or universals would make the job easier, while bigger the tendency groups (i.e. fewer groups) would also make algorithms more powerful.

6.4.3 Using Linguistic Universals to Translate Dead Languages

The possibility that universals and tendency groups could be of use in translating language is supported by recent translation of a long dead language. Benjamin Snyder, Regina Barzilay and Kevin Knight of MIT used a computer program of their own design to decipher and translate ancient Ugaritic (a Semitic language used in 14th–12th centuries BCE) in just a few hours. The researchers made several assumptions, the biggest one being that Ugaritic belonged to a known language family (Hebrew in this case), and that there were cognate words in the two languages. They programmed the algorithm to allow for a measure of fluctuation from the assumptions. The computer algorithm then bounced between the assumptions about Ugaritic and Hebrew as it maximized the translation accuracy (Snyder et al. 2010). Ugaritic had been deciphered through years of work in the 1930s, so we know that the computer algorithm got it right—the computer model just took a fraction of the time. Unfortunately, having a good model language may not be possible when dealing with a newly encountered alien species. What some linguistics hope is that if some universals or tendencies can be assumed, then a family might be found to help with the translation.

Another computer program was used to try and determine if a set of documents from the Indus civilization indeed represent a language. The program used universals in ordering words and symbols to predict if the texts were a linguistic language or more like non-linguistic symbols. Cave paintings would be an example of non-linguistic pictographs. The study results suggested that the sequences were most like those in natural languages (Rao et al. 2009). On the opposite side of the fence, linguist Richard Sproat (Google) contends that the Indus symbols are non-linguistic. His own study showed that the Indus pictographs fall into the non-linguistic category (Sproat 2014). Sproat correctly points out that computational analyses have a downfall in that biases can enter the picture based on which text is selected and what it might depict. Could someone learn English if all they had to study were shopping lists? Perhaps the brain can provide an unbiased view of language—*Star Trek's* UT was based on it.

6.4.4 Can Neurolinguistics Identify Universals

Neurolinguistics, the study of the activity of the brain by EEG during language tasks, demonstrates electrical responses to specific events of speech or language processing. If the event-related potentials (ERPs; EEG spikes associated with a specific stimulus) for understanding or forming speech can be grouped across languages, then neurolinguistic universals might be identifiable. This would support the Chomsky psycholinguistic theory and give credence to Kirk and his brainwave reading UT. Among the many research programs doing work on speech processing using EEG, Dr. Ming Xiang studies how the brain processes sentences and syntax. She and her group have looked at the ERPs that wh- questions (who, what, where, why) stimulate in the brain. The wh- usually appears at the beginning of sentences and far from the verb in English, but in languages like Mandarin Chinese the word order is the same as for a statement, with the wh- word close to the verb (What did that Vulcan Spock say? vs. That Vulcan Spock say what?).

A specifically located and timed ERP is seen in English speakers when presented with a wh- question, so Xiang presented Chinese speakers with one of two versions of a question. Both used their accustomed word order; one had a wh- and one did not. The wh- versions gave identical ERPs as in English, therefore, the word order must matter little. The brain comprehends wh- questions the same across languages (Xiang et al. 2014). An EEG sensor that detects the same spike in an unknown language might be able to decipher it based on the prediction that it is a wh-question.

Neurolinguistic studies also show that there is spatial separability in processing nouns and verbs—we store the meanings for nouns and verbs in different places in the brain. EEG spike locations could help a UT discern whether a speaker was using the word "love" as a verb or as a noun. The left inferior temporal cortex (straight in from where your glasses curve over your ear) play a role in naming and representing object knowledge, while the prefrontal cortex (right behind your forehead) is more involved in in action word processing (reviewed in Vigliocco et al. 2011). Studies of aphasia (inability to speak) in the 1980s support this notion. Some aphasics have damage in the left inferior temporal cortex and can't say the names of things, while those with damage to specific parts of the prefrontal cortex are unable to speak action words (Cotelli et al. 2006; Robinson et al. 2009). The separability between word classes is greatest and most reliable when semantics are taken into account, so a UT would garner clues about what kind of words were spoken, and would get clues as to context based on the degree of separability. Using statistical datasets to predict translations in association with the semantics clues given by ERP locations would ensure that a UT would not just translate a list of words.

6.5 Limitations in Developing a UT

Inherent in the Chomskyan theory of biologic and psychological commonalities for language is the idea that all groups use language to express similar ideas. The culture of the specific group is less crucial than the physiologic wiring that guides the process. However, not every linguist holds this view so strictly; they hypothesize that the culture of a people could greatly influence language development. A famous *Star Trek* episode and a native tribe in Brazil illustrate this idea.

6.5.1 Culture and the UT

T-shirts with phrases like "Shaka when the walls fell" and "Darmok and Jalad at Tanagra" sell briskly, and are recognized by those that understand the classic failure of the UT during Starfleet's dealings with the Tamarians (*TNG: Darmok*). The Tamarian language was based solely on cultural history, where metaphor was used to convey meaning through examples in their history or experience. The UT can translate the Tamarians' actual words just fine, *but not the meaning*. This suggests that language is more than words in an order; language is the collective mind of a group, connecting them to each other and the world.

Unless the UT had access to a cultural database for every species encountered, there would be gaps in the translation wherever the language used cultural references, abstractions, or idioms. There was such a database for the Tamarians, and it was used to finally break the code of the Tamarian language. Unfortunately, in true first encounters a database or cultural record would not be available. If culture plays a larger role in language development than the Chomskyans believe, this would be a significant challenge in developing a true UT.

The problem pointed out in *Darmok* is eerily similar to Dr. Dan Everett's (Bentley University, Waltham, MA) work with an Indian tribe in the Amazon Basin. The Piraha (pronounced pee-da-HAN) live on the Maici tributary of the Amazon River in Brazil. Their language is unrelated to any known language and has one of the simplest phonetic systems on Earth, with only eight consonants and three vowels. But using a small number of phonemes isn't what makes it unique. Piraha language uses a complex system of stresses, tones, and syllable lengths; speakers often dispense completely with vowels and consonants and express meaning completely through hums and whistles (Colapinto 2007).

The Piraha use this wonderfully odd system to express all the ideas that are important to their daily lives, yet Everett has found that at least one characteristic of language though to be universal is completely missing in the Piraha tongue. Chomsky holds that sentence recursion (placing one phrase inside another within the same sentence) is universal; it is what makes human language different from animal language. For example, a signing ape would never say, "The Andorian who has the broken antenna is named Shirley." However, Everett has shown that the Piraha do not use recursion. This does not make the Piraha language primitive, just more influenced by their culture. The Piraha live in the present and they never speak in abstractions. If they can't see it or touch it, it does not exist and they don't talk about it. In fact, they do not even have words for numbers above two. This does not mean they are backward—quite the opposite. They understand the concepts completely, they just choose to reject them.

Not all linguists accept the lack of recursion in Piraha language. It is recursion that permits a person to say infinite number of things; a particular sentence can always be made longer by including nested phrases, yet remain grammatically correct. If it were true for the Piraha, then it would be possible to list every possible sentence in their language—an unlikely possibility. But one cannot argue the crucial role Piraha culture has played in development of the Piraha language, just as with the Tamarians. The more culture plays into a language, the less useful a culturally neutral UT will be.

6.5.2 *The Lessons of Klingon*

Consider the genesis of the Vulcan and Klingon languages. They are designed languages, brought about first by the work of linguist Marc Okrand. He was initially called to produce only a few lines of Vulcan to match the lip movements of Leonard Nimoy and Kirstie Alley for a scene they had already shot in English for the original motion picture, but it was picked up by Mark Gardner and the "Vulcan Language Institute" and turned into a language. Okrand then wrote a few lines of

Klingon for *Star Trek III: The Search For Spock* and in 1992 he published his own full Klingon dictionary. To date, more than 250,000 copies of the Klingon dictionary have been sold, far outpacing the Vulcan book. Okrand explained to Reuters news service in 2012 that he thinks Klingon is more popular because, "Vulcans aren't as much fun as Klingons. They are much more serious. The Klingons let loose and the Vulcans don't." (Serjeant 2012)

One person who definitely favors Klingon is Dr. Tracy Canfield, a computational linguist and one of the few people on Earth who speaks both Klingon and Esperanto. She has been a science fiction fan all her life, yet didn't start learning Klingon until she was doing her PhD work in linguistics. One particular class required her to produce some machine translation code, so she chose Klingon because it can be written using a standard ASCII keyboard. In the process, Dr. Canfield became both an expert in Klingon and a thorn in her professor's side. Her hard work has paid off; she has been invited to Australia and other exotic destinations to help with Klingon translations and interpretations.

Canfield says Klingon can be thought of as the anti-Esperanto. While most engineered languages work to simplify rules and make the language easier, Okrand purposely picked some rarely used patterns and rules for Klingon. For example, English and Esperanto use a subject/verb/object (SVO) pattern in most writing and speaking. Klingon is OVS, a pattern seen in very few languages. Also, Klingon vocabulary is hard to learn because Okrand made sure it doesn't match any known language set, while Esperanto uses common roots and words from several languages and all the pronunciations are phonetic. All these factors would make Klingon a difficult subject for a UT.

Okrand's choice of word order based on his personal ideas of Klingon culture and development echo a recent study suggesting that word order is based more context (culture and history) within a language family than on linguistic universals. Many different sentence part combinations, like verb-object (or object-verb) or preposition-noun (or the reverse), are influenced by other structures within the sentence (Dunn et al. 2011). One word preceding another in some languages causes a reverse in other pairs, while the reverse might be true in different language families. Evolution of sentence structure via word ordering does not follow an inevitable course, hinting that languages are influenced by the collective experiences of the speakers and therefore aren't predictable.

Dr. Morten Christensen of Cornell is another computational linguist and Trek fan. He and his wife have passed on their love of *Star Trek* and all things science fiction to their children; however, even Dr. Christensen has a hard time imagining that enough linguistic universals are tendencies can be found to produce a UT that his children might one day use to talk to aliens. Dr. Christensen said, "Chomsky's notion that we all speak but a single language is only true at so high a level of abstraction that it would be useless in terms of creating a universal translator." While Dr. Canfield believes in the Chomskyan idea of a biologic basis of language, she is with Christsensen in contending that the abstraction inherent in language universals and the arbitrary evolution of languages based on culture and history will make a true UT difficult to realize.

On a more positive note, Dr. Christensen is more optimistic on a *human* UT being developed. He told me, "What may provide a better explanation of cross-linguistic similarities are similarities in basic cognition, culture, and communicative constraints, even if how to fit those into a universal translator is far from obvious." As to a translator that could function for all species, he did point out one possibility by adding, "However, it could be extended into space, if you adopt the *Next Generation* premise than an earlier progenitor species went around the universe and sowed the seeds of all species, all having a humanoid-like form—two legs, one head, etc." (*TNG: The Chase*) This sets up an interesting dilemma; would you trade the infinite possible diversity of life on the millions of planets in our Galaxy for the ability to chat right away with whomever we meet? Is communication that important?

6.6 Conclusion

The possibility of designing a true UT comes down to three issues; the hardware, the software, and the nature of language. Twenty-first century engineering is managing the hardware/software matters with skill and innovation, so this part of the process does not present a problem. Current technologies for recognizing speech and text and for reproducing speech are not just adequate, they're amazing. The time needed to learn speech patterns and recognize accents and dialects is going down with each new version released. Talking to a computer and getting it talk to you have already reached a *Star Trek* level. Even the exotic sensor technologies of the UT can be approximated in the present day. Solid-state recording provides a record of conversations on a computer that can be uploaded to the cloud. Wireless EEG headsets can help neurologists and neurolinguists "see" your language processing. And Microsoft's Kinect package can use cameras to interpret the gestures of American Sign Language and possibly other speech characteristics.

Programs like Kinect and the advances in language learning and translation indicate that software is not a hindrance to producing a UT. Machine learning is progressing at an increasing pace every year, and while true artificial intelligence might not be realized for decades, it might not even be necessary for producing a working UT. The software coding would be much more involved without AI, yet it could be accomplished—*if* the rules for specific language recognition, learning, and translation can be generalized to a finite number of patterns. And therein lies the rub; science is clearly at the mercy of language in producing a UT.

To what degree do languages follow a common pattern or patterns? There are undoubtedly many common features to language development, both in structure and in the brain, and new information can either strengthen or destroy theories on universal rules. Linguists thought that they had a fairly strong hold on recursion until the Piraha language was examined in greater detail; now the issue must be examined again. A newly discovered language, a revived dead language, or an alien language brought to us by visitors from outer space—any of them can negate a previously supported linguistic universal or tendency. It will be the linguists, asking pertinent questions about language processing and production, who will slowly group and partition language characteristics. They may use and be guided by technology, but technology itself will take a back seat to good old-fashioned thinking and number crunching. Just as Uhura and Sato are used as final arbiters and interpreters on *Star Trek*, language research is always going to come down to beings talking to beings. In a way, this personal involvement seems appropriate for language research; whatever universal language structure turns out to be, shouldn't communication always begin by being personal?

References

S Allen. New Questions, Old Tools. *Tableau* 12; Spring, 2011. https://tableau.uchicago.edu/articles/2011/03/new-tools-old-questions

ZM Bazarbayeva, AM Zhalalova, and YN Ormakhanova. Universal Properties of Intonation Components. *Review of European Studies* 7(6); 226-230, 2015. doi:10.5539/res.v7n6p226. http://www.ccsenet.org/journal/index.php/res/article/view/48079/25835

T Carter. Lights! Camera! China! Global Times. June, 19, 2015. Accessed November 18, 2015. http://www.globaltimes.cn/content/932769.shtml

J Colapinto. The Interpreter. *The New Yorker* April 16, 2007. http://www.newyorker.com/magazine/2007/04/16/the-interpreter-2

M Cotelli, B Borroni, R Manenti, A Alberici, M Calabria, C Agosti, A Arevalo, V Ginex, P Ortelli, G Binetti, O Zanetti, A Padovani, and SF Cappa. Action and object naming in frototemporal dementia, progressive supranuclear palsy, and cotricobasal degeneration. *Neuropsychology* 20(5); 558-565, 2006. doi:10.1037/0894-4105.20.5.558. https://www.researchgate.net/publication/6850707_Action_and_object_naming_in_frontotemporal_dementia_progressive_supranuclear_palsy_and_corticobasal_degeneration

S Dean. Konstrui pli bonan lingvon (To build a better language). *The Verge* 05/29/2015. Accessed November, 29, 2015. http://www.theverge.com/2015/5/29/8672371/learn-esperanto-language-duolingo-app-origin-history

C Dillow. How Microsoft's machine Learning is Breaking The Global Language Barrier. Popular Science Online. Posted December 19, 2014. Accessed on November 14, 2015. http://www.popsci.com/how-microsofts-machine-learning-breaking-language-barrier

M Dunn, J Simon. Greenhill, Stephen C. Levinson, Russell D. Gray. Evolved structure of language shows lineage-specific trends in word-order universals. *Nature* 473; 79-82 2011. doi:10.1038/nature09923. http://www.nature.com/nature/journal/v473/n7345/full/nature09923.html

J Greenberg. *Universals of Language*. Cambridge, MA, MIT Press, 1963.

G Hinton, L Den, D Yu, G Dahl, A Mohamed, N Jaitly, A Senior, V Vanhoucke, P Nguyen, T Sainath, and B Kingsbury. Deep neural networks for acoustic modeling of speech recognition. *Signal Processing Magazine, IEEE* 29(6); 82-97, 2012.doi:10.1109/MSP.2012.2205597. http://static.googleusercontent.com/media/research.google.com/en//pubs/archive/38131.pdf

L Hyman. Universals in phonology. *The linguistic Review* 25(1-2); 83-137, 2008. doi:10.1515/TLIR.2008.003. http://www.degruyter.com/view/j/tlir.2008.25.issue-1-2/tlir.2008.003/tlir.2008.003.xml

F Joseph. *Star Trek – Starfleet Technical Manual: Training Command, Starfleet Academy*. New York: Ballantine Books, 2006.

M Kontra, R Phillipson, T Skutnabb-Kangas, and T Varady. *Language, a Right and a Resource: Approaching Linguistic Human Rights.* Budapest, Hungary, Central European University Press, 1999.

CD Manning, P Raghavan, and H Schultze. *Introduction to Information retrieval.* Cambridge, UK; Cambridge University Press, 2008.

AF Martin and JS Garofolo. NIST speech processing evaluations: LVCSR, speaker recognition, language recognition. *Proceedings of the IEEE Workshop on Signal Processing, Applied Public Security Forensics.* Washington DC, USA, 1-7, 2007.

YK Muthusamy, N Jain, and RA Cole. Perceptual benchmarks for automatic language identification. In *Proceedings of the IEEE International Conference on Acoustics and Speech Signal Processing*, Adelaide, Australia 1; 333-336, 1994.

A Mehrabian. *Silent Messages.* Belmont, CA, Wadsworth, 1971.

RE Owens. *Language Development; An Introduction* 98th edition. Boston, Allyn & Bacon, 2011.

RP Rao, N Yadav, MN Vahia, H Joglekar, R Adhikari, I Mahadevan 2009 Entropic evidence for linguistic structure in the Indus script. *Science* 324(5931); 1165, 2009. doi:10.1126/science.1170391. https://homes.cs.washington.edu/~rao/ScienceIndus.pdf

J Robinson, J Druks, J Hodges, and P Garrard. The treatment of object naming, definition and object use in semantic dementia: the effect of errorless learning. *Aphasiology* 23(6); 749-775, 2009. doi:10.1080/02687030802235195. http://www.tandfonline.com/doi/abs/10.1080/02687030802235195

J Serjeant. Klingon goes boldly beyond "Star Trek" into pop culture. *Reuters Entertainment* November, 02, 2012. Accessed 12/03/2015. http://www.reuters.com/article/2012/11/02/entertainment-us-books-klingon-idUSBRE8A10HB20121102#3QjyyF5tVbHOmtOw.97

OP Sharma, MK Ghose, KB Shah, and BK Thakur. Recent trends and tools for feature extraction in OCR technology. *International Journal of Soft Computing and Engineering* 2(6); 220-223, 2013. http://www.ijsce.org/attachments/File/v2i6/F1158112612.pdf

B Snyder, R Barzilay, and K Knight. A statistical model for lost language decipherment. *Proceedings of the 48th Annual Meeting of the Association of Computational Linguistics*, pg. 1048-1057. Uppsala, Sweden, July 11-16, 2010.

R Sproat. A statistical comparison of written language and nonlinguistic symbol systems. *Language* 90(2); s1-s27, 2014. doi:10.1353/lan.2014.0042. http://www.linguisticsociety.org/document/language-vol-90-issue-2-june-2014-sproat

P Taylor. *Text-to-Speech Synthesis.* Cambridge, UK, Cambridge University Press, 2009.

DA van Leeuwen, M Boer, and R Orr. A human benchmark for the NIST language recognition evaluation 2005. Presented at the *Odyssey: Speaker Language Recognition Workshop*, Stellenbosch, South Africa. Paper 012, 2008.

G Vigliocco, DP Vinson, J Druks, H Barber, and SF Cappa. Nouns and verbs in the brain: A review of behavioural, electrophysiological, neuropsychological and imaging studies. *Neuroscience and Behavioral Reviews* 35: 407-426, 2011. doi: 10.1016/j.neubiorev.2010.04.007. http://neurocog.ull.es/wp-content/uploads/2012/04/Vigliocco-Vinson-Druks-Barber-Cappa-2011.pdf

M Xiang, B Dillon, M Wagers, F Liu, and T Guo. Processinf covert dependencies: an SAT study on Mandarin *wh*-in-situ questions. *Journal of East Asian Linguistics* 23: 207-232, 2014.

P Zhan, S Wegmann, and S Lowe. Dragon Systems' 1997 Mandarin Broadcast News System. *Proceedings of the Broadcast News transcription and Understanding Workshop*, Lansdowne, VA, pp. 25-27, 1998.

MA Zissman. Predicting, diagnosing and improving automatic language identification performance. In: *Proceedings of the Eurospeech Conference*, Rhodes, Greece. pg, 51-54, 1996.

Chapter 7
Geordi's Visor: A Vision of the Future

Our eyes reflect our lives, don't they?
And yours, so confident!

—Shinzon
Star Trek: Nemesis

7.1 Introduction

A man named George died from the complications of muscular dystrophy in 1975. He had been confined to a wheelchair for many years, but when he could, he had traveled to his favorite events—*Star Trek* conventions. George was well known amongst the trekkies, and was also a friend of *Star Trek* creator Gene Roddenberry. Gene must have been impressed by George; he was the inspiration for a *Star Trek: The Next Generation* (TNG) character even though the series didn't air for a dozen years after George's death. Roddenberry found George to be courageous, genial, a little irreverent, and a man passionate about the history and culture of his favorite things. He had a good sense of humor despite the constant pain and discomfort he felt, and he spoke of the universe with a sense of wonder and awe in spite of his personal challenges. George's last name was La Forge. Gene's memories of George were brought to the screen in the form of Geordi La Forge, the blind from birth ensign cum Chief Engineer cum Lt. Commander of the *USS Enterprise NC-1701-D*.

Whether the idea for a blind crewmember was Gene's or that of story editor and TNG bible writer David Gerrold is a matter of some dispute. In his book, *Gene Roddenberry: The Myth and the Man Behind Star* Trek, author Joel Engel contends that Gene merely parroted back an idea for a character who could "see" by way of visual implants that Gerrold had sent out in a memo just two days previous (p. 228). Regardless of who first suggested the character, Gene supervised the creation of an engineer for the second series of the franchise who embodied all of George's traits. Just as Lt. Uhura had broken barriers of race and gender for the original series in the sixties, this new character raised public consciousness of the abilities of those people usually described as handicapped.

M.E. Lasbury, *The Realization of Star Trek Technologies*,
DOI 10.1007/978-3-319-40914-6_7

The audience's first views of Geordi are from behind as the pilot progressed. We see that his skin is dark, we hear him be gregarious and outgoing, yet it isn't until we see him from the front that we realize that something is very different (TNG: Encounter at Farpoint). The *Star Trek* canon tells us that Geordi was blind due to a birth defect, but by the age of five had received a remarkable piece of technology that allowed him to "see." His VISOR (**V**isual **I**nstrument and **S**ensory **O**rgan **R**eplacement) was an electronic device that fit over his natural eyes and attached to his temples just in front of his ears. The prop for the series had many incarnations before filming began; if the version he wore in the early episodes looked a little like a hair barrette or a banana clip to you, it's because that was exactly what the final model was based on.

The VISOR was a great visual prop to draw attention and interest to Ensign La Forge, and it was one of the things that made him an instant fan favorite. For this reason, Geordi is the character most Trek fans think of when asked about physically challenged characters; however, he wasn't even the first blind Federation officer in the franchise. Even though Dr. Miranda Jones was completely blind, she could sense her environment through a sensory array or "web" worn over her clothing (*TOS: Is There No Truth In Beauty*?). The many sensors recreated her immediate environment and somehow transmitted it to her visual cortex, but the series never explained just how this was accomplished. The original series also gave us Fleet Captain Christopher Pike, a severely injured and mute former captain of the Enterprise. It is interesting to note that at the same the US space program was picking the most able-bodied men for the Gemini and Apollo programs, Pike made his debut just two days after Buzz Aldrin set an extravehicular activity record of five and a half hours and Dr. Jones appeared on *Star Trek* the same week that the Apollo 7 mission with Captain Wally Schirra started its ten day orbit of Earth. *Star Trek* led the movement for inclusion in space.

Of course, complete inclusion eluded the writers of the 1960s. Blindness was approached differently in the 1968 version of the show as compared to the late 1980s. Dr. McCoy stated that Miranda was unfit to pilot the Enterprise due to her blindness, even though she could compensate with elegant technology. Yet just nineteen years later, Geordi, also using technology that allowed him to "see," was the trusted pilot of a ship with the same name. The later series also gave us many more characters with physical or mental challenges to overcome; Julian Bashir was developmentally delayed (*DS9: Doctor Bashir, I Presume*), a famous mediator was deaf and mute (*TNG: Loud as a Whisper*), a new officer couldn't walk if away from her planet (*DS9: Melora*), and Christopher Plummer's Klingon character had an eye patch (*Star Trek VI: The Undiscovered Country*). Did their increased presence represent a change in society's attitude or in the technologies to help them?

Medical and engineering advances on *Star Trek* that help solve medical problems or compensate for disabilities are now becoming more common on Earth. Artificial limbs, neuromuscular controlled prostheses, and even exoskeletons are helping people to overcome physical disabilities. Expanded transplant programs are rendering once lethal or debilitating conditions moot; 2016 has seen the advent of uterus transplants (Johannesson and Jarvholm 2016) and the implantation of a 3D printed vertebra (Xu et al. 2016). The advances even include ways to compensate for blindness. Visual prostheses are transmitting light signals to the brain to help people who are losing their sight. And yet science is going one step further, the ability to match *Star Trek* in the power to see heat, X-rays, and tap into a WiFi signal is within our grasp.

7.2 Visual Processing

If an implant or other technological device has a chance of recreating visual signals in the human brain, the pathways through which light is converted to nerve impulses and those impulses are assigned meaning must be accessed. Like entrance ramps on a highway, the signals that are to be processed as sight travel along specific routes to specific places, and the electronic signals that are to be substituted for the original biochemically induced impulses have to enter that highway at some point. To do so, the signals have to be similar and in the same format the natural nervous system messages would take, or the brain will not know how to process them—one doesn't fly a plane on a highway, the rules and processes are different for planes and cars. Therefore, to develop technologies to restore sight using an individuals visual processing system, one must understand how photons of light are converted into messages that the brain can interpret and where and how this interpretation takes place.

7.2.1 Converting Light to Electricity

Perceiving your environment visually starts with light passing through the lens and the pupil of the eye and landing on the retina that lines the inside rear wall of the eyeball. The retina is made up of several layers of neurons, including about 95 million cells modified so that they can react to light. These photoreceptor cells come in two types, rods and cones (Fig. 7.1). Rods are located all over the retina and are more numerous than the cones in the periphery, while cones are mostly located around the center of the retina, and greatly outnumber the rods in an area called the fovea centralis, the area of most sensitive vision located where the light strikes as a person looks at an object directly in front them. Cones need high light levels to be activated and come in three types, each responding to a different color of light, red, blue or green. Rods are stimulated by any color light and will respond to lower levels of light intensity, but their output is only in black and white; they are for night vision.

The part of the rod or cone that faces the light is called the outer segment and has many layers or discs of membranes studded with thousands of pigment molecules that are stimulated by light. In cones, the pigment is photopsin, with slight variations in structure accounting for the versions that absorb the different colors of light. Rhodopsin, also called visual purple, is the pigment in the rods, yet both photopsin and rhodopsin act in much the same way. Both pigments are made up of opsin proteins connected a version of vitamin A called cis-retinal. When light strikes the pigment molecule, the cis retinal undergoes a structural change (to retinol) and is

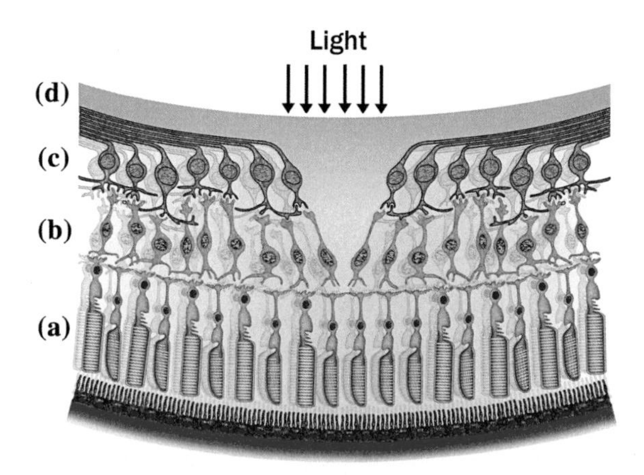

Fig. 7.1 The retina has a very interesting anatomy. The photoreceptor cells (*a*) are the farthest from the light source, and the outer segments that contain the responsive pigment in those receptors cells is farther from the light than are their cell bodies. The rods are represented by the rectangular outer segments, while those of the cones are curved to be conical at the end. The action potentials stimulated by light travel to the interneuron layer (*b*) that is a few layers of different cells that organize and collate signals from several photoreceptors. The signals then converge into the retinal ganglion cells (*c*) and their axons (*d*) join to form the optic nerve. (*Image credit* Alexilusmedical/Shutterstock)

released from the opsin. This allows the opsin to initiate a signal cascade that eventual moves to the base of the cell and stimulates an electrochemical neural impulse called an action potential. The retinal is retuned to it cis- form and re-associates with the opsin in a short time. This is why you can't see for a moment if you look at a bright light. All your photoreceptor molecules have been reduced to opsins and retinols and they haven't re-associated yet; your retinal cells can't respond to light until the photopsins or rhodopsins are remade.

Box 7.1 Electricity and Vision

Benjamin Franklin (1706–1790) developed a way of storing electricity in a series of Leyden jars in the 1760s; he called it a "battery" since it resembled an artillery battery of two rows of cannon.

Luigi Galvani (1737–1798) put a static charge through the legs of a dead frog in 1790. The muscles contracted and the legs twitched. He called this "animal electricity" and hypothesized that a special electrical energy controlled the life force.

Alessandro Volta (1745–1827) produced electricity de novo based on the electron transfer properties of two different metals. He connected two copper wires to a bimetallic pile he had constructed to produce the first bimetallic battery!

1792—Volta also hypothesized that the human body ran on a form of electricity, but surmised that electricity inside and outside the body were the same. He connected one wire of his battery to the corner of a person's eye, and touched the other to the roof of their mouth. His volunteer saw blobs of light, even in a darkened room, and the amount of electricity controlled what the eye would "see." Each spot of light the volunteer sensed is called a phosphene, and Volta could change the position and intensity of the phosphene by moving the contact points. The connections of lead to skin or mucosa completed a circuit and some of the electricity that passed through the body stimulated the optic nerve or retina.

Mary Shelley (1791–1851) based her novel, *Frankenstein*, on the work of Volta and Galvani. Her writings focused on the reanimation of dead tissue and were influenced strongly by work on the electrical energy in the body. However, nowhere in the book does Dr. Frankenstein use lightning or a zap of electricity to bring the monster to life. And he didn't have bolts as electrodes in his neck.

It is interesting to note that light must pass through two layers of neurons and their connections before reaching the photoreceptor cells. These cell processes that lie on top of the photoreceptors are the pathways that the stimulated bioelectrical neural impulses take to leave the eye on their way to the brain. The retinal ganglion cells are the top most layer of the retina and beneath them lay the interneurons. The interneurons act as collators of information between the photoreceptor cells and the retinal ganglion cells that make up the optic nerve that leaves the eyeball. Once the rods and cones transduce the light energy to spikes of electrical information (action potentials), the impulses are processed through three types of interneurons to the retinal ganglion cells. The axons (long processes) of the retinal ganglion cells lay across the top of the retina and all exit at one point. Because the exit point has no rods or cones, you don't see anything in that region—that's your blind spot.

7.2.2 Translating Visual Information

A neuron (individual nerve cell) sends messages along its very long cell process (the axon) and transfers it to the next neuron by electrochemical impulse. We think of electricity as a flow of electrons from an area of more negativity to one of more positivity, but electrochemical transmission can use other charges, namely chemical ions. The outside of the axon is more positively charged than the inside, since positive sodium ions (Na^+) are continuously and selectively pumped out of the cell. This builds up a resting electrical potential across the membrane. When a neuron is stimulated by an upstream neuron or by the action of some specialized receptor, like a photoreceptor for light, sodium is allowed to flow inside by the opening of a channel protein in the membrane in a very small area. This change in the electrical potential (depolarization) stimulates the membrane sodium channels immediately downstream to open, and they in turn stimulate their neighbors, etc., until the depolarization is sent down the entire length of the axon. This is the action potential. Shortly after the sodium ions leak in, they are pumped back out and the polarized state is re-established so that the neuron can fire another action potential when stimulated.

The axons of individual neurons are grouped together to form nerves. Some axons can be very long, for example the lower motor neuron axons of a human body can travel from the spinal cord all the way to the tip of the big toe. However, the axons of the retinal ganglion cells don't travel all the back to the part of your brain that processes the images. Instead, they transfer their information to different neurons in the lateral geniculate nucleus of the thalamus, a part of the brain that is very old and buried deep in the middle of the brain mass. There is some changing of the signals in the lateral geniculate and nearby nuclei that will tell you about the motion of objects and help you to track them by controlling your eye movements, but most of these downstream neural axons fan out to the back of the brain to be processed for vision. These fanned axons are called the optic radiations and they travel to the visual cortex in the occipital lobe (back) of your brain. Once the nerve signals reach the visual cortex, more processing takes place. Some cells sense shapes while others process orientation to show you where things are in your field. Some interpret color; others translate movement. Put all the information together, and you have your visual field at any given time. Of course, your brain has to repeat the entire process one millisecond later.

The visual system is even more intricate than the previous description lets on. The focusing of the rays through the lens causes a reversal and inversion of the image on the retina, so the signals have to be converted at some point to right side up again. Additionally, the signals that start in the retinal photoreceptor cells and travel through the optic nerve to different parts of the brain get split and put back together many times, each time a portion of the information may be extracted and processed, and then collated together with the additional information from other parts of the brain. Indeed, the visual cortex interacts with many other parts of the central nervous system, particularly your memory, so that you can recognize what it is you are seeing and assign meaning to it. In a general sense, visual processing comes down to two determinations, what is seen (shape and color recognition) and where it is seen (location, motion, and orientation). Each of these parameters uses several pathways and cues to put together all the information.

For instance, to process location, the signals from the part of your left eye's retina closest to your nose and those of the part of your right eye's retina closest to your right ear are combined in optic chiasma (crossing) as they travel back in the optic nerve. Therefore, the action potentials for images corresponding to objects to the left of you are all processed in your right lateral geniculate, optic radiations and visual cortex. The combining of information from the left and right visual cortices then places objects in your field of vision. Where the light from an object strikes on each retina is also mapped by the brain and compared from eye to eye. Your brain can calculate the angle for line of sight based on the portion of the retina responding. A wide angle from the object to the appropriate portion of each retina means the object is close, while a narrow angle occurs when objects are farther away. However, there are even more mechanisms at work for location; your brain recognizes perspective, relative size of objects, expected sizes of familiar objects, and even the degree to which angles change as an object moves past you (motion parallax). Your visual cortices and lateral geniculate nuclei process all this information in a fraction of a second, and that is *just* for location.

This has been a very rough description of the visual system, although it will serve the purposes of this discussion on visual prostheses. A term or two may be added later to clarify how a prosthesis works within the visual system, but if you want to learn more about how the brain converts light to visual images, I suggest a book entitled *A Natural History of Seeing* by Simon Ings (Ings 2008). However, for our purposes, pictures are turned into electrical impulses and then turned back into pictures. It sounds like television. We don't actually *see* pictures in our brain, the messages are much more complex and spread out. It isn't as if we hang a sheet in our brain and just project images based on the light that strikes the backs of our eyeballs. Having learned how complex vision actually is, Geordi's "bionic" vision certainly sounds far fetched, but it is much closer to real life than you might imagine. Science is showing us how to convert pictures into the language of the brain so that we can interpret camera images as if our eyes sensed them.

7.3 Camera Based Visual Prostheses

The stimulation to fire an action potential in a neuron is usually biochemical, although it can also be electrical. The electrical impulses from Geordi's VISOR start action potentials in the neurons to which his implants are connected. Electrical spikes are then translated into the language of the nervous system. From there, Geordi's own visual processing system takes over and interprets the firing of the nerves just as if they have come from his eyes. This is how Geordi's VISOR works and is exactly how instruments in real life are restoring some vision to people with two types of degenerative retinal disease.

7.3.1 Degenerative Retinal Diseases

Retinal degeneration accounts for millions of cases of blindness worldwide. There are several distinct diseases that can cause retinal function loss, with two of the most well known being retinitis pigmentosa (RP) and age-related macular degeneration (ARMD). Both of these diseases kill retinal cells and lead to progressive vision loss, but each does so in a different manner. The patients with these diseases lose vision over time, which makes them good subjects for testing visual prostheses that restore some sight. Since they often have partial vision and have been sighted during their lifetime, they can give good feedback on prosthesis function, as well as their wants and needs. Therefore, current implant technologies to restore vision have focused on these two conditions.

RP is a general term for any of a group of similar genetic diseases that affects 1 in 3500 people. The symptoms are caused by a mutation in the DNA that codes for controlling retinal pigment production or retinal cell survival. There are over 100 known mutations that can lead to the gradual loss of peripheral vision that is common to RP or RP-like diseases. Some mutations are inherited as autosomal dominant traits, meaning that only one copy of the defective gene is necessary to show the disease, so these patients usually show signs and symptoms beginning in childhood. Other types are inherited as recessive or X-linked recessive diseases, meaning that two copies are needed for disease or just one copy on the X chromosome since males are XY. These more common inheritance patterns give disease that first presents in the late teens to middle age.

RP begins with a loss of peripheral vision due to a degeneration of the photoreceptor cells of the retina. The earliest symptoms usually include a difficulty in seeing in dark rooms or after dusk, making it hard to drive at night. Later, the field of vision is reduced from the outside in, deteriorating to a central tunnel vision before complete blindness. There is no cure for RP and the current treatments are limited to some antioxidant vitamin supplements that only slow the timing of vision loss. The millions of people suffering with this disease make use of visual aid devices, yet bypassing the damaged retinal cells and see again might be a much better aid than a cane or service animal.

ARMD has a higher prevalence than RP; it is the most common cause of blindness in people over the age of 65, with over 200,000 cases every year in the United States alone. The macula (an area including and surrounding the fovea centralis) is a small portion of the retina onto which impinge the light rays of things that you are looking directly at. The photoreceptor cells, especially the color sensing cones, are most densely packed in this area, so it is the macula that has the greatest light transduction to neural impulses and therefore the most sensitive sight. As a person ages, there can be a build up of fatty deposits under the macula (called dry ARMD) which can, over time, cause that portion of the retina to thin out. A thin macula does not function well and the patient will begin to lose their most central, most sensitive vision. Fewer cases of ARMD are brought about by a pathologic overgrowth of blood vessels within the macula. This "wet MD" causes a faster, more severe loss of central vision and can develop in a person originally diagnosed with the dry version.

ARMD disease manifests itself in the exact opposite fashion as RP (Fig. 7.2). The central vision is the first to be affected, and as the disease progresses, a wider and wider field of vision is lost, ending in complete blindness. There are risk factors for ARMD, including smoking, some genetic predispositions, and increased body mass index, yet why some get the dry version, some the wet, and others suffer no ARMD is not understood. It may be that an age related decrease in the ability to prevent oxidative damage to cells could stimulate both fat production and blood

Fig. 7.2 These images are a vast simplification of the vision loss seen in retinitis pigmentosa (*RP*) and age related macular degeneration (*ARMD*). Tunnel vision develops slowly in most *RP* cases, and generally leads to a loss of visual field from the outside in. *ARMD* starts with a loss of central vision, although it need not be so circular, in many cases there are several small areas of loss in the central zone, merging to give full loss of vision in the most sensitive portion of the retina over time. (*Image credit* Library of Congress)

vessel formation (Nita and Grzybowski 2016), while changes in the immune system, specifically the control of the complement cascade (Lechner et al. 2016) could be the reason for so much neovascularization. Antioxidant treatments are the only way to possibly slow the dry version, while laser coagulation of blood vessels can be helpful in wet MD. Regardless, nothing can be done to completely stop or reverse the loss of vision, which is especially cruel since ARMD robs a person of their most central, most acute vision first. Therefore, ARMD patients are a population that is likely to be helped by neural implant visual prostheses.

7.3.2 Parameters for Making Visual Prostheses

To restore vision to ARMD or RP patients, the damaged or dead photoreceptor cells have to be bypassed. The electrical signal that represents light falling on the retina in specific patterns, movements, and colors has to be translated directly to the retinal neural cells that usually receive the signal *from* the photoreceptor cells. It is now possible to attach microelectrodes directly to the optic nerve or to interneuron cells. These are called visual or retinal prostheses and can stimulate neural firing using electricity. In newer prototypes of these devices, science is even going *Star Trek* one better. Researchers and engineers have learned to send the code to the optic nerve without even implanting microelectrodes.

One important feature of your vision that makes artificially replacing it possible is that a sense of position is maintained from when light strikes your retina until it is interpreted in the visual cortex of your brain. Photoreceptor cells that are near neighbors in the retina will give information that will be translated by near neighbor neurons in the visual cortex of the brain. The pattern that photon strikes on the retina is topically represented in the same way in the visual cortex, even if it isn't a one to one relationship. When you deliver artificial inputs to the retina in the "shape" of an object, that same shape can be grossly recreated in the visual cortex. This is what is meant by topographic or retinotopic vision. Put the right spike patterns in the right place, and the neurons will interpret them as visual images, even if they started out in a camera. This is how most current visual prostheses translate a picture into a neural pattern.

If you can determine which part of the visual system is defective or injured, you can input the signals downstream of the inactive portion and restore vision using the parts that still function since the visual system isn't one set of neurons. As we have discussed, there are several connections along the way to the brain; photoreceptors —interneurons—retinal ganglion cells and the optic nerve—lateral geniculate—optic radiation—visual cortex. It's a long road, but it offers some advantages for curing blindness. If the photoreceptor cells are the only damaged area, then inputting the signals to the interneurons or optic nerve will allow the rest of the visual system to work. If the optic nerve is damaged, then move the implant back to the lateral geniculate. However, if the defect is in the visual cortex of the brain that reconstructs the electrical patterns back into meaningful images, then it's a whole different ballgame. You can get on the highway to the cinema at many different places, but is there any point to going if you know the projector is broken? Even though scientists and engineers can't yet make the blind see as well as a normally sighted individual or as well as Geordi La Forge, they can help the almost blind be a little less blind. And science is only beginning to learn how to help. The next decade may indeed lead to the elimination of blindness, not just ways to ease the difficulties of the visually challenged.

7.3.3 The Argus II Epiretinal Implant

As stated earlier, degenerative retinal diseases are a significant cause of acquired blindness. The diseases render the retinal photoreceptor cells inactive, and in the case of RP, the interneurons *may* be damaged too. But the retinal ganglion cells are intact. One visual prosthetic device that came to the marketplace in 2014 takes advantage of the functional retinal ganglion cells by implanting a grid of micro-electrodes directly on the top layer of the retina where the retinal ganglion cells lie —an epiretinal implant. The Argus II is a camera and transducer device designed and produced by Second Sight of Sylmar, CA. The Argus II substitutes a pattern of electrical impulses that approximate the shape and contrast (light and dark) of objects as seen by a miniature camera housed in set of dark glasses. The visual field elements are converted to electrical impulses in the device's video processing unit (VPU), which is carried in one of the patient's pockets. The small, short duration electrical pulses generated by the VPU are returned to a transmitter on the temple of the glasses, and these impulses are sent wirelessly to the epiretinal implant. The implant itself has three main parts; a scleral band that wraps around the eyeball and keeps the implant in place; a grid of microstimulators that lies on the retinal ganglion cells; and the receiver coil that senses the electrical impulse code from the transmitter on temple of the goggles to the electrode array (Fig. 7.3).

Since the photoreceptor cells are non-functional, laying an implant over the top of part of the retina does not have any negative effect. The microelectrodes of the implant emit small electrical shocks that trigger action potentials in the retinal ganglion cells to which they make contact. The spots of light (phosphenes) induced by the electrical stimulation can be manipulated via the VPU to give different brightness levels. A brighter image on the camera will stimulate higher current stimulation via the electrode, and this produces a brighter phosphene. Patients can discriminate up to 10 levels of phosphene intensity (Humayun et al. 2003 and Zrenner et al. 2011) and this intensity scaling creates a grey scale image. Unfortunately, increased electrical stimulation also increases the phosphene size, as neighboring retinal ganglion cells are also stimulated by the same signal (Humayun et al. 2003 and Wilke et al. 2011). This reduces spatial resolution (Behrend et al. 2011).

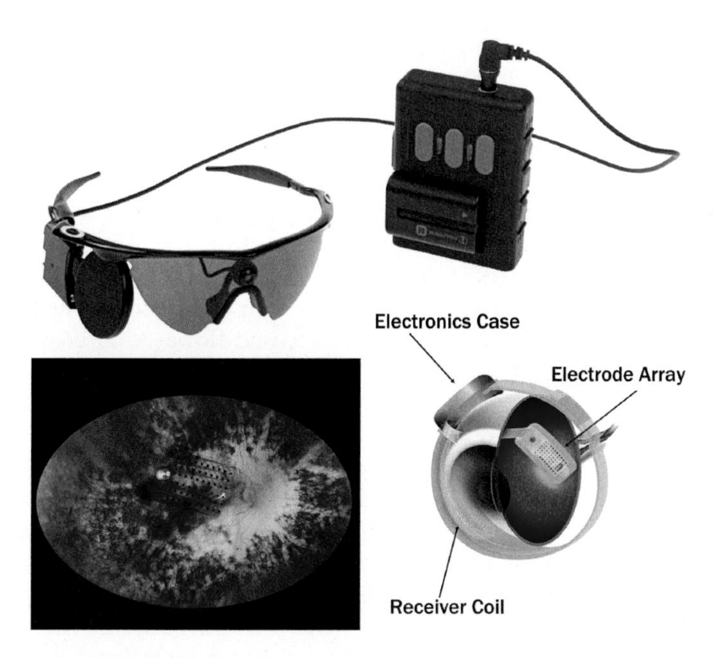

Fig. 7.3 The Argus II visual prosthetic system gathers information from the environment with a camera centrally located on a set of glasses. The signal is processed in the VPU (*top right*), and is then transmitted wireless by the transmitter on the temple of the glasses. On the eye (*bottom right*) the array is positioned on top of the retina, and held in place by band around the eyeball. The receiver coil captures the signal sent by the transmitter on the glasses. The electronics holds the wireless battery and the electronics that convert the signal into the electrode impulses. The photomicrograph (*bottom left*) shows the epiretinal implant and its electrode array. [*Image credit* Second Sight Medical Products, Sylmar, CA; (Lauritzen et al. 2012)]

The action potentials stimulated by the electrodes travel out of the eye via the optic nerve and through the visual system to the brain. Phosphenes are "seen" by the user within an apparent field of vision, about 8.9 cm by 16.5 cm (3.5 by 6.5 in.). The device doesn't offer a full visual field since the implant only has 55 electrodes (actually 60, with five held back as replacements for failed electrodes) that stimulate a small portion of the retina. After implantation and a period of training and adjustment, the VPU can be optimized for the individual based on their stimulatory thresholds. Each electrode can be adjusted independently to maximize signal strength and resolution using the external VPU alone.

Over 100 patients have been fitted with Argus II implants and camera glasses as of late 2015. Studies of their abilities show that users improve in daily tasks that require vision. More than 95 % of the patients can locate large objects (Humayun et al. 2012), and they are better at locating and grasping a target object (Kotecha et al. 2014). A majority can trace a maze or follow a moving bar of light (Humayun et al. 2012; Dorn et al. 2013; Barry and Dagnelie 2012); however, the system does not restore normal vision. The patient manual for the Argus II states that the device provides a type of "artificial vision," and the information it provides to the user must be incorporated with their existing assistive devices, such as seeing-eye dogs and canes. However, the results have been nothing short of life changing for the patients, says Dr. Robert Greenberg, MD, PhD, the CEO and President of Second Sight.

7.3.3.1 Training to Use the Argus II

The period of adjustment mentioned above is important, and here again Geordi's VISOR mimics real life. He said he had never seen a sunrise… at least not the way that Captain Picard had (*Star Trek: Insurrection*). The information his VISOR sent to his brain wasn't a view of the world that normal sight would provide; he had to interpret the world from the information he was given. However, in contrast to people who lose their sight and need to learn to interpret Argus II signals, Geordi was born blind, so maybe it was actually easier for Geordi than for real life patients.

Significant training is required for the users to learn to interpret what they are seeing. Objects will have consistent characteristics, but the patient must learn to recognize them. Think of it as written language—Americans read "dog" while the French see "chien." Even though they look different, they mean the same thing. If you have read English your entire life and the suddenly have to read French, it's going to take some time. Now imagine that the pen used to write "chien" is running out of ink. Only parts of each letter can be seen. This is the state of current implants; first one must figure out what the shapes are, then they can try to figure out what they mean. Fortunately, some Argus II users have learned with training to resolve letter shapes with more precision than was thought theoretically possible for the device (da Cruz et al. 2013); some people are just better at learning and recognizing the clues in this new visual field, just like some people are good test takers.

Patients with no vision at all can expect about 20/1400 vision; they can see at 20 ft. (6 m) what a person with normal vision can see at 1400 ft. (428 m). Legal blindness is considered to begin at 20/200, so at most the first generation of the Argus II gave a general idea of light and dark and the edges of bright object, but that doesn't keep the recipients from crying like babies when they first use the device, says Dr. Greenberg. A child of the 60s and 70s, Dr. Greenberg is a *Star Trek* fan, and he says that it certainly stimulated his youthful appetite for science. However, it was the 1970s action show *The Six Million Dollar Man* that was more of a guide for the Argus II than was Geordi LaForge. Steve Austin's bionic left eye, with its 20.2:1 zoom lens, infrared vision, and super slo-mo inspired Dr. Greenberg to find a way to help the blind, and Austin and Geordi still inspire him to extend what the Argus II can do.

Dr. Greenberg told me that the number of electrodes on the Argus II could be increased to strengthen acuity and/or field of vision, yet the increased numbers is not always a good thing. More and smaller electrodes will actually increase the chance of failure of the device. He likens the implantation of an electronic device on the retina to throwing a television in the ocean and expecting it to work. The liquid and salt environment of the eye is very hard on electronics, so keeping things simple is often the more prudent course. Another example of simplicity enhancing the overall device is the use of the epiretinal implant. The signal (20 μm diameter) is spread out over many retinal ganglion cells (10 μm diameter), resulting in lower resolution; however, it permits the use of the Argus II with patients who have lost both the photoreceptor layer of cells and the processing layers of cells that feed information to the retinal ganglion cells. Therefore, even though the Argus II has been tested mostly in RP patients, it is now being tested in people with dry age-related MD that destroys the photoreceptor cells and many of the interneurons but spares the retinal ganglion cells. This is important since many more people suffer from age-related MD than RP. In the patients tested so far, the centrally located implant has been able to restore some of the central vision of MD patients, and they have found that the individuals are able to integrate their natural peripheral vision with the induced central field signals.

7.3.4 The Alpha-IMS Subretinal Implant

There is a newer piece of technology that better mimics the ocular implants that Geordi received later in life (*Star Trek: Insurrection*). The Alpha-IMS device from the German company Retina Implant AG is now entering the European market. Instead of using an external camera attached to a pair of glasses to transduce the light signals to electrical impulses, the Alpha-IMS is a subretinal device that combines the camera, processor, and stimulator all in one implanted instrument (Fig. 7.4).

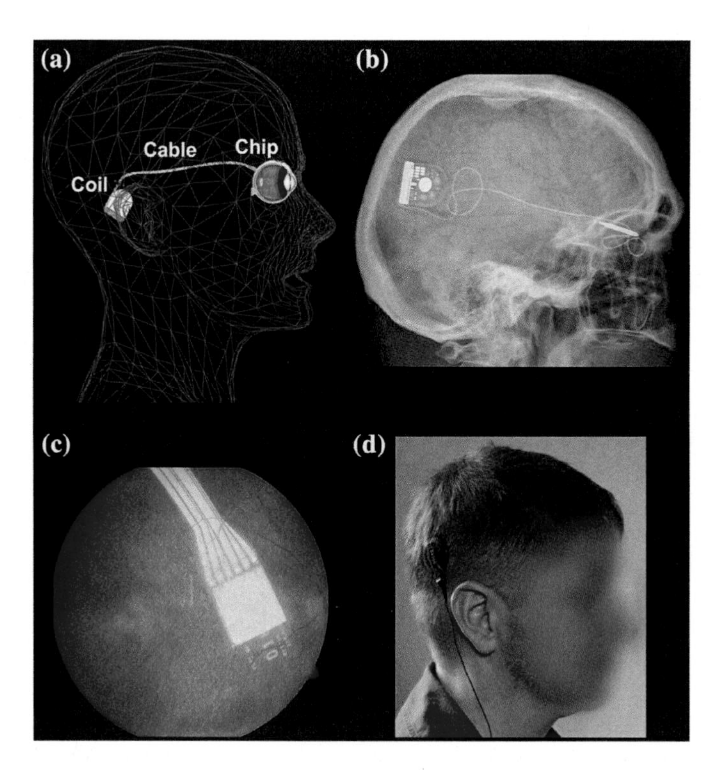

Fig. 7.4 The Alpha-IMS system uses a chip and camera implant (**a**) connected by a subepidermal cable to a wireless battery coil behind the ear. The set up can be visualized better by X-ray (**b**). The subretinal implant (**c**) has all the electronics (no software) to convert the camera on the implant to stimulate the 1500 electrodes. The subdermal coil is recharged by a wireless unit (**d**) held in place on the skin by magnets on the coil apparatus. [*Image credit* (Stingl 2013a)]

Box 7.2 Vision and Pain

In the TNG series premiere, Geordi tells Dr. Crusher that his VISOR comes with a price—constant pain. Treating the pain doesn't work or negatively affects the activity of the VISOR, so Geordi has to concentrate and endure discomfort in order to see (*TNG: Encounter at Farpoint*).

Pain in the Brain—The brain doesn't have pain neurons, and the optic nerve is part of the brain. This is why you can do brain surgery under local anesthesia and have the patient respond as the surgeon pokes around in his grey matter, so a VISOR shouldn't directly cause pain. However, there are a few syndromes that might account for his pain when using the VISOR:

Synesthesia—a crossing of sensory inputs. A sight may trigger a sound that only the observer hears; numbers may have a shape, color and texture. The numbers are seen as objects and make mental math much easier. Geordi tells Dr. Crusher that the source of his pain was in his sense systems, "They say

I use my natural sensors in different ways" (*TNG: Encounter at Farpoint*), which sounds a bit like synesthesia. Dr. Greenberg of Second Sight remarked that early on, some of the implant patients had pain when the system was turned on because the current was turned up too high and the discomfort was the same as staring at a bright light.

Pain synesthesia—there are rare examples of synesthetes who have pain accompanying a sensation, but usually this occurs when they see *someone else* have the experience. It's called mirror touch synesthesia—when someone else touches his or her own cheek, the synesthete feels it as pain (Ward and Banissy 2015). Phantom limb pain is the most common type of self-synesthesia for pain (Goller et al. 2013). People that sense phantom pain are also the people most likely to have empathy synesthesia pain (you literally feel someone else's pain; Banissy et al. 2007).

Blood vessel pain—blood vessels in the brain have pain neurons even if brain tissues doesn't. The constriction or dilatation of these vessels is the source of pain in most headaches. There are many visual triggers for migraine headaches, although what triggers a headache in individuals can vary greatly. A military helicopter pilot who only got migraines when he used his night vision goggles (Cho et al. 1995) suggests that Geordi's VISOR just might be giving him a headache. Surprisingly, colored filters can reduce the visual illusions associated with striped patterns (Harle et al. 2006 and Shepherd et al. 2013), so perhaps if Geordi had tried some color therapy he could have cured his headaches.

For patients using the Alpha-IMS, light passes into the eye and strikes the retina as it does in sighted individuals, so a user can just move their eyes to alter their visual field, not their entire head as they would to move the camera of the Argus II. Each photodiode (of which there are 1500) of the CMOS image sensor of the implant has its own amplifier and a 50 μm × 50 μm (1/510th of an inch square) iridium electrode that makes contact with several interneuron cells. The entire unit is 3 × 3 mm (0.12 × 0.12 in.) and uses only 10 mW of electricity, so there is little damaging heat (0.5 °C) generated. The implant has a flat cable that exits the eyeball and connects to a wirelessly charged power pack inserted beneath the skin behind the ear. All in all, it is a self-contained, wirelessly recharged, artificial system for vision (Stingl et al. 2013a).

The subretinal implant provides an additional layer of normal visual processing to the action potentials stimulated by the electrode array. Instead of shocking the retinal ganglion cells, it is the interneuron cells of the second retinal layer that are stimulated. These neurons process the signal and then pass it on to the correct retinal ganglion neurons in the language that those cells understand. By stimulating the interneurons, an added level of natural visual processing is integrated into the artificial system; this processing results in more accurate and refined images when the visual cortex interprets them. The signals are also more closely retinotopic because the electrodes stimulate interneurons which correspond to very nearby photoreceptor locations. In the Argus II, stimulation of the retinal ganglion cell axon could correspond to a photoreceptor cell a good distance away.

Test subjects described significant improvements in their vision while using the Alpha-IMS system. Almost half the individuals could distinguish alphabet letters with training; some could do so immediately after turning the device on for the first time (Stingl et al. 2013b). A more recent version of the Alpha-IMS achieved vision levels of 20/546 to 20/606, with several participants able to read a wall clock on at least one post surgical visit (Stingl et al. 2015). A not so apparent reason for the increased visual acuity using the Alpha-IMS relates to the tiny movements our eyes make when they naturally scan objects and refresh the detection of objects on a microsecond level. The movements include the very small horizontal jerk-like movements of the eyes (microsaccades) which correct drifting in eye position like the minor adjustments you make to your steering wheel while you drive in a straight line and a refocusing to prevent image fading. These "oculomotor fixational patterns" are only possible because the camera is part of the implant and moves as the eye moves (Hafed et al. 2016).

Implants using external cameras cannot include these movements and therefore are limited in the acuity of the signal transmitted to the visual cortex, not because the camera is inferior, but because they do not include the natural signals that brain expects from the eye to finely interpret a signal. Dr. Greenberg of Second Sight is hoping to incorporate eye tracking to move the camera instead of the head as is possible with the Alpha-IMS implanted camera. However, this may produce more than just more natural head movement. It could also help to incorporate the oculomotor fixational patterns into future versions of the Argus implant by tracking pupillary movements with a camera directed at the eye. This will track the minute movements of the eye as the patient hones in on the object(s) they *think* that they are looking at. The signal from the implant will look to them to be an object in the environment, and their eyes will move accordingly to provide the most acute image to the brain, including the microsaccades. By tracking these movements, the software can adjust the signal to mimic them in the stimulated retinal ganglion responses. Or perhaps the software could be altered to introduce these fluctuations in the signal and mimic the saccades without actually tracking them by camera.

For the Alpha-IMS, higher resolution is a result of the combination of increased electrode number, the ability to incorporate natural eye movements, and the increased natural processing that can take place via the interneurons, yet the interneuron stimulation helps in other ways as well. Dr. Eberhart Zrenner, Chairman of Retina Implant AG's Supervisory Board and Executive Director/Full Professor at the University Eye Hospital at Tubingen, told me that the subretinal placement of the implant eliminates the need for surgical fixation of the device, the neurons themselves hold the implant in place. Also, the stimulation of this intermediary layer with smaller and more numerous electrodes makes for a smaller phosphene signal; therefore, more definition and acuity in the perceived image is possible.

Unfortunately, stimulating the interneurons also limits the uses of the implant. The interneuron layer must be functional for the subretinal implant to work properly. Accordingly, the Alpha-IMS is being tested only in RP patients, not in those with age-related MD, a disease that often destroys the interneurons as well.

Another feature of the Alpha-IMS, which may be seen as an advantage or disadvantage, is that there is no software included in the device. The signals, based on intensity and position, are used to stimulate the neurons directly, with no interpretation or processing involved. This one-to-one correlation means that the instrument is simpler in design and production, and fewer things can go wrong, although it also means that no improvements can be made once the device is implanted. This is in contrast to the software upgrades on the Argus II system (see below). However, Dr. Zrenner says that Retina Implant AG is planning on expanding the Alpha-IMS system to include multiple implants in a single eye to expand the visual field. Increasing the field is not possible with the Argus II device unless the original implant is extracted and a larger one implanted.

7.3.5 Visual Cortex Implants—The Orion I

The integration of photoreceptor signals via the different retinal cell types is just one level of visual processing that occurs before the neural spikes arrive at the visual cortex. Most of the optic nerve inputs are processed through the lateral geniculate of the thalamus, but some spread to other nuclei. These signals in other areas of the brain are processed and feedback to the eye muscles so that you can track objects without having to consciously decide to move your eyeball.

Geordi's birth defect was somewhere in the front half of his visual system, since his implants were attached to his optic nerves or perhaps embedded in his thalamic nuclei. The temporal area where his VISOR connected to his head was in the general vicinity of the optic nerve, but Dr. Crusher was amazed in *Star Trek: Insurrection*, that the metaphasic (healing) radiation of the planet Baku was regenerating the cells around his optic nerves. This would imply that his optic nerves were included in his birth defect and the implants would had to have been downstream, perhaps in the lateral geniculates. Locating an implant further downstream in the system means that more processing will be lost and must be substituted artificially. There is a trade off here; artificial processing is more difficult as you go farther back, yet you can correct more types of blindness. You gain the natural processing of the lateral geniculate, optic radians and visual cortex if you use a retinal implant; however, the use is limited to retinal degenerative diseases. On the other hand, an implant directly in the visual cortex would help many more blind people, even if it would be technologically harder to generate a precise image.

The World Health Organization states there are 39 million people with blindness (Aug. 2014 estimates) that could be treated with an implant in the visual cortex. They have a disease, defect, or infection that has destroyed the optic nerve along

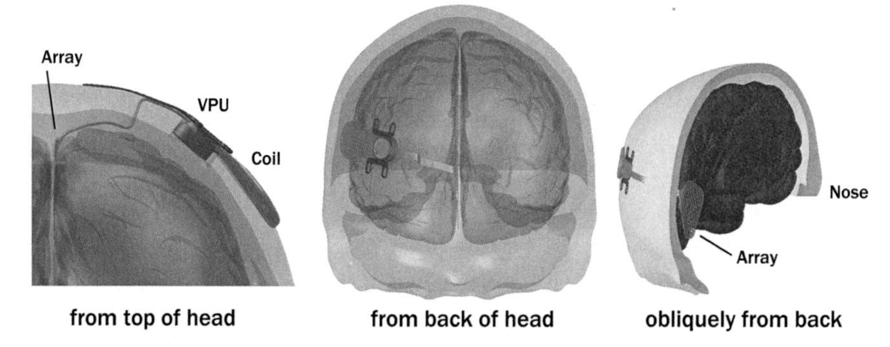

Fig. 7.5 The Orion I is a cortical visual prosthesis implant, plugged directly into the visual cortex of the brain. The *left image* shows the implant device in place from the top of the head, with the electronics package and the recharging coil outside the skull. The *middle* cartoon is from the back of the head, the electrode array can't be seen, as it lies on the visual cortex in the calcarine sulcus between the hemispheres. The *right image* shows the array against the top layer of the visual cortex, making the surgery not so difficult. Because there are right and left visual cortices but only one implant, the images offered by the Orion I will be limited to one half the visual field. To cover both halves of the visual field, two implants would be needed, and they would have to be placed at precisely the same place in both hemispheres. (*Image credit* Second Sight Medical Products, Sylmar, CA)

with the retinal cells. Fortunately, a visual cortex implant called the Orion I is being developed by Second Sight (Fig. 7.5). Mechanical models showed good tolerability and stability in 2015 animal tests, and human testing is scheduled for 2016–2017. The Orion I is based on technology similar to the Argus II, although the implant will be located directly in the visual cortex at the back of the brain. Because the visual cortex maintains the retinotopic layout, the implant will be larger but will not need to cover the entire visual processing area. The portion of the visual cortex that interprets images from the macula is located on the outer surface of the brain, so it will be the easiest to manipulate in surgery. Improvements in neural encoding and transducing will enable higher resolution and contrast, and software updates will improve the function as well (see below).

Dr. Greenberg told me that the biggest advantage of the Orion I is that it will be able to treat the six million or more people in the US with injuries to the optic nerve or lateral geniculate nucleus that cannot make use of the Argus II. Previous surgeries and research efforts in the 1960s through the 2000s to use implants in the visual cortex have helped Greenberg's team to hone in on placement of the implant electrodes, yet the progress made with the Argus II has also helped in developing the hardware and software. Dr. Greenberg says that if they were starting their work now, they might begin with cortical implants, but the Argus work has been of immense help in their current efforts and he knows that development of the Argus II first was the responsible way to proceed in this field.

7.4 Vision Based on the Retinal Code

While the achievements with the Argus II, Alpha-IMS and Orion I are impressive, there is a long way to go before scientists can create an artificial retina. The physical array of electrodes on the implant of both the Alpha-IMS and Argus II are much larger than individual neural axons of the retina, so each electrode stimulates many axons or neurons. The result is blobs of light rather than accurate pinpoints of light for each phosphene, and therefore objects are not seen sharply. More important, both current implant types provide electrical signals based on light or dark only (stimulation or no stimulation). This is not the language that the retina uses to talk to the brain. The Argus II and Alpha-IMS, while ground breaking and helpful, are like listening to a concert via telegraph; most of the information just isn't getting through. What we need is more precise localization of signals in the natural language of the brain.

Geordi's VISOR took the electromagnetic information it sensed and translated it into compressed delta-wavelengths (*The Mind's Eye*), something supposedly similar to what happens in the brain. There are delta waves in EEG tracings (Box 6.1), but these do not correlate strictly with visual input or processing; the description was most likely a contrivance for the series. Science must translate the artificial sensations into a language the brain could understand. This is where the current work of Dr. Sheila Nirenberg and her group at Cornell University's Weill College of Medicine in New York City becomes crucial.

7.4.1 The Retinal Code

The interaction of light with an individual photoreceptor will cause it to transmit its signal in a pointillist fashion. However, it does not stay this simple as we progress from the photoreceptor layer through the interneurons to the third set of neurons that make up the optic nerve. There are approximately 95 million photoreceptor cells in each retina, but only one million retinal ganglion axons leave the eye as the optic nerve. Each layer of neurons integrates the image as received by the previous layer. Some inputs are combined; some are used to suppress other signals. The message is turned into a series of electrical spikes; therefore, each layer has its own language.

There are roughly sixty different types and subtypes of neurons contained with in the retina. Their locations and functions are precise in that they develop small fields of acute vision that can overlap to give produce larger fields integrated with one another. For instance, some neurons fire when light strikes them and they just keep on firing whether or not the stimulus remains. These are the "on" cells. Other neurons act in exactly the opposite fashion. They do not fire when a photoreceptor responds to light striking it, but when that light *stops* activating the receptor, then the neuron fires. These are the so-called "off" cells.

The input to the photoreceptors is deconstructed and integrated according to twelve different characteristics—shape, motion, color, direction, orientation, etc. To add a further level of complexity, the different cell types are arranged in topical patterns, with "on" cells dense in a "center" and "off" cells more dense in the area encircling an "on center." This is a spatial type of processing called "center surround antagonism." The areas overlap and talk to one another so that an overall picture of shape, contrast, and motion can be developed. To talk amongst themselves and collate their information, the neurons do some spatial arithmetic based on the communications they have with the other retinal cell types in their vicinity and the stimulation those cells are receiving. They send on the results of their math to the next set of neurons as a pattern of action potentials. The coordination of the inputs from photoreceptors through the different types of ganglion cells is called the retinal code; this is their neural language.

7.4.2 *Visual Prostheses as Artificial Retinas*

Much work was done in the 2000s and early 2010s to derive the retinal code using isolated normal retinas (Simoncelli et al. 2004; Paninski et al. 2007; Nirenberg et al. 2010; Pillow et al. 2008; Nichols and Nirenberg 2010; Victor 1987; Famulare and Fairhall 2010). Images projected onto the retinas produced firing patterns that could be recorded. The recorded data was analyzed and many models were derived to try and approximate the coding of the signals. Dr. Nirenberg and her graduate student, Chethan Pandarinath, used the previous results and their own of series of mathematical equations to predict and model the firing of all types within a topical area of the retina based on which receptor cells were stimulated (Nirenberg and Pandarinath 2012). With this electronic "encoder" in hand, they then took the further step to use the encoder to translate images captured by a camera into the retinal code and then deliver them to blind mice.

Box 7.3 Optogenetics

Opto—means visible and often refers to light, and *genetics* refers to the fact that a gene is involved. Optogenetics is a method to study the response of cells to a changing environment or to monitor cells' downstream effects by enabling scientists to control the activities of the cell with light as an and off switch. Optogenetics was named "research technique of the year" in 2010.

The technique uses a gene product (protein) that responds to light. Use the response element of that gene as a switch. Scientists can add the response element to other genes to make them responsive to light.

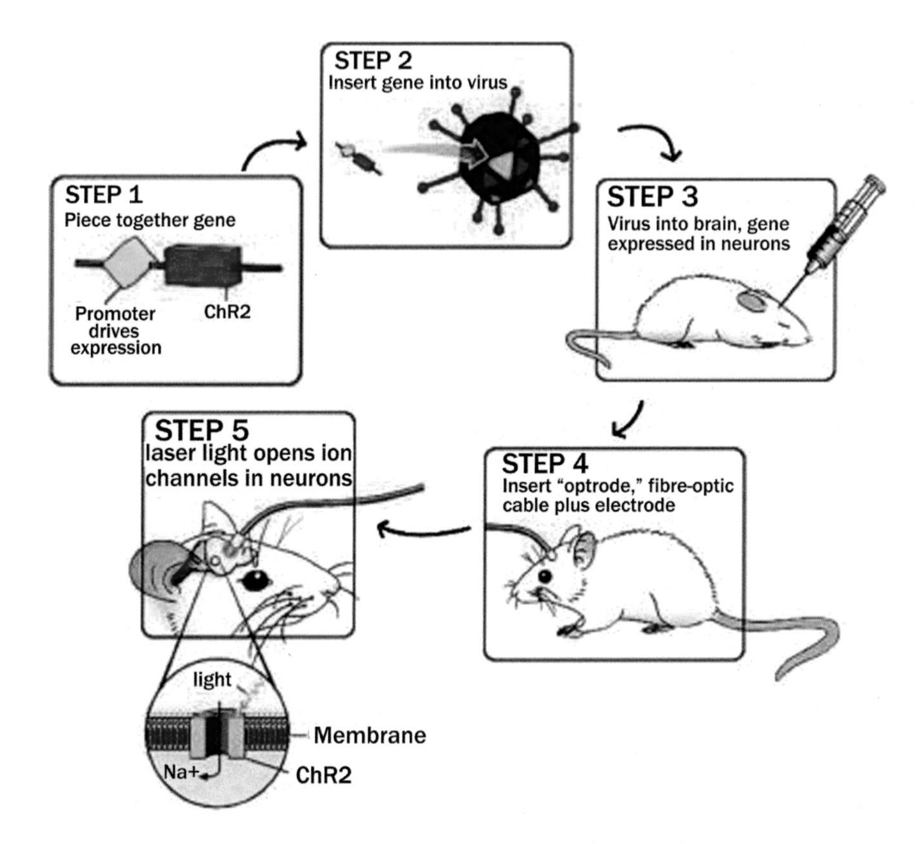

Fig. 7.6 The steps to create an optogenetically modified animal. The promoter is the piece of DNA that carries instructions for which cells will make the ChR2 photosensitive protein, in this case, the specific type of neurons targeted. Virus is an excellent way to transfer the DNA construct into the animals cells. Even if all cells are infected, only those that respond to the promoter will make the protein coded for by the gene. The implanted optrode shines a blue light on the neurons which causes the ChR2 to open a channel in the neuron and start a nerve impulse (action potential). (Reprinted from G et al., Channelrhodopsins: visual regeneration and neural activation by a light switch. *New Biotechnology* 30(5); 461–474, 2013 Copyright (2013), with permission from Elsevier)

Rhodopsin is a classic example of a gene product that responds to light. It spans the membrane and opens a pore temporarily when a photon of light strikes it. It can start a neuron action potential—it makes the neuron fire (Fig. 7.6).

Any neuron that expresses (makes) the protein that includes the rhodopsin switch will fire when exposed to the proper wavelength (blue) and intensity of light. Record the electrical responses downstream of the target neuron and you can map every neuron that the target gene activates or suppresses.

The key is to specifically express the channel rhodopsin switch in the neurons the researcher is interested in targeting. Genes have several components including the functional protein product and the control elements that determine where (in what cells) and when (under what environmental circumstances) the protein is made.

If he/she wants to express study neuron type A, then find a control element that promotes expression type A cells. Combine the DNA for the control element, the rhodopsin switch, and the functional protein together. If you want to express that construct all the time, find a control element that codes for constitutive (continuous) expression. With surgery to implant a blue light emitting diode, a researcher can now fire the neuron anytime he/she wishes.

Dr. Nirenberg's group used a new research technique called optogenetics to deliver the retina spike patterns derived from the mathematical algorithms of the encoder to the retinal ganglion cells. Very briefly, optogenetics uses light to bring changes in cell behavior. Nirenberg and Pandarinath introduced a gene into the retinal ganglion cells that coded for a protein called channel rhodopsin-2 (ChR2). When ChR2 is exposed to blue light, it changes shape and opens a channel in the membrane. This channel allows the sodium ions in and starts an action potential.

To create the spatial and temporal pattern of action potentials on the ChR2-expressing retinal ganglion axons, Nirenberg used a "transducer" made from a grid of incredibly small mirrors attached to motors. By controlling the angle of the individual mirrors they could precisely shine light on ganglion cells in the exact pattern they chose, both spatially and temporally. Spatially, the mirrors focused the stimulatory blue light signal more narrowly than do the electrodes of the Argus II or Alpha-IMS because they can be placed more closely together (Nirenberg and Pandarinath 2012).

Nirenberg's encoder-transducer device leads to a much more refined final image after interpretation of the signals by the visual cortex (Fig. 7.7). The encoder represents the retinal code for the stimulus, so if it were transduced with 100 % efficiency, it should reproduce normal sight within that part of the visual field. Mice in Nirenberg's study were able to track moving images well (Nirenberg and Pandarinath 2012), and successful tracking in this study was higher than in studies using intensity-coded prostheses (Argus II and Alpha-IMS). And it can still improve; the better the transducer, the closer to normal vision the entire device will come. Nirenberg's company, Bionic Sight, has the task of bringing the encoder-transducer technology to the market for human use. Dr. Greenberg of Second Sight and Dr. Nirenberg are old friends and colleagues and they have discussed the possibility of incorporating retinal coding into the Argus II implant (since the Orion I is a cortical implant, the code would be completely different). Unfortunately, since the signaling of the Argus II is less retinotopic and is not one-to-one with photoreceptor cell inputs, they say it would be hard to predict the results of retinal coding in an epiretinal implant.

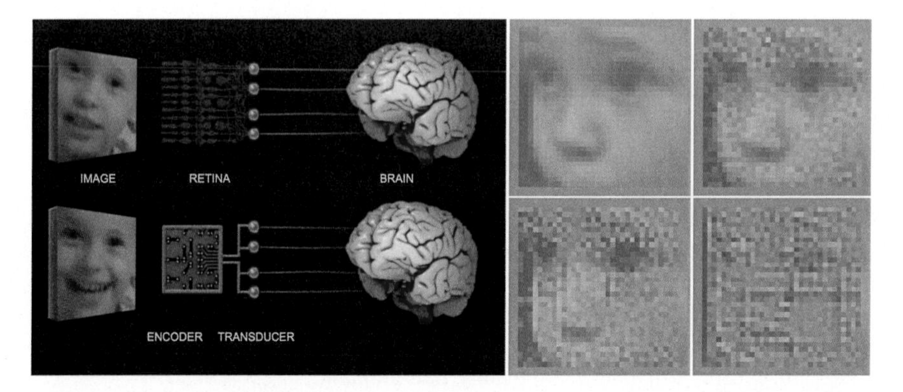

Fig. 7.7 The encoder-transducer system developed by Shelia Nirenberg's group at Cornell substitutes the encoder for the retinal photoreceptors cells and the interneurons. The transducer then collates and translates that signal into the retinal code and flashes the code retinotopically to retinal ganglion cells expressing the channel rhodopsin 2 receptor protein. Those neurons then fire, sending the signal along the optic nerve (*left image*). The *right images* shows the transduced image using natural vision (*top left*), the encoder alone (*top right*), the encoder and transducer (*bottom left*), or the standard method of electrical impulses from camera images on the retinal ganglion cells (*bottom right*). The encoder alone would provide a more life like image if the signal could be conveyed to the retina by the transducer in a more precise manner. Research is being conducted to improve the function of the transducer. [*Image credit* (Nirenberg and Pandarinath 2012)]

7.5 Future Visions for Visual Prostheses

The retinal code work of Nirenberg shows us that just increasing the number of electrodes isn't enough. Increasing the number of pixels has a very limited effect on improving resolution; the key is to talk to the brain in its own language. Much work needs to be done to improve the vision provided by implants; science lacks the specificity and precision to mimic natural vision as of now. Case in point, all work up to this point has been with light and dark. A few shades of grey can be generated by artificial phosphenes, but color vision cannot yet be replicated—or can it?

7.5.1 Software Updates

Despite the limitations of the hardware and encoding in current implants, improvements are coming in the manner in which the devices help users interpret what they can see. Methods to sharpen the signals from visual neural prostheses might include more electrodes for greater acuity or field of vision, better training for patient's to interpret what they sense, or better electrode placement and attachment. But they may also include better software. More like the android Lt. Commander Data than Geordi, people with the Argus II wireless prostheses will undergo software upgrades. This is particularly helpful since software upgrades alone do not require a new FDA approval of the entire device. Software enhancements certainly outstrip the pace of hardware or biologic upgrades, so this will be an integral part of improving prosthesis accuracy and efficiency in the near future.

An immediate example is a software upgrade to the Argus II prosthesis called Accuboost. It is a firmware update that allows for higher resolution and focus, as well as a zoom mode controlled by the user. Once again Geordi's VISOR shows us the way; he has a microscopy mode on his VISOR that he uses to look at surfaces more closely (*TNG: Encounter at Farpoint*). With Accuboost, Dr. Greenberg stated that sight levels of 20/200 are being achieved, the threshold for legal blindness. Accuboost is a standard part of new installations, along with a new VPU that is 25× faster than the original model. However, a new VPU *will* require FDA approval, although with a lower bar than a completely new implant.

Most amazingly, by manipulating implants *already in place* with software changes, Dr. Greenberg's team has been able to restore some color vision as well. He relates it to the old optical illusion from the 1890s called Benham's Top, a circle that is half black and half white with some patterns of black lines. When spun at different speeds (frequencies), the line segments start to take on different colors, reds to blues, depending on the speed and the relative position of the lines from the center to the edge. The black half of the circle serves to break the line segments with a flicker effect, so this is also called "pattern-induced flicker color."

Scientists are not exactly sure why the effect occurs, although one hypothesis is that the black area creates and on and off stimulation of the all the rods and cones that the light falls upon. The different cones (three types) and rods have different activation rates and lags, so the fast rotation may induce a timing imbalance across the different retinal ganglion cells that could be interpreted by the brain as colored light. Using the same manipulation of the on/off stimulation speed of the retinal ganglion cells by the implant, colors can be perceived by the user even when black (off) and white (phosphene induction) are the only inputs. This system has been included in the Argus II software updates and the different colors might be used to represent distance to an object or some other additional data.

7.5.2 Virtual Electrodes and Current Focusing

The strategies for improving prosthesis function via software could focus on making the low-resolution phosphene images more defined, mostly by manipulating brightness. Reducing noise and enhancing contrast can be easily accomplished, often by calculating neighboring pixel values, while highlighting edges can be improved by enhancing light/dark changes that make object edges stand out.

One amazing advance is the use of *virtual* electrodes. Dr. Greenberg likens it to throwing pebbles in a pond. Where two sets of ripples cross one another the energy is the highest. It's the same with the energy from the electrodes. When two electrodes are stimulated, the signals ripple outward and can interact with one another. Where they meet is the highest energy, so this position halfway between the electrodes may also be able to stimulate an action potential. Manipulating the electrodes stimulated can create the virtual electrodes and produce more phosphenes. The virtual electrodes increase resolution without replacing the original implant. In contrast to this, Second Sight is also investigating the possibility of increasing image acuity by steering the electrical current to specific axons in specific situations. Both the virtual electrodes and the current steering can be achieved by using a phased array, similar to the laser optical phased arrays (Sect. 1.3.2.1). The interference patterns will direct the electrical stimulation in different directions without changing the position of the electrodes or the manner in which the individual electrodes fire. The process has already been successful in cochlear implants (Kalkman et al. 2015) and has been simulated with optic nerve stimulation (Li et al. 2013). Current focusing will make the same number of electrodes more versatile and it may be possible for the patient to provide feedback to the implant as to what patterns of firing increase acuity in different situations. This along with machine learning algorithms (Sect. 6.3.1.1) in the software could then provide better and better vision over time.

7.5.3 Algorithms to Improve Recognition

A different sort of improvement would be to increase the meaning of certain portions of the processed image; an attention grabbing device for things that the user needs to know. For instance, it would be helpful to know if you were about to run into something that is close to you. Geordi had an implant in each optic nerve; therefore, he had depth perception. Depth of field and distance measuring in the brain uses input from both eyes. Normally, the brain calculates the difference in retinal locations for the pattern of the object from one eye to the other. A greater difference between the locations of the object on each retina, the closer the object is.

Unfortunately, depth of field would be incredibly hard to reproduce using two retinal implants. The placement of each implant would have to be precise enough to produce a perfect match when the inputs from each eye were integrated in the visual cortex to reproduce the 3D field. Otherwise, it would be like viewing everything with one's eyes crossed. However, there are several ways to determine the distance from the camera to an object electronically—radio, ultrasound, or laser emitters would do the trick. In that case, a visual prosthesis would also be a range finder, great for golf. But these would be costly and require more power. Therefore, software solutions to the problem are being investigated using image processing alone.

One method is to take multiple images of a scene over time. In real time, the size of objects in the foreground are assessed using the change in size for two objects, one close and the other farther away. The changing size permits determination of the distance to the closer object. This *size perspective method* has been successful in locating and distancing objects within 7.5 m (25 ft.) of a visual prosthetic user (Mohammadi et al. 2012). However, the algorithm does not project distance information to the user in feet or meters, it merely subtracts out the cluttering information of objects farther away, and the closest object is *the only thing* the user "sees."

Other algorithms seek to convey information about a region of interest in the visual field, without reducing the rest of the field to blackness. Saliency (the degree to which something stands out against the background) cues are a method to highlight areas of the field that represent the most prominent object. There are many models of saliency detection in robotics. One experiment used one of several bottom-up models that simulate primate vision to locate a region of interest (Parikh et al. 2013). Bottom-up models (Zhaoping and May 2007) work on the hypothesis that you see what is most prominent in the field based on what is there, not on a specific object that your brain interprets you to be looking for. Blinking phosphenes then direct the user to the detected region of interest directionally. In their modeled experiments using hooded individuals and a 6×10 pixel phosphene readout (to simulate the Argus II), participants significantly improved their performance in a mobility task. The directional cueing to a region of interest reduced time to finish the task, the number of errors, and the number of head movements needed to find the target (Parikh et al. 2013). This algorithm would be helpful for finding a person and shaking hands, but would you know who it is?

This is an important point, as users of the current prostheses rate seeing their loved ones a central benefit of the technology. Consequently, a considerable amount of user training with current prostheses is geared to facial recognition. Future software changes may improve this experience. Simulation experiments have been used to determine if face extraction strategies improve participant recognition from a database of three dozen faces (Wang et al. 2014). Three different strategies were tried, and the *statistical face region strategy* (enlarging and highlighting three facial measurements; nose to crown of head, nose to chin and nose to edge of hair) was found to be most helpful. Interestingly, all three strategies improved recognition of female faces more than male faces (Wang et al. 2014).

Second Sight's Robert Greenberg stated that they have ongoing research in all these areas, but in his estimation, those that declutter the background will be more helpful than the "intelligent" algorithms that try to identify and highlight certain features in the field. His patients have told him that their number one priority is to be able to discern a person's face, not so much to recognize the person; it is more important more to them to see if the face is looking at them. So much of communication is non-verbal, and people with vision losses often are frustrated that they can't tell when someone has turned away from them or is speaking to them in particular as opposed to someone else nearby. Seeing the face makes their interactions with other much more informative. The second greatest desire of Dr. Greenberg's patients is to be able to increase mobility without aid. Decluttering of the visual field will enable them to move about with more confidence. His example was the street crosswalk. Crossing is no problem with a cane, but moving straight across the street within the crosswalk is. Veering into traffic in the middle of the intersection is a constant danger (for example, the beeps or verbal "walk" cues at some crosswalks are more for keeping people moving straight than for telling them when to cross), so decluttering will allow them to move more efficiently.

7.5.4 Expanding Our Vision Beyond Visible Light

Using software to maximize the information provided by his prosthesis wasn't a problem for Geordi. In fact, he received *too much* input from his VISOR; he had to concentrate to maintain focus on just a portion of the data that his VISOR was blasting at him (*TNG: The Masterpiece Society*). The reason?—Geordi's VISOR showed him much more than just visible light.

Box 7.4 The Electromagnetic Spectrum
Electromagnetic radiation waves are a form of energy; some have a wave particle duality yet have no mass. They generate an electrical field in the same direction as the wave and a magnetic field perpendicular to the direction of the wave. They do not need a medium through which to propagate, so they can travel through space.

Wave—the peaks and troughs through which EM radiation cycles.
Amplitude—the height of the peak from the zero value; a measure of intensity.
Wavelength—distance between same point on two adjacent peaks or troughs.
Frequency—number of wavelengths completed in one second, measured in Hertz (cycles/sec).
Inverse relationship—wavelength = 1/frequency, so as wavelength decreases, frequency increases.

Energy—measured in electron volts (eV); the amount of energy it takes to move one electron through a one volt potential. High-energy EM waves have short wavelengths and high frequencies. Ultraviolet waves have higher energy than radio waves (Fig. 7.8).

Ionizing radiation—EM waves with energy high enough to remove electrons from the atoms to which they are bound, forming ions. Includes EM waves with energies higher than 10 eV, so high frequency ultraviolet, X-ray and gamma ray are ionizing. High-energy particulate radiation is also ionizing radiation.

Particulate radiation—completely different from EM radiation; this radiation is made of particles with mass. Alpha particles (helium nuclei), beta particles (high-energy electrons), and other cosmic radiation such as high-energy heavy nuclei. They present a physical danger to astronauts in space.

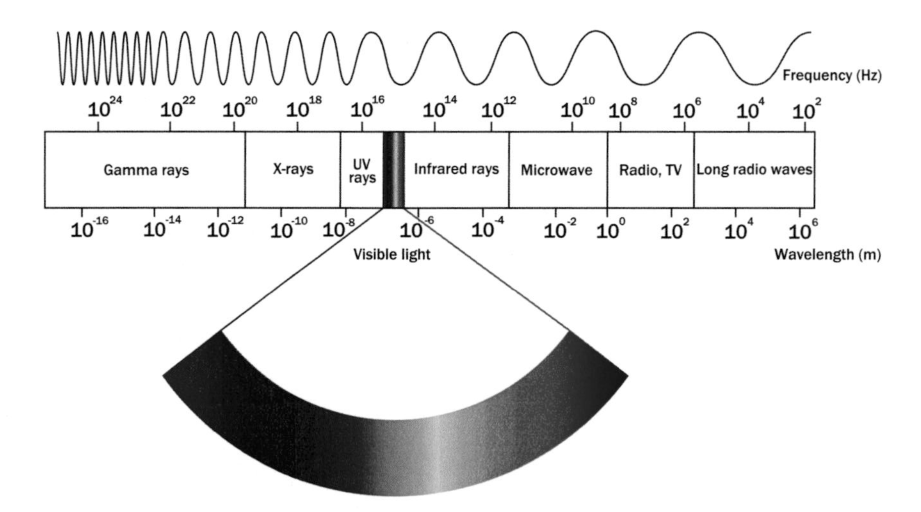

Fig. 7.8 Visible light makes up a small portion of the electromagnetic (EM) spectrum. You can see that frequency varies inversely with wavelength. As the frequency increases and the wavelength decreases the waves have more energy. The longer wavelengths are harder to classify; the boundaries of far infrared, microwave and ultra high frequency radio are often blurred. This image does not even show terahertz radiation, which is located between the infrared and microwave bands. (*Image credit* Biro Emoke/Shutterstock)

Radiation damage—waves propagate until they strike and atom, when they are scattered or absorbed and transmit some of their energy to the atom. X-rays and gamma rays wavelengths are smaller than most atoms, so they are more penetrating. They are high energy waves, so much energy can be transmitted to the atoms they strike. If the atoms are in a living organism, they can damage cells and perhaps kill the organism.

Geordi's VISOR—could detect and transmit interpretable signals to his brain all waves from radio through X-rays.

Geordi didn't see just the small part of the electromagnetic spectrum that represents visible light—he saw nearly all of it. In the *The Masterpiece Society*, Geordi stated that his VISOR could image that part of the EM spectrum from one Hertz (Hz, waves that have a peak that occurs once each second, or one cycle/second) to 100,000 THz (equal to 100,000,000,000,000,000 Hz). The only electromagnetic waves that aren't included in this range are hard X-rays and gamma rays. Radio waves, microwaves, infrared heat waves, visible light, ultraviolet light, and soft X-rays—Geordi could see them all. The higher the energy, the more likely they would be to penetrate an object instead of bouncing off and into his sensors, but still, he must have been inundated by information. No wonder his head hurt.

Brian Mech, a former vice president of Second Sight, stated in an interview that there is no reason why visual prostheses today couldn't be modified to pick up all these sources of EM radiation. It is just a matter of including a detector in the external camera, or the implant CMOS detector of the Alpha-IMS, and then converting the signal into electrical spikes that will trigger phosphenes. Night vision goggles detect infrared heat waves; and in truth, Second Sight has already included an infrared detector in the Argus II. Dr. Greenberg says that the infrared detector is included mostly to pick up heat signatures around the home. The users appreciate this because they can tell if a pan on the stove is hot and can detect other sources of heat that might otherwise injure them.

Higher energy waves, like X-rays are detected in an X-ray machine. This isn't to say Geordi has X-ray vision; an X-ray machine has an emitter to produce rays which are then captured on film or by a digital X-ray detector after passing through some object, like your broken arm. Geordi can only detect the naturally occurring X-rays that find their way to his VISOR. Stars and the like produce soft X-rays, but they can't penetrate our atmosphere. So Geordi only displays decent X-ray vision in space; on planets with a substantial atmosphere he is X-ray blind. Dr. Greenberg sees no reason why X-ray, radio, or ultraviolet detectors could not be included on the Argus II; however, they will likely wait until specific needs for each sensor can be identified. Picking up all the stray signals from the spectrum at once could make Geordi a very nosey person if he chose to be, just as he tapped into the telemetry data coming from an experimental probe (*TNG: Interface*). It would be overwhelming to have a camera and implant that could detect radio, micro, IR, visible, UV and soft X-ray waves all at once. With only 60 electrodes in the Argus II, or even 1500 in the Alpha-IMS, the receiver would be maxed out with incoming data every second of every day. Every electrode would be firing all the time and the image would be continuously white. To filter the information, one would need a switching mechanism to allow the user to pick which frequencies would be detected and translated at any point in time. This will probably be the strategy pursued by future visual prostheses unless a mechanism to separate the outputs for each part of the spectrum in the cortex can be developed.

And yet, why stop even there? Geordi recorded and transmitted what his visor sensed while he was on the freighter *Batris* to the main screen of the Enterprise (*TNG: Heart of Glory*). To do so, he had to use a visual acuity transmitter of his own design and he could only broadcast over a range of a few kilometers. Heck, science could do better than that today. Merely install a WiFi hotspot on the prosthesis and you could stream your vision live around the entire world via the Internet. But be careful of being hacked. Geordi's VISOR was modified by Romulans to transmit E-band messages directly to his brain after his abduction. Using this back door into his brain, they mentally conditioned him into becoming an assassin (*TNG: The Mind's Eye*).

Geordi even used his VISOR to analyze molecular structures, including the form and function of the world-destroying crystalline entity (*TNG: Datalore*) and the walls of the passages below Farpoint Station (*TNG: Encounter at Farpoint*). O.K., this is a bit harder, though who is to say that researchers couldn't miniaturize an electron microscope, a field ion microscope or mass spectrometer so that they fit into a set of goggles. Look at how far miniaturization has come in terms of analyzing tissue samples for the tricorders (Sect. 9.5 and 9.6). Will there come a time when Geordi's VISOR, Superman's X-ray vision, Cyclops' energy rays, or Neo's ability to see the Matrix will be regarded as ho-hum? All we'll need is a rad pair of glasses, an implant, and Geordi's integrity to use our powers for the good of the Federation.

7.6 Conclusion

La Forge's VISOR was an important part of *Star Trek*. It represented a tangible example of the inclusion built into the franchise and the societal forward thinking that the writers demonstrated when telling stories of space exploration by a truly united Earth. Entire books have been written about the social implications of *Star Trek* and the risks the writers, creator, and producers took to put forth their vision of a society without prejudice, yet I don't think enough attention is paid to the inclusion of physically or mentally challenged (as in Reginald Barclay and his undiagnosed Asperger's syndrome, social anxiety disorder, and stuttering) characters in all the series of the franchise. Geordi have been the most central of those characters, but he was hardly the only person with challenges to overcome.

The VISOR is also important as a piece of technology that helps the individual, not the crew or universe as a whole. It is a practical instrument meant to do one thing for one person, as are the real world visual prostheses that are so important to a growing segment of our population. Geordi's implants and detector are one of the most individual and personal pieces of tech on the show. It is also one of the most mature technologies in real life. Tractor beams are real, although we can only move miniscule objects at the present time. Cloaking devices exist as well, but science is nowhere near being able to cloak a vessel from all EM radiation in all directions. Science has all the pieces to put together a universal translator, yet we still don't know if languages have enough in common to make it worth building one. These are all amazing advances, yet none is near as mature as the visual prostheses that Second Sight, Retina AG and Dr. Shelia Nirenberg have produced. Partial vision has been restored to hundreds of patients, with the numbers climbing every day. If one ignores the more modest accomplishments of talking computers and cell phone communicators, this has to be the most stunning real life translation of a *Star Trek* technology that we have encountered. It's importance and genius cannot be overestimated.

However, there is a long way to go with visual implants before seamless, natural vision is restored. Second Sight's founder Robert Greenberg is of the opinion that bionic eyes will be possible within the next ten to twenty years, although he is partial to keeping most of the hardware of the system external to the body so that upgrades and repairs will be easier. This is just the same as Geordi's VISOR. *Star Trek's* writers had the ability to place the hardware completely inside his head and only show the bionic eyes as they did in *Star Trek: First Contact*, but they chose to keep the hardware external most of the time. This may have been for story telling purposes and as a visual cue to the audience about the specialness of Commander

La Forge's situation, or maybe it's just that they knew future technology specialists would hold the same opinion as Dr. Greenberg. Could *Star Trek* see that well into the future?

References

MJ Banissy and J Ward. Mirror-touch synesthesia is linked with empathy. *Nature Neuroscience* 10(7); 815-6, 2007. doi:10.1038/nn1926. http://www.nature.com/neuro/journal/v10/n7/full/nn1926.html

MP Barry and G Dagnelie. Use of the Argus II retinal prosthesis to improve visual guidance of fine hand movements. *Invest Ophthalmol Vis Sci* 53(9); 5095–5101, 2012. doi:10.1167/iovs.12-9536. http://iovs.arvojournals.org/article.aspx?articleid=2165826

MR Behrend, AK Ahuja, MS Humayun, RH Chow, and JD Weiland. Resolution of the epiretinal prosthesis is not limited by electrode size. *IEEE Trans Neural Syst Rehabil Eng* 19(4); 436–442, 2011. Doi:10.1109/TNSRE.2011.2140132. http://ieeexplore.ieee.org/xpl/articleDetails.jsp?arnumber=5752253

AA Cho, JB Clark, and AH Rupert. Visually triggered migraine headaches affect spatial orientations and balance in a helicopter pilot. *Aviat Space Environ Med* 66(4); 353-358, 1995.

L da Cruz, BF Coley, J Dorn, F Merlini, E Filley, P Christopher FK Chen, V Wuyyuru, J Sahel, P Stanga, M Humayun, RJ Greenberg, G Dagnelle, for the Argus Study Group. The Argus II epiretinal prosthesis system allows letter and word reading and long-term function in patients with profound vision loss. *Br J Ophthalmol* 97(5); 632–636, 2013. doi:10.1136/bjophthalmol-2012-301525. http://bjo.bmj.com/content/97/5/632.long

JD Dorn, AK Ahuja, A Caspi, L d Cruz, G Dagnelle, J Sahel, RJ Greenberg MJ McMahon, for the Argus II Study Group. The Detection of Motion by Blind Subjects With the Epiretinal 60-Electrode (Argus II) Retinal Prosthesis. *JAMA Ophthalmol* 131(2); 183–189, 2013. doi:10.1001/2013.jamaophthalmol.221. http://archopht.jamanetwork.com/article.aspx?articleid=1375731

M Famulare and A Fairhall. Feature selection in simple neurons: How coding depends on the spiking dynamics. *Neural Comput* 22(3); 581-598, 2010. doi:10.1162/neco.2009.02-09-956. http://www.mitpressjournals.org/doi/full/10.1162/neco.2009.02-09-956#.VpGnssArLfQ

N G, A Tan, Y Farhatnia, J Rajadas, MR Hamblin, PT Khaw, and AM Seifalian. Channelrhodopsins: visual regeneration and neural activation by a light switch. *New Biotechnology* 30(5); 461-474, 2013. doi:10.1016/j.nbt.2013.04.007. http://www.sciencedirect.com/science/article/pii/S1871678413000599

AI Goller, K Richards, S Novak, and J Ward. Mirror-touch synaesthesia in the phantom limbs of amputees. *Cortex* 49(1); 243-51, 2013. doi:10.1016/j.cortex.2011.05.002. http://www.sciencedirect.com/science/article/pii/S0010945211001468

ZM Hafed, K Stingl, K-U Bartz-Schmidt, F Gekeler, and E Zrenner. Oculomotor behavior of blind patients seeing with a subretinal visual implant. *Vision Research* 118; 119-131, 2016. doi:10.1016/j.visionres.2015.04.006. http://www.sciencedirect.com/science/article/pii/S0042698915001510

DE Harle, AJ Shepherd and BJW Evans. Visual stimuli are common triggers of migraine and are associated with pattern glare. *Headache* 46(9); 1431-1440, 2006. doi:10.1111/j.1526-4610.2006.00585.x. http://onlinelibrary.wiley.com/enhanced/doi/10.1111/j.1526-4610.2006.00585.x/

MS Humayun, JD Dorn, L da Cruz L, G Dagnelle, J Sahel, PE Stagna, AV Cideciyan, JL Duncan, D Eliott, E Filley, AC Ho, A Santos, AB Safran, A Arditi, LV del Priore, RJ Greenberg, the Argus Study Group. Interim results from the international trial of Second Sight's visual prosthesis. *Ophthalmology* 119(4); 779–788, 2012. doi:10.1016/j.ophtha.2011.09.028. http://www.aaojournal.org/article/S0161-6420(11)00884-0/abstract

MS Humayun, JD Weiland, GY Fujii 2003, R Greenberg, R Williamson, J Little, B Mech, V Cimmarusti, G van Boemel, G Dagnelle, and E de Juan. Visual perception in a blind subject with a chronic microelectronic retinal prosthesis. *Vision Res* 43(24); 2573–2581, 2003. doi:10.1016/S0042-6989(03)00457-7. http://www.sciencedirect.com/science/article/pii/S0042698903004577

S Ings. *A Natural History of Seeing: The Art and Science of Vision*. New York: WW Norton, 2008.

L Johannesson, and S Jarvholm. Uterus transplantation: current progress and future prospects. *International Journal of Womens Health* 8: 43-51, 2016. doi:10.2147/IJWH.S75635. https://www.dovepress.com/uterus-transplantation-current-progress-and-future-prospects-peer-reviewed-article-IJWH

RK Kalkman, JJ Briaire, and JH Frijins. Current focusing in cochlear implants: an analysis of neural recruitment in a computational model. *Hearing Research* 322; 89-98, 2015. doi:10.1016/j.hearres.2014.12.004. http://www.sciencedirect.com/science/article/pii/S0378595514002032

A Kotecha, J Zhong, D Stewart, and L da Cruz. The Argus II prosthesis facilitates reaching and grasping tasks: a case series. *BMC Opthalmol* 14; 71, 2014. 10.1186/1471-2415-14-71. http://bmcophthalmol.biomedcentral.com/articles/10.1186/1471-2415-14-71

TZ Lauritzen, J Harris, S Mohand-Said, JA Sahel, JD Dorn, K McClure, and RJ Greenberg. Reading visual braille with a retinal prosthesis. *Front Neurosci* 6; 168, 2012. doi:10.3389/fnins.2012.00168. http://journal.frontiersin.org/article/10.3389/fnins.2012.00168/abstract

J Lechner, M Chen, RE Hogg, L Toth, G Silvestri, U Chakravarthy, and H Xu. Higher plasma levels of complement C3a, C4a, and C5a increase the risk of subretinal fibrosis in neovascularization age-related macular degeneration: Complement active in AMD. *Immun Ageing* 13; 4, 2016. doi:10.1186/s12979-016-0060-5. http://immunityageing.biomedcentral.com/articles/10.1186/s12979-016-0060-5

M Li, Y Yan, Q Wang, H Zhao, X Chai, X Sui, Q Ren, and L Li. A simulation of current focusing and steering with penetrating optic nerve electrodes. *Journal of Nueral Engineering* 10(6); 06607, 2013. doi:10.1088/1741-2560/10/6/066007. http://iopscience.iop.org/article/10.1088/1741-2560/10/6/066007

HM Mohammadi, E Ghafar-Zedah, and M Sawan. An image processing approach for blind mobility facilitated through visual intracortical stimulation. *Artificial Organs* 36(7); 616-628, 2012. 10.111/j.1525-1594.2011.01421.x. http://onlinelibrary.wiley.com/doi/10.1111/j.1525-1594.2011.01421.x/abstract

Z Nichols, and S Nirenberg. Correlations in complete retinal populations play a very small role in coding natural scene stimuli. Proceedings of the 40th Annual Meeting of the Society for Neuroscience, San Diego. Prog. No. 891.3, 2010.

S Nirenberg, I Bomash, JW Pillow, and JD Victor. Heterogeneous response dynamics in retinal ganglion cells: The interplay of predictive coding and adaptation. *J Neurophysiol* 103 (6); 3184-3194, 2010. doi:10.1152/jn.00878.2009. http://jn.physiology.org/content/103/6/3184.long

S Nirenberg and C Pandarinath. Retinal prosthetic strategy with the capacity to restore normal vision. *Proc Natl Acad Sci, USA* 109(37); 15012-15107, 2012. doi:10.1073/pnas.1207035109. http://www.pnas.org/content/109/37/15012

M Nita, and A Grzybowski. The role of reactive oxygen species and oxidative stress in the pathomechanism of the age-related ocular diseases and other pathologies of the anterior and posterior eye segments in adults. *Oxid Med Cell Longev* 2016; 3164734, 2016. doi:10.1155/2016/3164734. http://www.hindawi.com/journals/omcl/2016/3164734/

L Paninski, J Pillow, and J Lewi. Statistical models for neural encoding, decoding, and optimal stimulus design. *Prog Brain Res* 165; 493-507, 2007. doi:10.1016/S0079-6123(06)65031-0. http://www.sciencedirect.com/science/article/pii/S0079612306650310

JW Pillow, J Shlens, L Paninski, A Sher, AM Litke, EJ Chichilinisky, and EP Simoncelli. Spatio-temporal correlations and visual signaling in a complete neuronal population. *Nature* 454; 995-999, 2008. doi:10.1038/nature07140. http://www.nature.com/nature/journal/v454/n7207/full/nature07140.html

N Prikh, L Itti, M Humayun, and J Weiland. Performance of visually guided tasks using simulated prosthetic vision and saliency-based cues. *J Neural Engineering* 10(2):026017, 2013. doi:10.1088/1741-2560/10/2/026017. http://iopscience.iop.org/1741-2552/10/2/026017/

EP Simoncelli, L Paninski, J Pillow, and O Schwartz. Characterization of neural responses with stochastic stimuli. *The Cognitive Neurosciences*, ed. Gazzaniga, M (MIT Press, Cambridge, MA), pp327-338, 2004.

AJ Shepherd, TJ Hine, and HM Beaumont. Color and spatial frequency are related to visual pattern sensitivity in migraine. *Headache*. 53(7); 1087-1103, 2013. doi:10.1111/head.12062. http://onlinelibrary.wiley.com/doi/10.1111/head.12062/abstract

K Stingl, KU Bartz-Schmidt, D Besch, A Braun, A Bruckmann, F Gekeler, U Greppmaier, S Hipp, G Hortdorfer, C Kernstock, A Koitschev, A Kusnyerik, H Sachs, A Schatz, KT Stingl, T Peters, B Wilhelm, and E Zrenner. Artificial vision with wirelessly powered subretinal electronic implant alpha-IMS. *Proceedings of the Royal Society B* 280(1757); 20130077, 2013a. doi:10.1098/rspb.2013.0077. http://rspb.royalsocietypublishing.org/content/280/1757/20130077

K Stingl, KU Bartz-Schmidt, F Gekeler, A Kusnyerik, H Sachs, and E Zrenner. Functional outcome in subretinal electronic implants depends on foveal eccentricity. *Investigative Ophthalmology & Visual Science* 54(12); 7658–7665, 2013b. doi:10.1167/iovs.13-12835. http://iovs.arvojournals.org/article.aspx?articleid=2127911

K Stingl, KU Bartz-Schmidt, D Besch, CK Chee, CL Cottriall, F Gekeler, M Groppe, TL Jackson, RE McLaren, A Koitschev, A Kusnyerik, J Neffendorf, J Nemeth, MAN Naeem, T Peters, JD Ramsden, H Sachs, A Simpson, MS Singh, B Wilhelm, D Wong, and E Zrenner. Subretinal visual implant Alpha IMS – clinical trial interim report. *Vision Research* 111(Part B); 149-160, 2015. 10.1016/j.visres.2015.03.001. http://www.sciencedirect.com/science/article/pii/S0042698915000784

JD Victor. The dynamics of the cat retinal X cell centre. *J Physiol* 386(1); 219-246, 1987. doi:10.1113/jphysiol.1987.sp016531. http://onlinelibrary.wiley.com/doi/10.1113/jphysiol.1987.sp016531/abstract

J Wang, X Wu, Y Lu, H Wu, H Kan, and X Chai. Face recognition in simulated prosthetic vision: face detection-based image processing strategies. *J Neural Eng* 11(4); 046009, 2014. doi:10.1088/1741-2560/11/4/046009. http://iopscience.iop.org/1741-2552/11/4/046009/

J Ward, and MJ Banissy. Explaining mirror-touch synesthesia. *Cogn Neurosci* 6(2-3); 118-133, 2015. 10.1080/17588928.2015.1042444. http://www.tandfonline.com/doi/full/10.1080/17588928.2015.1042444

R Wilke, VP Gabel, H Sachs, KU Bartz-Schmidt, F Gekeler, D Besch, P Szurman, A Stett, B Wilhelm, T Peters, A Harscher, U Greppmaier, S Kibbel, H Benav, A Bruckmann, K Stingl, A Kusnyerik, and E Zrenner. Spatial resolution and perception of patterns mediated by a subretinal 16-electrode array in patients blinded by hereditary retinal dystrophies. *Invest Ophthalmol Vis Sci* 52(8);5995–6003, 2011. doi:10.1167/iovs.10-6946. http://iovs.arvojournals.org/article.aspx?articleid=2188619

N Xu, F Wei, X Liu, L Jiang, H Cai, Z Li, M Yu, F Mu, and Z Liu. Reconstruction of the upper cervical spine using a personalized 3D-printed vertebral body in an adolescent with Ewing sarcoma. *Spine (Phila Pa 1976)* 41(1); E50-E54, 2016. doi:10.1097/BRS.0000000000001179. http://journals.lww.com/spinejournal/Abstract/2016/01000/Reconstruction_of_the_Upper_Cervical_Spine_Using_a.21.aspx

L Zhaoping and KA May. Psychophysical tests of the hypothesis of a bottom-saliency map in primary visual cortex. *PLoS Computational Biology* 3(4); e62, 2007. doi:10.1371/journalpcbi.0030062. http://journals.plos.org/ploscompbiol/article?id=10.1371/journal.pcbi.0030062

E Zrenner, KU Bartz Schmidt, H Benav, D Besch, A Bruckmann, VP Gabel, F Gekeler, U Greppmaier, A Harscher, S Kibbel, J Koch, A Kusnyerik, T Peters, K Stingl, H Sachs, A Stett, P Szurman, B Wilhelm, and R Wilke. Subretinal electronic chips allow blind patients to read letters and combine them to words. *Proc R Soc B* 278:1489–1497, 2011. 10.1098/rpsb.2010.1747. http://rspb.royalsocietypublishing.org/content/278/1711/1489.long

Chapter 8
The Transporter: Are We There yet?

I signed on to this ship to practice medicine, not to have my atoms scattered back and forth across space by this gadget.
—Dr. Leonard McCoy
TOS: Space Seed

8.1 Introduction

"Beam me up Scotty," is one of the most memorable of all *Star Trek* quotes, right up there with, "I'm a doctor not a (fill in the blank)." The phrase has been used on T-shirts, greeting cards, as a verbal exit after an embarrassing situation, as well as a dig when followed by, "There's no intelligent life here." In my youth, I uttered it countless times. So I was interested to find out that no one ever said that specific that line in all of *Star Trek*. Captain Kirk did once say, "Scotty, beam me up,"—once (*Star Trek IV: The Voyage Home*), and the heart of the quote, "beam me up" was used only three other times by the original cast (*TOS: This Side of Paradise, The Squire of Gothos, Time's Arrow*). For a piece of equipment that was so integral to the franchise, that's precious few mentions in the most remembered parlance. Of course, later series became fond of the verb "energize," when the original series only used that word once. Believing he has killed Kirk during his first experience with the Vulcan mating ritual *pon farr*, Spock asks to be beamed back to the ship with Kirk's body using that one word command, "Energize." (*TOS: Amok Time*)

The transporter, like the terms for its use, is the most iconic of the *Star Trek* technologies. Even *The Big Bang Theory* acknowledges the fact, when Penny gives both Sheldon and Leonard mint-in-box transporter pad collectibles (*The Transporter Malfunction*). The idea of traveling anywhere one wishes in an instant is attention grabbing and thought provoking, making it all the more amazing that the transporter was an original series compromise; one that changed the order of the first season. Like most things concerning television production, it was a matter of money. Gene Roddenberry originally scripted for the *USS Enterprise* or a shuttle to land on the surface of a new planet each week, which would have been a budget buster (sets, models, etc.), especially for a show that didn't have a lot of backing to begin with.

M.E. Lasbury, *The Realization of Star Trek Technologies*,
DOI 10.1007/978-3-319-40914-6_8

The production team had to think of a cheaper way of getting crew members down to a planet and back the ship. Voila—the transporter (Okuda et al. 1995, p. 519).

Later on, Roddenberry deemed this budget limitation a creative benefit. He said that the compromise to use a transporter forced the writers and directors to be inventive; coming up with the idea of the transporter and the sci-fi technology to back it up (sort of) wasn't a simple undertaking. In the end, the technology itself seemed plausible, although not needing to land a fourteen story tall starship on a planet every week also helped the series' credibility (Whitfield and Roddenberry 1968, pp. 43–44). And just how did the transporter change the order of the first season? The third episode (*The Man Trap*) began with Kirk and cohorts transporting down to the planet surface; by showing this first, the producers and writers didn't have to go to the time and effort of explaining the transporter—the audience understood its purpose when they saw it perform its job (Kooper 2013, p. 150).

Indeed, the original series wasted little time explaining the workings of the transporter. The *Enterprise* series took more time to investigate the early transporters, especially the ambivalence or outright hostility that some of the population had towards the prospect of beaming people here and there (*ENT: Daedalus*). However, it was *The Next Generation* that tried hardest to explain how transporting takes place, always within the narrative of the story and always with enough scientific validity to make it believable. They, with *DS9* and *Voyager*, introduced more parts of the apparatus; the annular confinement beam, the multiplex pattern buffer, the molecular imaging scanners, the phase discriminator, the site-to-site interlock, and the Heisenberg compensator. They were all devised and tinkered with over the seasons as an indication of how much the writers wanted the transporter to be seen as a plausible piece of technology. Now it is the job of science to prove the franchise correct; researchers and engineers need to come up with a feasible mechanism of teleportation in reality.

To describe the work being done on the subjects surrounding teleportation, this chapter is set up a bit differently than the others. There are so many working parts to the process—none of which is scientifically mature, or even immature just yet. Some of the most basic tenets of physics and engineering have to be met head on or maneuvered around in order to move an object from point A to point B without ever existing anywhere in between. While some fantastic research is being carried out, just how these discoveries might be used to move a person around in space is still a project in a pre-embryonic state. While the research described in most chapters is demonstrating why those technologies are possible/probable, think of the studies of this chapter as suggestions why the transporter might not be *impossible*.

The issues will be addressed in the order that teleportation takes place: (1) targeting a destination, (2) defining a quantum level pattern of the individual or object; (3) dematerializing the target to a matter stream; 4) sending the matter stream and pattern information to the destination; and (5) rematerializing the matter stream according to the pattern. Each issue has its own problems and possible solutions, with some of those solutions coming from research being done for completely different purposes. For instance, the search for faster and more powerful computing architectures (quantum computing) could solve problems in patterning a person's entire quantum state. The overall process may sound daunting, yet one may draw hope from the knowledge that science has already achieved some degree of teleportation. As a scientific concept teleportation is more than plausible, it's science fact.

8.2 Issue: Targeting a Spot in Space

When you determine that you need to mail a package, the first bit of information you need is an address. You can't send anything to anyone until you know exactly where you want the package to go. The same is true in teleportation. First you need an address in space. For UPS or Fed Ex, a manmade coordinate and tracking system works fine; people have defined areas as cities, grids of those cities as named streets, and positions on those streets as numbered addresses. This is adequate for what it does, although sometimes more precision is needed. Certainly, transporting to an alien planet surface is going to require a bit more information; a specific point in space needs to be defined as accurately as possible. On Earth, satellite navigational systems like the global positioning satellite (GPS) system are good. Would this also work for outer space? In reality, there are actually many different systems in play for finding an address in our galaxy.

8.2.1 Celestial Reference Frames and Coordinate Systems

The 30 or so satellites United States military GPS system are fine for locating a point on Earth, but defining a point in space would need a few more to get adequate coverage—try billions. Even with a sea of cubesats (Sect. 5.4.1) this is an untenable idea at best and most likely impossible, so a different method is needed. Each solar system could have a different coordinate system, using the barycenter (gravitational center) as the 0,0,0 point. Then any point at a given time could be defined by directions x, and y, and a distance z from the origin, as long as one defines the plane of the X and Y axes, i.e. a point on at least one side of the origin through which the axis runs. In addition, when McCoy, Kirk, and Spock beam down to the red soil surface of Vulcan (*TOS: Amok Time*), the precise distance is very important; too low and they end up buried in the dirt, too high and they appear in mid-atmosphere and plummet to the surface at a rate determined by Vulcan gravity. Yet the away team always arrives just where they ought. How can this be done if every planet isn't mapped exactly or uses a coordinate system based on the planet itself? The success implies that the *Enterprise* works with a very precise mapping system that is general to wherever it might find itself. On the other hand, away teams often beam from ship to ship. This might be best accomplished using a frame of reference and coordinate system that is centered on the *Enterprise*, a spacecraft-fixed frame of reference. Therefore, the reference frame and the coordinate system used may be specific to the situation. This being the case, many reference frames and coordinate systems would probably be developed—and they have been.

To define a system in which one wants to locate an object, there is a need for both a frame of reference (or just frame) and a coordinate system. The frame is the ordered set of three vectors, meaning it gives the coordinate system a point to which it is pinned; as the pinned object moves, so does the frame. The coordinate system is then the mechanism used to locate an object within a frame of reference. Therefore, any defined point is only meaningful if one also knows both the frame and coordinate system being used. In planetary science, there are several coordinate systems used, just as there are several reference frames used. The frames can be determined by Earth as it moves and rotates, so the frame moves with the Earth and doesn't appear to change for an observer on Earth (Earth-centered, Earth-fixed frame, ECEF) or it might be fixed, it does not rotate as the Earth does (Earth inertial frame, EIC). There is also the International Celestial Reference Frame (ICRF) that is based on the positions of 1500 extragalactic radio sources (quasars). On the other hand, a frame of reference might be a single spacecraft, with the frame following it wherever it goes. All other objects would be measured by some coordinate system as its position relative to the spacecraft.

Coordinate systems are just as numerous as reference frames, with the axes of the system either in rectangular (Cartesian, X, Y, Z) or spherical (angles from a plane on a reference ellipsoid) arrangement. Coordinate systems have an origin through which at least one axis passes, and the origin need not be at the center of the system. The ecliptic coordinate system is based on plane of the Earth's orbit around the Sun (called it's ecliptic). It can have the Earth or the barycenter of the solar system as its origin, may use an EIC or ECEF frame, and can use Cartesian or spherical coordinates. However, the ecliptic wobbles over time, so an additional parameter is needed to define a point—time. Most ecliptic coordinates are assigned to the year 1950 or 2000, so that the coordinates are pegged to the ecliptic at that point in time. For this reason, the ICRS (International Celestial reference System) uses the ICRF because extragalactic quasars are fixed over much longer periods of time.

The *Star Trek* franchise has spent all its time exploring the Milky Way galaxy with the exceptions of when Kirk and company briefly move just beyond the limits of the Milky Way (*TOS: Where No Man has Gone Before*) and when Picard's crew travels to galaxy M-33 (*TNG: Where No One has Gone Before*). When the *USS Voyager* is knocked 70,000 light years away into the the Delta quadrant by a massive displacement wave (*VOY: Caretaker*), it is still in the Milky Way, just in a different quarter of the galaxy as defined by an X and Y axis through the galactic center and along the galactic plane. Most of the stories of the *USS Enterprise* take place in the Alpha and Beta Quadrants, where Earth lies on the Alpha side of the border between the two (Star Trek Encyclopedia, p. 393). Therefore, a galactic coordinate would seem in order for this degree of exploration, and is just the system described in the Star Fleet Academy Training Manual (p. TO:02:06:10). However, the author doesn't say which reference frame is being used. In real life, a galactic

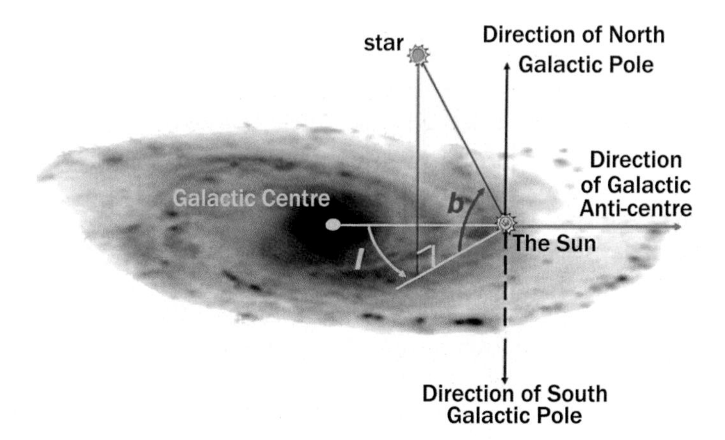

Fig. 8.1 There are several methods to chart a point in space; many reference frames and coordinate systems apply to the solar system, galaxy and intergalactic space. One common way of charting the Milky Way uses our Sun as the origin, and runs through the galactic center in the galactic ecliptic. To locate a target star or object, there are angles from the galactic center through the origin (*i*) and to the star out of the galactic ecliptic (*b*). These two angles and the distance from the Sun define the location of the target star. (*Image credit* Swinburne University of Technology, Centre for Astrophysics and Supercomputing)

coordinate system is often used; it is more encompassing than the ecliptic system, yet smaller than any supergalactic system.

The galactic coordinate system as used by many astronomers employs our Sun as the origin (Fig. 8.1). The primary axis is through the center of the Milky Way and the fundamental plane is through the galactic plane (similar to the ecliptic for Earth). It uses spherical coordinates; there is a galactic latitude equal to the angle above or below the galactic plane, and a galactic longitude described by the angle eastward from the galactic center as marked by the radio source Sagittarius A. The third point needed to locate an object is the distance from the Sun to the object. The galactic system would be fine for traveling throughout all four quadrants, although communicating locations with other species might be difficult since this system uses a minor, medium sized star as its origin. A rendezvous with a Vulcan ship would probably require transposing the coordinates to a mutual system. The take home message is that everyone needs to be on the same page, and to beam someone to within a few centimeters of an exact spot is going to take some major math skills.

8.3 Issue: Quantum Level Patterning

In order to build a building, one needs a blueprint, showing what goes where and how each part contributes to the whole. To sew a dress exactly like the one seen on the model, someone has to make a pattern of the pieces and write down how to put them together. In *Star Trek*, transporting a crewmember begins by making a pattern of the person right down to the last detail. That way, when all the parts arrive at the destination, the person can be put back together again. The detail to which the pattern must be made is excruciating. When the construction company builds a house, it doesn't matter which brick goes in the wall next, but for patterning a human, every piece of information matters. The water molecule in the top cell of your brain is likely to be slightly different on a quantum mechanics level than the water molecule in your big toe and it has relationships with the other molecules in that brain cell that must be preserved. It's going to matter.

Unfortunately, patterning a human or other intricate object of any size on a quantum level is fraught with problems that must be overcome. Some were recognized by *Star Trek* and cheats were employed in the writing. Others weren't contemplated because they didn't know how daunting the task actually is. Finally, some parts of the transporter just plain break the laws of quantum mechanics. This would seem to lead to a conclusion that teleporting a person is not possible, yet there are some advances that give one hope. Maybe, just maybe, it will all work out.

8.3.1 Quantum Information

The *NX-01 Enterprise* was one of the first starships equipped with a transporter for biological materials (*ENT: Broken Bow*). To do this, the pattern has to be much more detailed than that for inanimate objects, just as with the replicator (Sect. 3.2.2). However, there are cargo transporters that are larger (*TNG: Symbiosis; Power* Play) and are not directly purposed for beaming people from place to place. The cargo transporters pattern on a molecular level, just as the cargo replicators do. Luckily, the precision of the cargo transporters can be increased for personnel evacuation purposes (*TNG: 11001001*). In both cases, patterning for the transporter is carried out by the *molecular image scanners*. There are four molecular image scanners, at each of 90°, so one of four can be ignored if its results do not agree with those from the three others. So what data are the image scanners collecting?

Patterning for transport consists of recording the quantum level identity, location, state, and relationships of every particle in an object. Objects are made of matter, matter is made of molecules, molecules are made of atoms, and atoms are made of elementary subatomic particles, some of which are combined into composite particles. Holding those particles together are forces that are mediated by yet other subatomic particles (bosons and mesons), some of which have no mass. A few parameters of the information are relatively simple to deduce, given detailed enough scanners. If each atom can be detected, as with emerging types of microscopy, then the identity and location can be deduced and stored. Since the elementary particles that make up protons and neutrons are known (quarks), and the bosons that hold them together and give them force are known (Box 8.1 and Fig. 8.2), then the identity and location of every subatomic particle can be inferred. Please note that science cannot do this as of today, but the prospect of superlenses

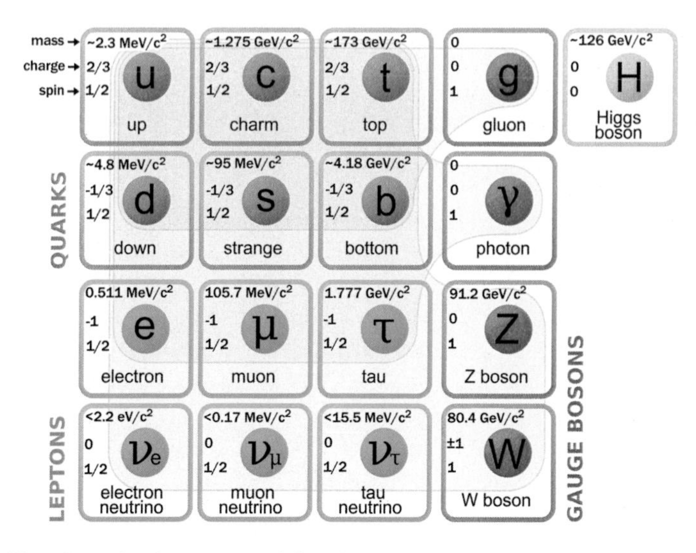

Fig. 8.2 The subatomic, elementary particles of matter can be arranged and organized neatly. It is in part the beauty of this system that suggests it must be true. As indicated *top left*, the columns of numbers indicate mass (in terms of energy equivalent), charge, and spin. Note that the photon and gluon bosons are massless, and that the fermions (quarks and letpons) have ½ integer spins, while the bosons have integer spin. Although not indicated, the fermions are all subject to the Pauli exclusion principle, while the bosons are not. (*Image credit* By Miss MJ [CC BY 3.0 (http://creativecommons.org/licenses/by/3.0)], via Wikimedia Commons)

to resolve images to the atomic or subatomic levels are coming, making use of transformation optics metamaterials (Sect. 3.5.2.1)

Box 8.1 Components of Matter and Forces

Hadrons — particles made of quarks, antiquarks or gluons, include mesons and baryons

Fermions — particles that obey Pauli exclusion principle and have ½ integer spin, include the baryons and the leptons

Bosons — particles that don't obey Pauli exclusion and have integer spin, a group unto themselves

Baryons — hadrons that have mass, composite particles made of three quarks, include protons and neutrons, as well as lambda, sigma, xi, and omega particles

Leptons — low mass, elementary particles, include electron, electron neutrino, muon, muon neutrino, tau particle, tau neutrino and their anti versions

Mesons — hadrons that have mass, composite particles made up of a quark/anti-quark pair, function with gluons strong force interactions, includes many types of particles.

Quarks	have mass, make up baryons and mesons, include up, down, strange, charmed, top and bottom and their anti-versions	
Bosons	massless elementary particles that confer some characteristic to baryons, mesons, and leptons	
Bosons include:	photon	confers electromagnetic field
Gauge Bosons	Z&W	confer weak nuclear force, important in radioactive decay
	Gluon	confers strong nuclear force
	Graviton	hypothetical, confers gravity, probably a gauge
	Higgs Boson	confer mass

The harder information to glean and record will be the state of each particle and the relationships each has with other particles. The state of a particle is reflected in its energy and direction, as well as it spin and its relative symmetry with other particles (called parity). Each characteristic is assigned a name and symbol, and the defined possible values of that characteristic are denoted by numbers. Everyone remembers learning about electron state from their days in high school chemistry. The principal quantum number (n) will be an integer (1, 2, 3 etc.) and equates to the orbital of the electron; a higher number meaning a larger orbital. The angular momentum (l, 0 to [n − 1]) represents the shape of the orbital. The magnetic quantum number (m) denotes the energy levels available to the electron (−l to +l) and the spin number for electrons will be either −½ or +½.

What they didn't teach in high school is that each nucleus (protons + neutrons) has its own set of quantum numbers, and every elementary particle that makes up or holds those nuclei together have their quantum state numbers as well. Elementary particles have many states, relating to both space-time factors and to internal factors. There are rules to how the particles can behave, as reflected in their state numbers. For fermions (quarks and leptons), no two particles in the same molecule can have the same set of quantum numbers. This is called the Pauli exclusion principle, and is crucial to the current methods of quantum teleportation. Boson particles do not have to obey Pauli, although they do have to conserve their states. Elementary boson particles must have the same sum of quantum numbers after they react with another particle, except for their symmetry numbers, for which the product must be conserved. Somehow, with some piece of instrumentation that science has not invented yet, the quantum state of every particle must be discerned instantaneously and recorded as the pattern to follow for reassembly. It must be instantaneous because changes are always occurring in objects, especially live objects. There are approximately ten million chemical reactions that take place in every cell every second, and even the movement of one molecule from one location to another will change its quantum state and the relations it has with other particles. Considering that there are about 7×10^{27} atoms in a 70 kg human (Brian Krauss, Chief Detector Engineer, Jefferson National Accelerator Facility, Newport News, VA), that's a lot of information to collect in no time what so ever. And the problems just get worse from there.

8.3.2 Heisenberg Uncertainty

During any effort to define and record the quantum state of a particle, or of all particles in an object, the detector itself will be a problematic issue. In 1927, Werner Heisenberg voiced what he called *uncertainty relations* in quantum characteristics. There are canonically conjugate pairs of properties of a particle that share an unusual relationship; the more you know about one parameter, the less you know about another. One conjugate pair is location and momentum (mass x velocity); Heisenberg said that the more precisely one is able to measure the location of a particle, the less precisely the same person can know the particle's momentum. Carried to its logically conclusion, knowing the exact location of a particle would mean that one could know nothing of its momentum. The other conjugate pair is energy level and time. You can measure how much energy it has, but not when it has it, or vice versa. The mathematics of these relations extends to the observation; just by observing a particle, you change its state—measuring one parameter alters the others. This expansion of the relations led Heisenberg to call it the *uncertainty principle*—it is more than just a couple of math relations.

The writers and scientific consultants of *Star Trek* were aware of the Heisenberg uncertainty principle and problem it brings up for the explanation of how the transporter operates. By stating that the molecular image scanners pattern an object at the quantum level, it implies that all quantum information about all particles can be recorded at the same time. Heisenberg says this can't be done. To overcome it, the writers introduce the "Heisenberg compensator," a device to make assumptions or otherwise mathematically overcome the lack of information one can have about every particle. By doing so, the matter stream can remain "coherent" until rematerialization takes place. Coherence is a matter of the relationships that exist between the particles (quantum entanglement), is defined by the state of each, and will be discussed below. The Heisenberg compensator is a nifty piece of "wuantum mechanics" in the parlance of sci-fi writer Benford (2002)—a fictional piece of technology springing out of nowhere to solve some intractable problem. For making this type of teleporting work, such a piece of technology to overcome the indefiniteness of the patterning will be crucial. The obvious issue is whether there is truly a way to get around the uncertainty principle. In other words, is science so certain about the uncertainty principle?

Certain questions started popping up in the early 2000s when a Japanese mathematician named Ozawa refined the mathematical formulation about how measuring one part of a conjugate pair would increase the fluctuation in the other member of the pair (position and momentum or energy and time). By adding in terms for disturbance and fluctuation as well as error, Ozawa concluded that as the source of error in one measurement is reduced, the other does not rise necessarily to infinity (Ozawa 2003). Perhaps Heisenberg's principle is not quite as stringent as once thought.

Recently, Ozawa's refinement has been shown experimentally when Hasegawa in Vienna simultaneously measured multiple spin components of a stream of neutrons (Erhart et al. 2012). In general, what this demonstration did was separate the two parts of Heisenberg. While trying to measure one might increase error in the other, the idea that the mere observation or measurement of a particle disturbs it so that other values will change was proven moot. In truth, the two had never actually been linked, even if most scientists and laypersons believed that that they were two parts of the same axiom. Just a year later, researchers in Toronto made a series of observations stringing together chains of precise and imprecise measurements to simultaneously describe the polarization of two photons in different planes (Rozema et al. 2012). Their results were more in line with Ozawa's version than Heisenberg's original definition. By measuring the system before and after it interacted with the detector, they showed that an observation doesn't necessarily alter the system.

It is true that Ozawa made some assumptions in math for his reconstruction and assigned some values where others might not, so the validity of the reformulation and the observations confirming it are in no way universally accepted. Nonetheless, the implications are large in terms of quantum mechanics, especially for something called quantum computing that we will discuss below. If the observation doesn't change the system, then codes built on quantum information which ought to turn information to mush if someone tries to read it, may not work as intended. On the other hand, a limit on the error introduced by making a measurement and the ability to detect and record without altering a system might just make a "Heisenberg compensator" just a bit less fiction and a bit more science. If it were to make a billion or so chained weak and strong observations, perhaps enough data could be gleaned to pattern an individual. Maybe. One might just as easily argue that the Heisenberg uncertainty principle isn't a practical problem to be overcome, it is more of a math law that cannot be gotten around. In a practical sense, it doesn't have anything to do with adding energy to a system when observing it; take the example of lasers. The narrower a slit is for a laser, the more it will spread, simply because it can act like a wave, so the velocity (speed and direction) becomes less precise as the position become more precise. This might be closer to what Ozawa was reformulating; the fluctuation is independent of any disturbance you make by observation.

So while the question remains as to whether observations can be made to quantify the state of particles as a whole, it may be that lasers could some day lend a hand in patterning at the quantum level. Attosecond laser pulses emit light for very short periods; it is accurate to say that an attosecond (10^{-18} s) is to the second what a second is to the age of the universe (4.4×10^{17} s since the Big Bang). Now that 67 attosecond lasers have been achieved (Zhao et al. 2012), researchers are looking to turn them toward viewing electron activities in atoms. Even longer attosecond lasers have been applied to looking at electrons as they change energy state

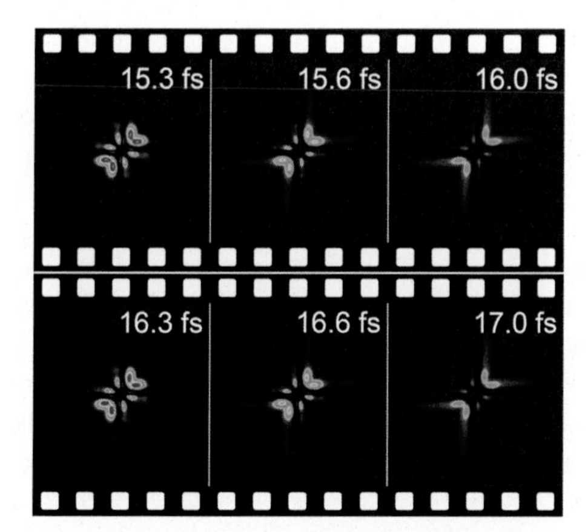

Fig. 8.3 Single electrons have been imaged before, but the work of Ott et al. (2014) demonstrates the first time an electron pair in an atom captured in movement. The probability density is shown in color for the position of one electron along the vertical axis and the other along the horizontal. Red indicates a high probability that both electrons are in that region, while black represents the lowest chance. By exciting two quantum states with an attosecond-pulsed laser-like flash of light at an extreme ultraviolet frequency, the superposition of the electrons reflects itself in a beating pattern of probabilities. The electrons are close to the central atom at 15.3 fs and then farther away at 15.6 and 16.0 fs. By 16.3 fs, the pattern is ready to repeat. Using tunable visible lasers, the authors controlled the movement of the electrons and hope to use this method to produce new chemistries on demand in the future. (*Image credit* Max Planck Institute für Nuclear Physics)

(Kirchner et al. 2014), and similar efforts have been made to produce electron movies (Fig. 8.3) with internal attosecond pulses stimulated by femtosecond lasers (Ott et al. 2014). Unfortunately, the lasers themselves are subject to matters of Heisenberg; the shorter the pulse, the broader the frequency spectrum, which will make precise measurements of photons or electrons more difficult. In addition, classical Heisenberg would state that observing the electrons would alter them, although the Ozawa's math might negate this problem. It is even hoped that a single light wave might be imaged using these induced laser pulses, observing for the first time both the energy and direction of the electric field as it evolves through a waveform. Ten years ago, making a movie of an electron cloud would have seemed fantastical; who knows what technological advance in the next ten years could tell us about the state of the particles in an atom.

8.3.3 Quantum Entanglement and Coherence

The particles that make up matter aren't loners. Every particle has relationships with other particles with which it has come into contact or traded energy. They are "entangled" with each other's states. The two or more particles become a system, where the state of one affects the state of the others. The quantum state of any of the objects can only be described with reference to the other particles in the system—even if they become separated by great distance at a later point. The Pauli exclusion rule is important when dealing with quantum entanglement of states, it says that no two identical fermions in a system can have the same quantum state. For two entangled electrons are in the same orbital and subshell, if the spin of one is, while the other will be −½. If the spin of one changes, then so shall the spin of the other. This relationship does not change with distance, the two electrons could be end up at opposite ends of the universe and the spin of one will flip if the other is changed—no wonder Einstein called it "spooky action at a distance." The communication takes place at least 10,000 times faster than the speed of light and may very well be instantaneous (Hensen et al. 2015). This example of the electron is correct, yet incomplete, as photons, particles that do not obey Pauli exclusion, can also become entangled and changing the state of one will also change the state of the other—even spookier.

Larger systems can become entangled—a planet and satellite, once in semi-stable orbit, are a system. The position of one will predict the position of the other around the barycenter of the system. Crush one and the other will be affected. The degree to which one will affect the other if separated over time and distance drops off quickly, but can be measured in some cases, as with tiny diamonds (Lee et al. 2011). Two nanodiamonds (about 10^{16} atoms each) were placed 15 cm (6 in.) apart; one vibrated and the other didn't, and it couldn't be determined which was which—they shared all possible quantum states. The key to seeing a change in one when you manipulate the other lies in getting all the atoms lined up correctly and staying that way. This is called coherence. Two waves (particles) are coherent if the have a constant phase difference at the same frequency—so they may interfere constructively, destructively or somewhere in between, with the relationship being constant over time or distance. In the case of Lee et al., a femtosecond laser pulse was split to strike both diamonds, starting a vibration and emitting a photon from one diamond. That photon is shone upon the second diamond, linking the vibration between the two, yet disguising which one was actually vibrating—coherence.

Entanglement is meshed with coherence via something called superpositioning. One hundred percent coherence does not imply entanglement and vice versa, yet they are connected if the degree of either is less than 100 % ($\varepsilon^2 + \kappa^2 = 1$). For example, if two electrons or other particles are in perfect coherence, they vibrate in unison, and will remain doing so no matter how far apart they are. They each exist in multiple states all at once (superposition), although as soon as you observe one, its wave function collapse to one state and the other must select the opposite state. That is the demonstration of entanglement. The relationship goes further according to a recent experiment, if two incoherent systems (or particles) are put through an incoherent operation, the outputs will be completely separable. However, if one of the systems enters with a degree of coherence (yes, a particle can have internal coherence), then when both systems are put through the same incoherent operation, they will emerge entangled (Streltsov et al. 2015). In essence, any nonzero degree of coherence can be converted to an equal degree of entanglement.

So we have entanglement and coherence that can join the particles of an object together into a quantum system, not just parts that are pieced together to make a whole. The product is more than just the sum of the parts, and everything known about entanglement can now be applied to coherence. Can this be useful for science for patterning a human prior to beaming down to a planet? For quantum patterning of an object, the knowledge that certain particle systems are coherent and/or entangled theoretically reduces the amount of information that must be gleaned, of course you have to crash the wave function to put one particle in a specific state. The speed of entanglement–induced changes is what allows for teleportation of the information and energy of the particles from one place to another, while never existing in between. Most importantly, entanglement, coherence, and superpositioning are the keys to the coming computing power revolution that might allow for patterning an entire person.

8.3.4 Quantum Computing Uses Qubits

If the problems with defining a location and state for every particle in an object simply vanished and science had a tool that was capable, attosecond laser-based or otherwise, there would still be a major sticking point for quantum patterning a human or any object of that size—it's just too much information being detected, analyzed and stored in too short a period of time. The 7×10^{27} or so particles that make up a decent sized human, each with significant amounts of information attached, would consume more than all the world's computing power and storage. A paradigm shift in the mechanisms of computing speed, power and memory is needed to drive the ability to obtain, manipulate and store data on this kind of scale. Luckily, there are more pressing problems that also require a new type of computer, so the problem is already being worked on intensely. The irony is that quantum mechanics could be the solution to the quantum patterning dilemma; new computers based on quantum properties for solving practical problems might be the solution to teleportation issues as well.

The amount of computing power and speed needed to process a human's quantum pattern in an instant is almost unfathomable. The memory requirement alone is said to be in the billions of kiloquads (*TNG: Realm of Fear*), although how a kiloquad compares to a kilobyte is purely conjecture; the writers consciously kept the terms from sounding too much like current measurement system for computers, "since reality frequently outstrips fiction when it comes to computer science." (Memory Alpha website) Memory systems continue to improve and evolve, yet this may not be enough to increase memory sufficiently, especially in terms of needed space. In addition, the issues with patterning may require a different type computing, one that can solve problems that current computers find unsolvable. Quantum computing uses the power of superpositioning, coherence and entanglement to define a completely new way to process, analyze, and store information.

It is difficult to give examples of how quantum computers work compared to conventional computers since most of what we perceive in the world can be simulated by those same conventional computers. If a traditional computer could simulate a quantum computer, then there would be no difference between the two and no reason to try and build a quantum computer. Where a conventional computer uses bits that exist as a 1 or 0 depending on whether a transistor is open or closed, the parallelism of a quantum computer allows for elementary units that can exist as a 0, 1, and anything between—*all at the same time*. These versions of bits, called qubits, exist in superpositioned quantum states. Qubits can be ions or atoms trapped by magnetic fields or laser beams, photons, electron or nuclear spins states in atoms of liquids or solids with dopants (impurities) controlled by radio or optical pulses, superconductors, or even quantum dots (nanoparticles of semiconductors with an electron that can be manipulated, Sect. 9.6.2). Each added qubit doubles the computing capacity of the device. In fact, the parallelism inherent in superpositioning means that a computer of thirty qubits would have the same processing power of a 10 teraflop (trillions of operations per second) classical computer. Your laptop runs about 10,000–100,000 times slower than that.

Qubits can be manipulated in many ways, switching some property that then affects the properties of all the other qubits. This is different than a conventional computer where changing one bit from 1 to a 0 does not affect the other bits, as long as they are not linked in an algorithm or program. Conventional bits are manipulated by strings of transistors called logic gates; quantum computers also have logic gates of a similar function but are of a different design. For example, once research group recently found a way to produce a difficult quantum gate, a Fredkin gate, wherein two qubits are swapped based on the state of a third. Without the Fredkin this type of operation would require a circuit of five logic operations; with it, the operation is reduced to one action (Patel et al. 2016). The Fredkin also allows for comparing two qubits to see if they are the same, which will be crucial for maintaining secure quantum communications. Similar advances are being made in devising various other quantum gates, making quantum calculations more powerful and quantum programs easier to design.

8.3.5 Coherence and Entanglement in Quantum Computing

The parallelism afforded by superpositioning allows for multiple problems to be worked on at once, and multiple solutions considered at the same time, although this strength is also one of quantum computing's weakness. Picking the correct solution from the parallel series is difficult, if not impossible in some cases. In fact, a calculation based on quantum parallelism has many solutions, all confined to different universes, defining different possible outcomes. Not being in our universe means that to obtain a solution, one must observe the interference between all the possible results. And once the result is observed, the computer is prevented from any additional computations until an initial, entangled state is re-established. The key is entanglement. Entangle the qubits (make them coherent) to a third particle and you can observe what is happening to the qubits indirectly by looking for the changes in the other entangled particle. It's not quite that simple, but it is an approximation of how to maintain parallelism and still observe outputs. Therefore, the key is to maintain entanglement (remember that this is functionally equivalent to coherence) as long as possible. The entanglement for diamond only lasted a few picoseconds (10^{-12} s) (Lee et al. 2011). In many instances, cooling the quantum computer to within a thousandth of a degree of absolute zero is done to increase coherence time. The cooling reduces the effects of the outside world on the qubits, although it has limited success.

The decoherence problem is different and must be solved differently for each type of qubit. For instance, silicon-based phosphorous donors qubits are able to maintain coherence for 39 min, and if cooled like other systems, for over three hours (Saeedi et al. 2013). For magnetic quibits, outside magnetic fields disturb them and break coherence, so a new type of shielding has increased coherence time for them to 8.4 microseconds in one system (Shiddiq et al. 2016), milliseconds in a similar system (Zadrozny et al. 2015), and 28 ms for a quantum dot-based semiconductor qubit (Veldhorst et al. 2014). These examples of increased coherence may not seem like much compared to 39 min, but they are records for their type of quibit. Scientists have tried avoiding the problem by using a sea of molecules in a magnetic field to give all the nuclei the same spin. Using radio waves to flip the spins and nuclear magnetic resonance to read the spins, a drop of liquid as a computer can maintain coherence for long enough to do several thousand quantum operations (Golze et al. 2012).

This is where the difference between coherence and entanglement can be discerned. It isn't enough to keep the qubits coherent for a sufficiently long time, they must also have a degree of entanglement that allows them to "communicate" at longer than local distances (to remain spooky), so the correlation between the particles has to be strong. This has been hard to produce in semiconductors; in 2015 the same group in Australia that produced a 28 ms coherence in semiconductor qubits also tied to atoms together with near perfect correlation—a first (Dehollain et al. 2016). It was a small step, since they entangled the nucleus and the electron of a single atom, but it is an important first step; the degree of fidelity in the connection was the amazing advance. Furthermore, once you can entangle two particles strongly, you should be able to entangle bunches of them, and since this work is done in semiconductors, it suggests that quantum computers could be made with standard computer materials. The group's four-year goal is to produce a 10 qubit computer in silicon with entanglement across a centimeter of space and lasting long periods of time (Borghino 2016).

If it seems as that achieving commercially viable quantum computing is plagued by large problems, then please note that there are a few quantum devices currently on the market. A company called D-Wave Systems (Burnaby, British Columbia) already sells quantum computing devices to research groups and companies with

Fig. 8.4 D-Wave 2X quantum computer The *left image* shows the complete D-Wave 2X quantum computer. Most of the device is dedicated to reducing the temperature of the qubit semiconductor to 0.015 K and the reporter mechanisms for the output. The entire system is capable of supporting a 1000 qubit processor, although this is much more than the current number of qubits. The *right image* shows the cooling tower with process or at the bottom. For each level of the tower, the temperature drops sequential 50 K, 4 K, 1 K, 100 mK, and 15 mK. The processor is usually shielded from magnetic fields but has been uncovered for the photograph. The qubits consider all possibilities simultaneously, determine lowest energy required to form relationships, and returns multiple solutions scaled to show optimals. (photographs courtesy of D-Wave)

large quantum computing research arms, like Microsoft and Google (Fig. 8.4). Despite this push, it still isn't known just how helpful quantum computers will turn out to be on an everyday basis. As of 2016, quantum computers have proven their worth in only a couple of areas, including factoring large numbers. Prime numbers and their factors are the basis for most cryptography today. The realization that factoring large prime numbers is too hard and takes a great deal of time for conventional computers is what makes codes secure. Quantum computers could factor large numbers very quickly, destroying many current security algorithms for credit cards and banks. However, quantum processing would also allow for a new method of securing data via the polarization of light photons and the superposition inherent in them. Anyone trying to observe the quantum processing would collapse the wave and the outputs to the observer would be scrambled.

For tasks like simple multiplication, quantum computers would function much like traditional computers and would take about the same amount of time. In general, the more complex the problem, the more quantum parallelism will be of use. This is especially true when calculations can be expressed in multiple permutations at once. Therefore, the uncertainty and complexity of quantum patterning (if possible) might be better managed using quantum computing. With so many pieces of data for each particle, and each bit of information will be inextricably linked to every other particle, it may be that only quantum computing could minimize the solution profiles. The research examples given above are just a drop in the bucket compared to all the research being done on quantum computing issues, although the difficulties are daunting enough that quantum laptops are still decades away. Thinking optimistically, the problems being solved indicate that most of the issues with creating quantum computing are technical, not fundamental problems with the physics; therefore, time and effort should overcome them.

8.4 Issue: Reducing You to Atoms or Less

Dr. McCoy often states that he doesn't want his atoms scattered across the galaxy. He must know what he is talking about because the next step in the transporter process is to create a matter stream made up of the atoms of the target object or person. Or is it? The evolution of the franchise has brought several explanations for the dematerialization that takes place during transport. The original series writer's guide (season 3) states that the phase transition coils convert matter to energy and energy to matter (p. 20), suggesting that the material sent from one place to another is energy. If so, why is it called a matter stream? The *Star Trek* Memory alpha website states that the phase transition coils converted matter to subatomic particles, and then to energy for transport in a three step process. On the other hand, the TNG: Technical Manual says that antimatter can't be transported without significant modification because the matter stream is for matter, not antimatter (p. 68). Later it says that the matter stream is subatomic particles; still matter and not antimatter, just in the form of the quarks that make up baryons (protons and neutrons) and fermions (electrons) instead of anti-quarks and positrons (p. 102). In the real world parlance, phase transition, as in phase transition coils, means changing from one form of matter to another. Melting ice is a phase transition, as is gas changing to plasma or even molecules to atoms or subatomic particles. So, is the matter stream made of matter or energy, and if it is matter, what form of matter? Whatever the answer is, it isn't easy to accomplish.

8.4.1 Breaking Down Matter

Matter is made up of elementary particles like quarks and leptons, some of which can be joined to form composite particles like protons and neutrons. The elementary and composite particles are held together or kept apart by forces mediated by other elementary particles called bosons (see Box 8.1). When the composite particles and the leptons (particularly electrons) are joined together in various configurations, they form the atoms of the different elements. Atoms are then joined together using different kinds of chemical bonds (ionic, covalent, metallic, hydrogen) to form molecules of various and sundry shapes, sizes, and properties. These make up the matter with which we interact everyday, including our own bodies and the air we breathe.

To build matter up, energy must be put into the system. Nuclear fusion in the stars is how the elements we interact are initially created, and the stars are both consuming and producing a lot of energy to accomplish this. Chemical reactions that build bonds between atoms also require some energy input. Therefore, if one wants to reverse the process and break matter down into atoms, it takes energy to break the stable chemical bonds, but the energy held in the bond will be released when it is broken. To break atoms down into composite or elementary particles, it will take even more energy, though the breaking of the atoms up will also produce even more energy (think atomic bombs). To break the composite particles down to elementary particles and bosons it takes even more energy, and even more energy will be released. Finally, if you want to convert all those elementary particles to energy according to Einstein's $E = mc^2$, it will take even more energy input and the energy released will be almost unimaginable. For the transporter system to work, the device will need a mechanism to input enough energy to break the molecular, atomic, and subatomic connections so the matter is reduced to the proper level. The transporter device will also need a method of dissipating or storing the released energy; perhaps it saves and transports the energy for use rematerializing the object at the destination.

It is the residual strong force (also called the nuclear force) that holds protons and neutrons together in the nucleus (this is still stronger than the electromagnetic force because it can hold two positively charge protons together over the short distance of a nuclear diameter). Therefore, it takes much less energy to liberate baryons from a nucleus, i.e. break the nuclear force, than it does to liberate quarks from a single baryon or meson, i.e. overcome the strong force. The nuclear force acts on very short distance (between two adjacent protons) while the electromagnetic force that would push them apart works on a longer distance, across the entire nucleus. As a result, the nuclei of larger atoms have a growing electromagnetic repulsion but a constant strong nuclear force. Above the atomic number 82 (lead), the electromagnetic force can win and it force a nucleus to fission by losing baryons and some energy—this is radioactive decay. When the nuclear force is overcome, either by natural electrostatic forces or by pumping enough energy in artificially (speed or heat), then the nuclear force is released in the form of high-energy particles and gamma rays. This is also something the transporter would have to deal with, and that isn't even taking into account that it requires a million times more energy to break a nucleus down to elementary quarks and bosons as it does to break a molecule down to a group of separate nuclei and electrons (a plasma). The energy to convert all that matter to energy—perhaps ten million times or more than that.

8.4.2 *Chemistry, Light and Plasma to Break Down Molecules*

There are many ways to break matter down. Chemical reactions can break bonds and produce smaller molecules or atoms. The trouble with chemical dissociation or atomization is that not every process works on every target molecule. The transporter would need millions of chemicals and reaction conditions to reduce all the molecules of an object down to atoms. For instance, one of the most stable chemical bonds is between two nitrogen atoms, (N_2). The N–N triple bond is not polar and has no charge, so it takes a lot of energy to break it. Whole programs of research are devoted to finding easier ways to break this bond for cheaper production of nylon, fertilizers, and proteins. One group recently found a room temperature method to break N_2 into two nitrogen atoms that requires much less energy input. However, the reaction takes two steps and uses carbon monoxide and a rare metal as catalysts (Knobloch et al. 2009). So, to break this one molecule apart, you would be adding new molecules and products into your mix. Therefore, chemical breakdown may not be the answer to the dematerialization problem. However, plasma is another story. Plasma converters can reduce matter to small molecules and individual atoms by blasting at them with a plasma torch in a process called pyrolysis. Combustion requires oxygen and oxidizes the substrate, but plasma converters can pyrolyze organic molecules to volatile small molecules in airtight containers so that they can be later be combusted for energy. On the downside, plasma converters vitrify (turn to glass) the non-organic garbage and the contaminants in organics, so it doesn't come close to reducing all matter to atoms.

Light energy can also be used to break down molecules, a process called photolysis. Light of sufficient energy can strip electrons from atoms to create a plasma if energetic enough, as with LIPC (Sect. 1.7.2), and if higher frequency light and more wattage is used, the heat and increased energy of motion can break molecules and perhaps nuclei apart. Photolysis of water occurs in the photosynthetic pathway of plants, so visible light is energetic enough to break water (H_2O) into its constituent hydrogens and oxygen atoms. Likewise, ozone (O^3) in the upper atmosphere can be photolysed into O_2, and O, just as carbon dioxide (CO_2) can be broken into C and 2O. Higher energy light is used in multiple photon dissociation (MPD) (Jiang et al. 2009), often using a free electron laser (Sect. 1.7.1) to break molecules and atoms down for analysis.

Laser induced breakdown spectroscopy (LIBS) ablates a surface using a high power laser (Hahn and Omenetto 2010), wherein some of the atoms released can be assessed by spectroscopic methods (Box 9.1). The Mars rover, Curiosity, uses LIBS to study the constituents of rocks on the surface of the red planet (Clegg et al. 2014) (Fig. 8.5). Photolysis and techniques like LIBS were not designed for the sole purpose of breaking down matter; aside from teleportation there just isn't much cause for that, but nothing says they couldn't be adapted for the purpose. These techniques are just some of the ways that matter can be broken down; all require large energy inputs, with the energy output being generally small because a very

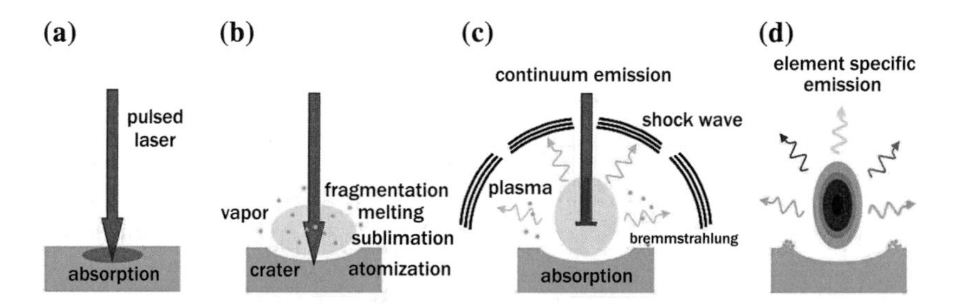

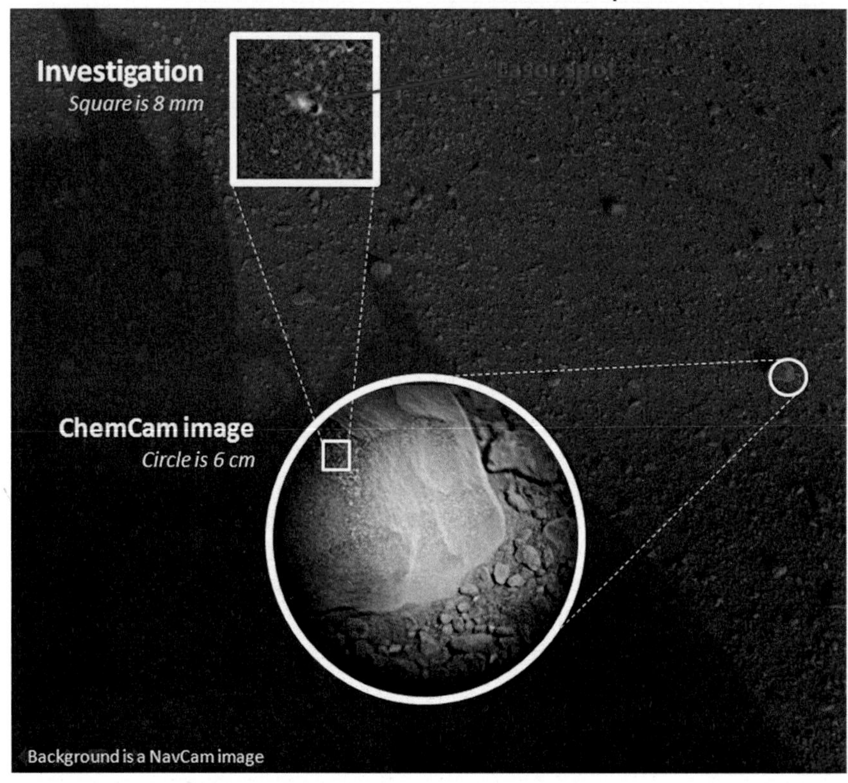

Fig. 8.5 Laser induced breakdown spectroscopy (LIBS) is an important tool on the Mars rover, Curiosity. The *top diagram* shows how LIBS uses the energy of a laser (*a*) to ablate the surface (*b*) and release radiation from stimulation of the individual atoms and small molecules that are released (*c*), which are then detected and identified by a the spectrometer (*d*). The *bottom image* has a background of the Mars surface, including the first rock chosen for LIBS analysis. The close up is the same rock, and the *inset* shows the 0.5 mm laser ablated surface. (*Image credit* NASA; Steven Rehse, University of Windsor)

small number of molecules or atoms are broken down. For teleportation, the efficiency would have to be 100 %, requiring more energy input, and resulting in more energy output.

8.4.3 Colliders Can Break Down Matter Further

Enough energy, in the form of heat, light, or the motion of other particles can strip electrons away to give plasma. It can also break energy that holds nuclei together to produce composite particles (protons and neutrons). Breaking the nuclear force that holds protons and neutrons together is a million times more work than stripping electrons away or breaking molecular bonds—and much more energy is released when they are overcome—like a nuclear bomb. The binding energy to break one tritium atom (^{3}H) into three protons is about 10 meV, mathematically expressed as the binding energy required to hold one nuclear particle times the number of nuclear particles in the atom. This is slightly more energy than it takes to break up a helium nucleus (with two protons and a neutron, about 8 meV). Even though helium and tritium have the same number of nucleons, it takes less energy to knock out a neutron than it does a proton. Multiplied by billions of atoms, the amount of energy needed to break a mass of atoms down to nuclear elements is staggering. One mechanism for doing this is an atom smasher, or more appropriately, a particle accelerator if there is one beam, or a particle collider if there are two beams directed at each other.

There are several different kinds of particle accelerators/colliders. We discussed a few in Chap. 1 when dealing with phaser technology, including the circular particle accelerator called the Large Hadron Collider (LHC) at CERN in Europe. Another circular collider is the RHIC (Relativistic Heavy Ion Collider) at the Brookhaven National Laboratory on Long Island in New York. Both RHIC and LHC, (as well as other colliders like the BEPC in China and the VEPP-2000 in Russia) accelerate particles to 99.9995 % the speed of light and then crash them into one another. For the RHIC, 111 particle bunches, each containing billions of specific nuclei, are accelerated and crashed into each other or target nuclei. The elementary particles released by the collisions (quarks and bosons) are tracked and identified by detectors (Fig. 8.6). To accomplish this task, huge amounts of energy are required; the temperature for the collisions must reach 4 trillion degrees, 250,000 times hotter than the center of the Sun. For instance, in Run 14 of the RHIC (Harrington 2014), the scientists produced 600 million collisions in a six-week period. Yes, it took six weeks to breakdown 0.0000000000000000016 % of the atoms in a human body, and they had the luxury of using a single type of atom for every collision. To reduce a 70 kg person to elementary particles would require *at least* 10^{14} times more energy to overcome the gluon-mediated strong force between all the elementary particles. This is to say nothing of the energy that would be released and would then need to be stored or dissipated. The typical RHIC collisions (still just fractions of the collisions needed for entire human) release plasma with so much energy that a cube that has sides 25 μm (0.001 in.) long of this plasma would be equivalent to all the electricity used by the USA in a year (Muzzin 2010).

Though *Star Trek* would need an RHIC on steroids to reduce an entire human to quarks and gluons, the heavy collider serves well to explore questions of the early

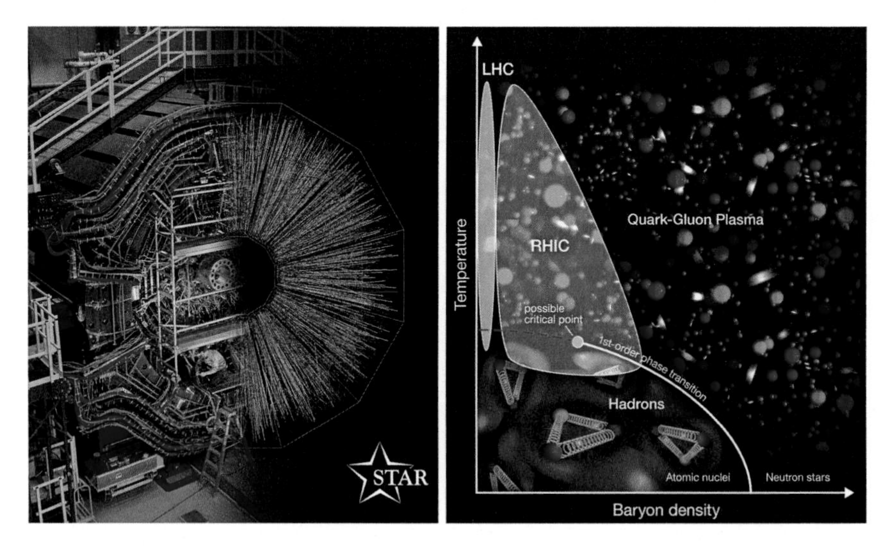

Fig. 8.6 The STAR experiment at the RHIC (Relativistic Heavy Ion Collider) uses the detector shown in the left half of the *left image*. When multiple particles strike each other at sufficient energies, the released particles produce traces that can be followed and identified as to energy and mass (right half of *left image*). The RHIC is particularly suited to investigating the region of energy where hadrons transition to a quark gluon plasma that is reminiscent of the early universe as matter coalesced from energy. Various temperature/baryon densities are needed to illustrate the transition boundary, and the RHIC can cover a larger portion of the phase transition map as compared to the LHC. (*Image credit* Courtesy Brookhaven National Laboratory)

universe and the fundamental nature of elementary particles. Because of the versatility of energy levels and nuclei that can be collided, the RHIC can specialize in large ion collisions, something other colliders are less able to do. For instance, LHC can explore the transfer from nuclei to quarks, but only over a very narrow range of densities and conditions. The RHIC can explore the phase transition to the "quark-gluon plasma" over a wide range of nuclei densities (Fig. 8.6), and this helps them better recreate the conditions of the very early universe, just fractions of a second after the Big Bang. Using this ability, researchers hypothesize that the early universe existed in the form of a perfect liquid, meaning that the superhot, super dense material formed by the collisions was a zero viscosity fluid (Song et al. 2011). It had no resistance to flow—our universe might have been liquid for a very short time and then converted to solid matter (Adamczyk and STAR Collaboration 2014). Other investigations are shedding light on why some elementary particles have integer spins and others have ½ integer spins (Lyuboshitz and Lyuboshitz 2010), and how the early universe might have broken a law of physics. Just after the Big Bang there may have been bubbles of *parity*, a left or right handedness; this might account for why there was more matter than antimatter in the first second of the universe and therefore why we now live in a matter, not antimatter, universe (Adare and PHENIX Collaboration 2011). So even if the RHIC won't be useful for transporting, it is certainly going where no person has gone before.

8.4.4 Converting Matter Directly to Energy

Some sources in the *Star Trek* world state that the transporter converts matter to energy and back. If this is the case, can it be done in real life? A nuclear bomb converts only a small percentage of the fissionable matter to energy, the rest is scattered as radioactive decay products like thorium, strontium and technetium, right down to the final decay product, lead (for uranium or radium series). In order to transport the energy, 100 % efficiency would be needed in the conversion of the matter. According to Lawrence Krauss in *The Physics of Star Trek*, simply heating the individual to 1000 billion degrees (a million times hotter than the center of the Sun) would provide enough energy to reduce a person to quarks and then convert the quarks to pure radiation; that's more than 1000 one-megaton H bombs of energy (for a 50 kg individual, Krauss 1995, p. 91). This could be difficult, but there are some additional hypothetical mechanisms that could convert matter to energy. Most have to do with black holes and each has its own issues, yet they aren't out of the realm of possibility. Please note that the word was possibility, not probability.

In theory, if one could create a black hole and then feed matter across its event horizon, much of the matter would be crushed, heated, and energized by the astrophysical black hole to the point of being converted to energy. This energy would then be spewed forth from the black hole as something called a particle jet (Fig. 8.7). If the radiant energy were collected, it could then be used as the substance that is transported to the destination for rematerialization according to the quantum pattern. Could this reach 100 % efficiency—probably not. Theoretical yields predict that about 40 % of the mass is converted to energy in a rotating black hole (Genzel 2010), so the lost energy would need to be recovered somehow. Granted, there are several "ifs" in that scenario, each with issues that might either

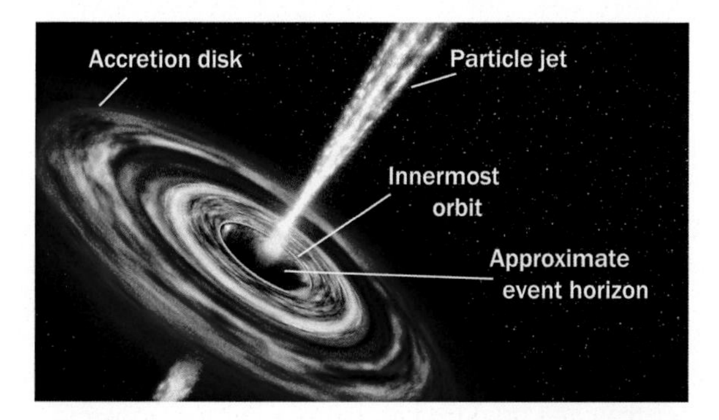

Fig. 8.7 An astrophysical black hole is huge. In 2012, the diameter of the innermost orbit (in *yellow*) around the event horizon of a supermassive black hole was measured at 113 billion km (60 billion mi.), where the distance from the Sun to the Earth is about 150 million km (93 million mi.). Just imagine the size of the accretion disk when it is feeding. As big as this is, the black holes that could be generated in colliders or with lasers would have an event horizon in the billionth of a meter range. Yet they may behave in much the same manner. (*Image credit* NASA)

require too much energy to overcome, or be of a size or nature to be unusable for this purpose. However, the first question, can we produce a black hole, is answerable—and that answer is yes.

8.4.4.1 Producing a Black Hole in a Collider

A 2013 paper suggests that particle accelerators could form black holes at lower energy inputs (East and Pretorius 2013) than previously thought (Choptuik and Pretorius 2010). The particles moving near the speed of light warp space time and can focus energy like a lens. Two colliding particles focus the energy of each other, so in this situation, it takes only 1/3 as much energy to form a black hole. Of course, that might still be a million times more energy than the LHC can currently muster. Another reason for a lack of black holes induced by particle collisions at LHC is that certain theories (gravity's rainbow) predict that the black holes will be found at only specific energies, with gravity leaking out into the other dimensions (Ali et al. 2015), so they may not be detected despite the increased energy of the LHC after the 2015 upgrade (Sect. 1.2.1.3). Regardless, any black hole that might be created by the LHC would be small, transient, and travel at nearly the speed of light, so scientists would have to learn to modulate the holes if they are to be used for practical purposes. However, they would be based on gravitational and particulate origins so they should, in theory, behave much like astrophysical black holes in deep space. They would be capable of feeding on matter and expelling flares of energy, very nearly the energy equivalent of the matter introduced across the even horizon.

The smaller a black hole is, the more stable and higher energy it would be, so a happy medium would need to be achieved for a black hole to produce energy as either an energy source for practical use or for transporting people as energy beams. For collider-induced black holes, the part about being small is a definite issue. In this case, small means that is mouth might be half the diameter of a proton, making it hard for a black hole of that size to consume matter. Perhaps the collider could first be used to reduce matter to elementary particles as speculated above (Sect. 8.4.4), and then these smaller particles fed to the black hole. In late 2015, an intense X-ray flare was detected from a supermassive black hole as it pulled a star about the size of our Sun into its gravitational field (van Velzen et al. 2016). As matter crosses the apparent horizon, it is heated and pulled and stretched, and energy escapes as it is consumed by the hole. The remnants of flares had been detected previous to this, but this was the first time that the flare was observed as it began grew, and ebbed. The tremendous amount of energy released by the flare would be much larger than any produced by introducing matter across the horizon of an artificial black hole in the LHC (if it can be created at all), yet the point is made, this would be a direct method of converting matter to energy. The issues to overcome are many; the energy to create a black hole, the energy and efficiency to reduce matter to a stream of elementary particles, the ability to hold and maintain the black hole as it fed. And of course, the unbelievably large of energy released from converting a human to energy would appear in the span of a nanosecond, so trying to control and store it for transport would be like trying to contain 1000 H-bombs in a trash can. You'd better have a tight lid.

8.4.4.2 Producing a Photonic Black Hole

In 2008, Ulf Leonhardt and his team at the University of St. Andrews in the UK created a photonic black hole. Basically, they built a black hole in a telephone line. Because light travels at different speeds through an optical fiber based on the frequency, they could send a slow wave down the fiber, only to be caught by a faster wave sent later (Philbin et al. 2008). The fast wave can't pass the slow wave, so it becomes trapped; it can't escape the energy of the slow wave. A couple of years later, Daniele Faccio's team in Italy also produced a black hole by shining two pulses of laser light into a small silica rod (Belgiorno et al. 2010). The first pulse warped the glass, and the second bumped against the warping, slowing it down to a stand still (slow light, Sect. 2.6.2). The point was really to create more of an event horizon than a black hole. In the Italian situation, it was more like a white hole, where nothing could get in instead of out. By studying the event horizon, they believe they could learn more about how black holes work. As a happy accident, they may also have gotten Stephen Hawking his Nobel Prize. Hawking proposed in the 1970s that black holes would emit low-level radiation, even when not feeding.

This "Hawking radiation" has to do with the effects of the gravitational and energy environment just at the event horizon. The interaction causes a particle pair to be emitted; one matter particle and one antimatter particle. In the situations where the matter particle was produced outside the event horizon and the antimatter particle was produced across the horizon, the pair would not self-annihilate (matter + antimatter goes boom, Box 1.1), and the matter particle would escape as Hawking radiation. Hawking had predicted it, but none had ever been measured because it is of such low energy and the events are so far away from us, so Hawking radiation was theory right up to the analysis of Faccio's artificial hole. A detector placed 90° to the side picked up the rare photon emission—something was escaping (Belgiorno et al. 2010). This could very well be the Hawking radiation, although there are problems. A white hole should eject the antimatter particle as well, and an annihilation of two photons should be detected; however, they only saw single photons.

For our discussion, it is enough to consider that these photonic black holes seem to behave like astrophysical black holes, at least in that they have a horizon and may emit Hawking radiation. There is even speculation that a metamaterial black hole (Sect. 4.6.1) might emit some radiation from a zone called a Rindler horizon, based on the way the photons interact with the metamaterial (Smolyaninov et al. 2012), but the question remains, would black holes such as these act in a way that would make them hypothetically useful for converting matter to energy for teleportation? Sir John Pendry says, no; they do not have a gravitational/particulate origin, so they would not feed on matter and emit flares such as astrophysical black holes do and collider-produced black holes might. Therefore, the ways to transduce matter to energy are still fairly limited and difficult to accomplish, and high efficiency would be harder still.

8.5 Issue: Transmitting Atoms or Information

In *Star Trek*, the matter transmission stream, whether it be matter or energy, as well as the quantum pattern information, are sent from one of the seventeen emitters on the dorsal surface of the *USS Enterprise's* saucer section (TNG Technical Manual, p. 8). The matter stream is sent almost instantaneously via a subspace domain, a portion of space-time within space-time. Perhaps subspace is another dimension, but it is ill-defined to say the least. The range of beaming has altered over time within the franchise, anything from 10,000 km (*ENT: Rajiin*) to 16,000 miles (Original Series Writer's Guide, p. 20), to light years (*TNG: A Matter of Honor*). Given the mixed messages that *Star Trek* has put out over the years about transporter function, it is hard to tell whether it is energy or matter that is being beamed from place to place. Energy can carry information, and this is probably how the pattern is beamed; the issue is whether the matter stream includes the energy of the person being transported or their mass as well. However, one subject is clear, for it to be true teleportation, the object has to be here….. and then there. And believe it or not, science can do that.

Box 8.2 The Problem of *The Fly*

22nd century—Transporter accidents merge extraneous matter with the person during patterning and beaming (*ENT: Strange New World*).

23rd century—Original series transporters decontaminate in two stages, first transport, then decontaminate (*TOS: The Naked Time*).

24th century—Biofilters are included on all transporters to decontaminate transported objects (*VOY: Macrocosm, Night*). Biofilters remove pathogens and theta radiation to keep the ship from being contaminated.

Problem One: In the 1958 film, *The Fly*, the scientist (Andre) teleported himself from point A to point B, not realizing that he had let a fly into the beam with him. The transporter didn't know how to deal with the contaminant, so it merged the fly and the main character. Andre ended up with a fly's head and forelimb.

This couldn't happen given how *Star Trek* says the transporter works. The molecular image scanners gather information on the state of each molecule and the *location* of that particle. At the destination, each particle is rematerialized exactly where it was at the origin. A fly to the left of the person will rematerialize to the left of the person at the destination. In essence, anything above the transporter pad will be moved to point B in its same relative position, including the air.

Problem Two: What exactly is a pathogen or a contaminant?
90 % of cells in and on human body are not human—they are bacteria, fungi, and parasites.
Only 9.5 trillion of the 75 trillion cells contain human DNA

Normal flora—most of the 90 % of microbial cells in human body are helpful.

(1) consume resources so pathogens cannot grow
(2) Produce toxins that kill pathogens
(3) Produce vitamins we need
(4) Crucial to digestion of food

Overgrowth of normal flora—too many or wrong locations can bring disease

(1) if immune system is not functioning efficiently
(2) *Candida albicans* fungus causes disease in many parts of body if overgrows
(3) Skin bacteria will cause rheumatic fever if on leaflets of heart valves
(4) Numbers that are normal in one person will cause disease in other people

So how does the transporter choose which organisms to filter out and which to allow to be beamed? In addition—unknown pathogens will not be filtered (*TNG: The Lonely Among Us*), while unknown species might be seen as contaminants.

Biofilters may need more thought.

8.5.1 Entanglement Allows Teleportation of Information

Can science teleport energy and matter over long distances? Yes and not yet; yes for information and energy, not yet for matter. What researchers and engineers have been able to send is information about certain electrons, photons of light, or atoms. Once again, entanglement and Pauli exclusion (if using hadrons) play a big role. The fact that information can be shared and communicated between particles at long distances means that teleportation and observing results from quantum computations (Sect. 8.3.5) are carried out in a very similar way to how solutions are presented in quantum computing. Don't ask me why, but when referring to teleportation, physicists call the particles and their observers in question Alice, Bob, and Charlie. We could change them to Kirk, Bones and Spock, although this might confuse the physicists so we will stick with the traditional names. For instance, you have a particle Alice with a certain initial state. This is the particle you want to teleport in that initial state from one place to another. You also have a pair of entangled particles, Bob, at the destination, and Charlie, next to Alice. If you take a measurement of Alice and Charlie that entangles them, the shift in state of Alice can be projected through Charlie to Bob. Bob now displays Alice's initial state. Of course, by entangling Alice and Charlie in the process, the initial state of Alice is lost. Alice's information can only be recovered with the help of Charlie, so this makes the teleportation a form of secret sharing.

This isn't the only way to teleport quantum information. The information could be part of the direct entanglement, thereby by teleporting the entanglement as well. There is also the technique where Charlie receives the measurement signal as well, known as telecloning (van Loock and Braunstein 2001); i.e. sending information from one sender to two or more recipients. Telecloning has been demonstrated for photonic qubits (Zhao et al. 2005) and coherent optical states (Koike et al. 2006) and should be possible for any number of recipients. This brings up an important point—one can achieve teleportation using any of the classes of qubits used in quantum computing, and more than one mechanism is currently know for each type of qubit. This reflects the interrelationship between quantum computing and quantum teleportation, and also extends to plans for quantum communication and the quantum internet. However, the same problems of decoherence and fidelity that plague quantum computing also enters into teleportation. Therefore, hybrid systems using different qubit and detection methods together may be the wave of the future (Kurizki et al. 2015). For example, there is a system that use semiconductor qubits hybridized to electron spin traps in a diamond (Kubo et al. 2011) (Fig. 8.8). This allows for short retrieval times, while increasing the overall fidelity of the system.

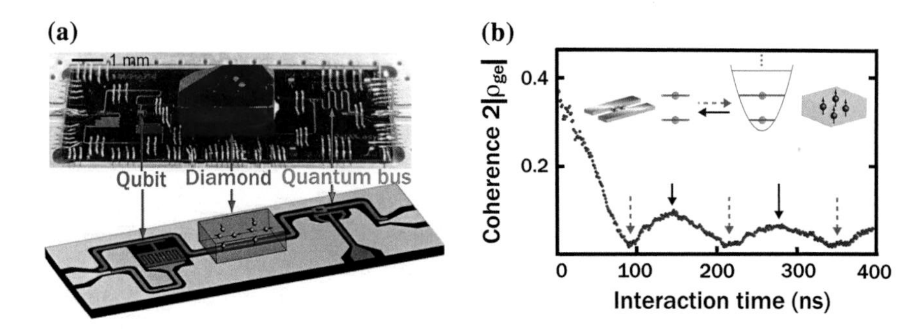

Fig. 8.8 Photograph and diagram **a** shows a semiconductor—diamond hybrid teleporting and quantum computing chip. This assembly is especially good for quantum memory storage. The qubit is a semiconductor; it is coupled to a nanodiamond to entangle electrons spins. The bus is s resonator that is tunable by frequency. The **b** chart shows cycles of storage of information (*red arrows*) and retrieval (*black arrows*) of information from the diamond spin ensemble. (*Image credit* Kurizki et al. 2015)

8.5.2 *Examples of Teleportation*

Papers from just before the turn of the millennium (Boschi et al. 1998) right through to the present demonstrate quantum teleportation of information states across space or through optical fibers. Each new study describes some small or large advance in the field, from distance (not) traveled to degree of fidelity of the information to the number of states teleported at once. The current record in distance (for any system) is 143 km (Ma et al. 2012), between La Palma and Tenerife in the Canary Islands. The record distance is significant because it is very close to the distance from low Earth orbiting satellites to the planet's surface. This bodes well for quantum communication with space in the near future. The current record for teleportation of states is two (Wang et al. 2015), wherein a single photon was teleported with respect to both its spin and its orbital angular momentum. The simple move from one to two quantum parameters is an important step for *Star Trek*-like transporter function; beaming anything or anyone will require that all quantum information for every particle is recorded and transported efficiently. If science couldn't find a way to transport more than one characteristic, they could never beam them all.

One proposal for a teleportation device smacks of *Star Trek* like no other. Tongcang Li at Purdue University in Lafayette, Indiana and Zhang-qi Yin of Tsinghua University in Beijing, China think that they can teleport the entire quantum information, the memory, as it were, of one microorganism to another microbe at a distance. Their idea is to place a single organism on a mechanical oscillator that can create a superposition of states in the molecules of the organism (Li and Yin 2016). Since the bacterium is so light compared to the oscillating membrane, it can sit on the membrane without affecting the quality of the oscillation. Both the membrane and the organism could be cooled to induce the quantum ground state, and then, using a superconducting microwave circuit, the states of its molecules could be entangled with its center of mass motion and be teleported to a distant organism on another oscillator (Fig. 8.9). The internal state contains information on that organism's molecular history, so it would, in essence, be like transporting the organism's memory to another member of its species. There have been microbe-like nanite species on *Star Trek* (*TNG: Evolution*), so perhaps beaming individuals around is closer than we thought.

The limiting factor for quantum teleportation, besides the fact that only information can be sent as yet, is that one needs an entangled particle at the destination, such as the organism and oscillator in the memory transport example above. There may not be a transporter pad at the target spot, but there must be something there to receive the information. What is more, it can't be a passive receiver; the entanglement must occur *before* the teleportation. If there is no particle ready to respond to a change at the origin, then there is no teleportation conduit through which to send information. This implies that for every particle you want to teleport, there must be a coherent particle at the destination to switch states to match the original target particle. If a human is made of 7×10^{27} particles, then 7×10^{27} entangled particles would be needed at the endpoint. Each would be acting as a tiny receiver. However,

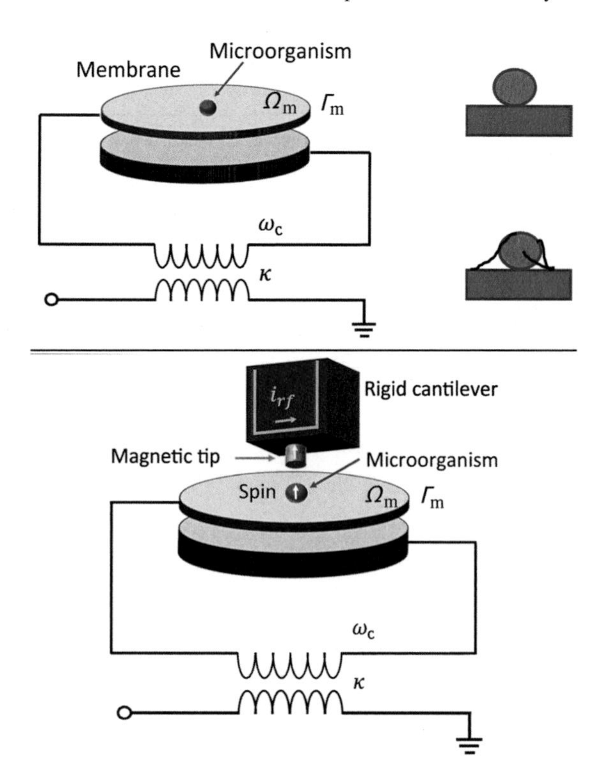

Fig. 8.9 On top is a schematic of the mechanical oscillator membrane connected to the superconducting inductor-capacitor that acts as the qubit. The organism, either with (lower red and blue cartoon) or without (upper cartoon) pili can be cooled cryogenically without ice crystal formation. The *bottom image* shows the cantilever with magnetic wire that will induce oscillation of the organisms spin states and the bacterial membrane as a system. This will be transmitted to the superconductor circuit for teleportation (electron spins, etc.) of the organism. (*Image credit* © (Li and Yin 2016) with permission of Springer)

Star Trek is silent on the point of how matter or energy is received at the point of rematerialization. It is basically assumed that this is receiver-less transmission. That's just one more issue to overcome in order to transport *Star Trek* style.

8.6 Issue: Rematerialization

Now that the individual or object has been patterned, dematerialized, and sent to the destination, the only job left is to reassemble the target according to the instructions in the quantum pattern. It should be a simple process given that everything the transporter needs to build the target is right there, and it has a detailed set of plans that says what goes next to what and what the quantum sate of every particle should be. There are no decisions to make, just follow the instructions. However, there is that nasty matter of not having any device (receiver) at the destination to interpret the quantum pattern or assemble the pieces. Here we will limit ourselves to investigating how one might build Kirk or another human from parts. The first question to ask is, what parts are we talking about? Did the transporter send you as atoms, subatomic particles, or energy? Rebuilding the target will be depend on the building blocks included in the kit that was received.

8.6.1 Building Macroscopic Objects from Atoms

If the matter stream sent by the transporter consists of the atoms that made up the person or thing on the transporter pad, then all that is needed is to reestablish every molecular bond present in the original body without creating any additional bonds. Of course, the quantum state of every particle within each bond would have to be perfectly replicated, but this would be theoretically possible because of the presence of the quantum pattern information. Just how the pattern would be implemented is more difficult to envision; for now we will assume that it could be.

The chemistry to put together a human being is fairly well understood. Biochemistry is the study of the building up, breaking down, and shuffling of the biomolecules that make up living things; proteins, carbohydrates, lipids or fats, and nucleic acids like RNA and DNA. The regulation of just when things are built up or broken down is also part of biochemistry, including the portions of DNA needed to code for and control each process. At the present time, people like Craig Venter and others have such a handle on what processes are necessary and how they come about that an entire new field has been created: synthetic biology. In this discipline, scientists use bioengineering, molecular biology and biochemistry to put together biomolecules and the systems they work in, absent the natural process that brings them about. In addition, synthetic biology can use existing or engineered biologic processes to yield biologic systems that do not exist in nature. Amongst a host of other achievements, Venter's team at the J. Craig Venter Institute have engineered a completely new, minimal genome bacterium (JCVI-syn3.0) in early 2016 (Hutchison et al. 2016), and in 2014, four new enzymes to manipulate nucleic acids were synthesized (Taylor et al. 2015) giving clues to how life on other planets might evolve. The ability to engineer an organism out of whole cloth favors the establishment of future programs to build specific molecules and systems outside of a living organism and then generate living organisms from the assembled parts. Do that in the blink of an eye, and you have molecular reassembly for teleportation.

In truth, building a human from atoms isn't so horribly difficult; it has been done almost 110 billion times since the dawn of man. Every biologic mother has managed to build a person from scratch; it just takes nine months to accomplish the feat. This might be too long for someone waiting to be rematerialized after transporting. Under normal conditions, most biochemical reactions require biomolecule reactants, protein enzymes to catalyze those reactions, and a source of biochemical energy (adenosine triphosphate, ATP). Every reaction would have to take place in a precise order, using the products of the previous reaction as reactants in the next one. It would also require the ability to place atoms and molecules in precise locations at an instant's notice. This is also hard given how molecules and atoms diffuse through space. Diffusion limits are the reason that cells have to be so small —diffusion through the cytoplasm is the rate-limiting step for all reactions, so keeping cells small limits the distance the molecules need to diffuse.

This seems like a lot of work to reassemble atoms into a human, and of course it isn't always people that are transported. Many times, it is inanimate objects that are sent around from place to place. Perhaps a more direct method of molecular bonding would be easier. Unfortunately, just like using chemistry to reduce an object to atoms (Sect. 8.4.2), building the object back up would require thousands of different reactions, each with different reaction conditions, cofactors and energy inputs. Therefore, it might actually be easier to build molecules from subatomic particles, just include the molecular bonds in the building process. If one builds the final molecules directly, the issue of chemistry is moot. Then again, perhaps it is just adding an additional step—first atoms, then bonds.

8.6.2 Building Macroscopic Matter from Subatomic Particles

Atoms are made up of protons, neutrons and electrons, held together by electromagnetic forces of photons, the strong force via gluons, and the weak force of Z and W bosons (Fig. 8.2). Three quarks (two up and one down) and appropriate gluons make up protons, while three different quarks (two down and one up) and gluons make up a neutron. Using one quark and one antiquark will give a meson, which are used in mediating the strong force of the gluons. For example a B_S meson is made up of one bottom quark and one strange antiquark, therefore, they are not opposing particles and do not annihilate one another when brought together. All these possible particles, together with all the bosons in right places, and the proper number of electrons (lepton particles) make up all of matter. If one has the correct recipe and materials, as well as the energy to bring them together, then one could build any atom one wished.

However, molecules aren't merely atoms merged together to form new atoms, they are different atoms joined together in space; each nucleus still maintains it's identity even if electrons are shared (covalent bonds) or swapped (ionic bonds). Therefore, transporting matter as subatomic particles just add two more steps to the process, reducing atoms to particles at the origin and then building atoms from particles at the destination. The reason for doing that would have to be some limitation on what can be transported, as our example above shows that photons, leptons, and very simple atoms (or at least their quantum information) has been transported, but not larger atoms. There may still be a limit to what can be transported in the 24th century, so reducing to and building from subatomic particles may be the only way to accomplish the task.

Does this mean that building up subatomic particles is limited to three quarks or a quark and an antiquark; are those the only two possibilities? Science is now learning that quarks can be put together in some additional numbers, with each new particle telling us more about the basic building blocks of the universe. The study of how forces hold and limit the shape and constituents of composite particles is called chromodynamics. There are some pretty good theories about how the system works, yet there isn't a firm idea about whether baryons (three quarks) and mesons (quark and antiquark) are the only hadrons that can be produced. It is evident now from experiments conducted at the LHC, RHIC, and the Tevatron at Fermilab that several other hadrons are possible.

A strange particle called X(4140) was reported at the Tevatron collider in 2009 (Shen and Belle Collaboration 2010). The researchers weren't quite sure what it was, it didn't fit the size or movement pattern for a typical baryon or meson. Experiments are being carried out to determine what it actually was, although additional observations have grabbed more attention in recent years. In 2013, the Japanese collider Belle experiment and the Chinese BESIII confirmed that they had observed a tetraquark called Z_c([3900]), an exotic type of meson (Ablikim and BESIII Collaboration 2013; Liu Belle Collaboration 2013). Then, in 2016, a different construction of four quarks (X[5568]) was confirmed at the Tevatron (Do Collaboration 2016). About this same time, a pentaquark was confirmed (LHCb Collaboration 2015) at the LHC. The exotic matter is made up of two up quarks, one down quark, one charm quark and one anti-charm quark. As of now, the type of association between the constituents is not known (Fig. 8.10). They may all be held tightly together, or it might be a loose association of a meson with a baryon held together by a completely unfamiliar force, with the charm quark and antiquark separated so they do not self-annihilate. Even more interesting, the pentaparticle can exist in more than one energy state.

This multiplicity of energy states applies to the two newest baryons discovered as well. In 2015, the LHCb team (the name of the experiment is LHCb; the team isn't the back-up squad) described two new baryons (Aaij 2015). They are made up of three quarks, just like protons and neutrons, but they are three different quarks;

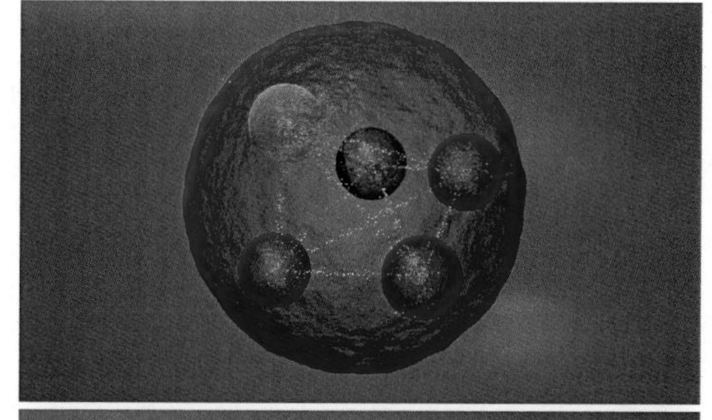

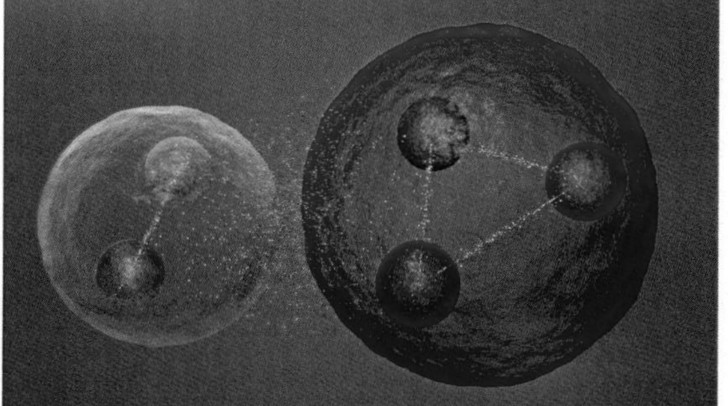

Fig. 8.10 The pentaquark is a completely new form of matter, made like nothing science has seen before. The association is still not understood. The five elementary particles might make one heavy particle, with forces mediated between all the particles. Or, it may represent a meson (quark/antiquark) and a baryon (three quarks) held together by some completely new force (the speckled portion between the two composite particles in the lower image). (*Image credit* © 2015–2016 CERN)

one bottom, one strange and one down. The bottom quark is much more massive than any other type of quark used in baryons, so these new three-quark particles are six times more massive than a proton. What is more, even though both the new baryons are made of the same three quarks, they have different spins in each, making one slightly more energetic (heavier) than the other. The point to describing these new and varying composite particles is hope-hope for a transporter. Research is showing that there are more paths to building matter from subatomic particles than science once imagined, so who knows how they might one day help to reassemble matter from elementary particles. As of now, they exist for short periods of time after running particles together at high energies. Yet, who is to say that tomorrow they might not be controllable and provide a way to build atoms without building individual protons and neutrons one at a time. Perhaps huge particles can be generated in one step and allowed to decay to a nucleus of many protons and neutrons in a second step. There is absolutely no evidence to suggest this possible, just like how we thought there were only two ways to build hadrons and baryons until 2003 or so.

8.6.3 Fusion Builds Matter

If it isn't possible to build large atoms in some exotic way, the standard mechanism the universe uses for constructing large elements will have to do—fusion. Colliding quarks of sufficient energy that find each other in the proper proportions will produce protons, i.e. hydrogen nuclei. Or, protons can be made from energetic protons in a process called pair production (Sect. 8.6.4). Given enough energy, as in a collider, fusion reactor, or star, nuclear fusion can occur. Fusion reactions releases energy, more energy than nuclear fission in most cases. The joining of two hydrogen atoms (protons) into a helium atom and two more protons (it is actually a four step process, Fig. 8.11) is the initial stage of the nuclear fusion that builds every natural element. The positively charge protons will repel each other and not fuse, but with sufficient energy, the electromagnetic repulsion can be overcome. This is why fusion takes place at such high temperatures. The final product is an atom with a smaller surface area:volume (SA:V) than the two starting hydrogens, so the nuclear strong force needed is less for helium than for the reacting hydrogens; i.e. smaller SA:V atoms have higher average nuclear forces. The excess strong force energy is released as kinetic energy or heat.

The fusion process continues as small nuclei are fused together to form larger ones, until you reach iron (atomic number 26). Beyond that, stars larger than our Sun produce most of their energy through a fusion cycle that involves carbon (6), nitrogen (7) and oxygen (8) (CNO cycle) or starting with the triple alpha process that converts three heliums to a carbon with a beryllium intermediate. Higher elements up to and beyond iron are formed in these larger stars and through the energy of nova . For atoms with larger atomic numbers than iron (26), the electrostatic forces between the protons in a nucleus are so repulsive that adding more

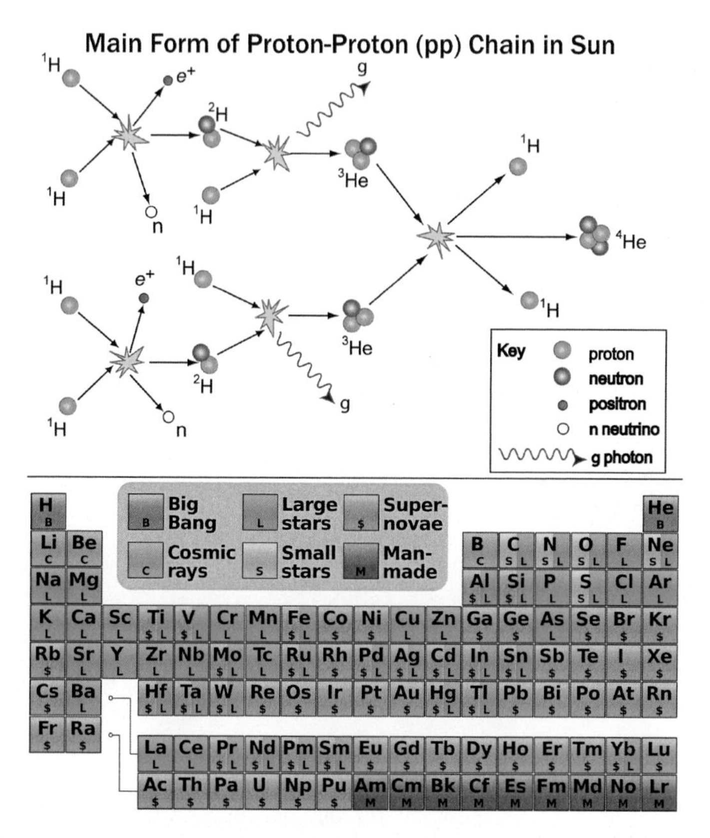

Fig. 8.11 The *top image* shows the classic version of nuclear fusion building helium from hydrogen. Even though helium is twice the number of as protons as two hydrogens, note how the two neutrons are products. Note also that it takes six hydrogens to produce one helium (*Image credit* CSIRO, http://www.atnf.csiro.au/outreach/education/senior/astrophysics/stellarevolution_mainsequence.html). The *bottom diagram* shows the natural source of fusion products. In general, the larger the element, the more energetic the collisions must be to produce it. This is true for all elements except those that are manmade, and exist for only fractions of a second in a laboratory. (*Image credit* Cmglee, by CC BY-SA 3.0, https://creativecommons.org/licenses/by-sa/3.0/legalcode)

protons or neutrons actually costs energy, although it is true that starting a nuclear fusion reaction is going to cost a tremendous amount of energy no matter what size of atoms are being produced. In fact, to start a fusion reaction, a deuterium plasma (ionized hydrogen gas made of hydrogen atoms with one proton and one neutron) must be held at 100 million degrees, have a density of at least 10^{22} ions/m^3, and a confinement time of longer than a second. These are the minimum requirements, not the factors that would make the production of energy by nuclear fusion cost effective. Across the globe, hundreds of researchers are trying to do just that, make nuclear fusion a cost effective mechanism for energy production. The biggest cost is in starting up a fusion reactor, so the longer an engineer can keep one going, the more efficient the energy production will be.

Science is well on the way to producing fusion energy using reactors called tokamaks (Fig. 8.12). A tokamak is a confined version of the free space spheromak plasma toroids discussed in reference to LIPC (Sect. 1.7.2) and the magnetic fields that can hold plasma in place as a cosmic radiation shield (Sect. 4.5.2). Tokamaks are instruments that use supermagnet coils to hold plasmas in a toroid (doughnut-shaped) container without the plasma ever touching the walls of the container. The original function of tokamaks was for the study of plasmas, and now they are important in the development of fusion energy reactors. Nuclear fusion can be a very efficient way to produce energy, as long as you have the initial energy input to produce a plasma stream and keep it at 100 million degrees Celsius. Current tokamaks have a limit of about 7 min; that's 420 s, just like the multiplex pattern buffer on the Enterprise (TNG Technical Manual, p. 108). Fortunately, some old ideas are being adapted and helping scientists to design new confinement systems that will confine the plasma longer.

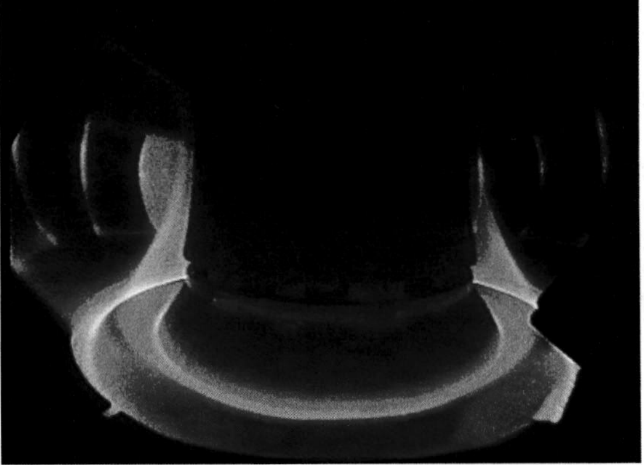

Fig. 8.12 The *top image* shows the inside of the plasma confinement chamber at the ASDEX tokamak at the Max Planck Institute during a facilities upgrade. The *bottom image* shows a plasma discharge in the same tokamak. It is hoped that this facility, and others like it, will usher in an age of fusion energy production. (*Image credit* Max-Planck-Institut fur Plasmaphysik, Volker Rohde)

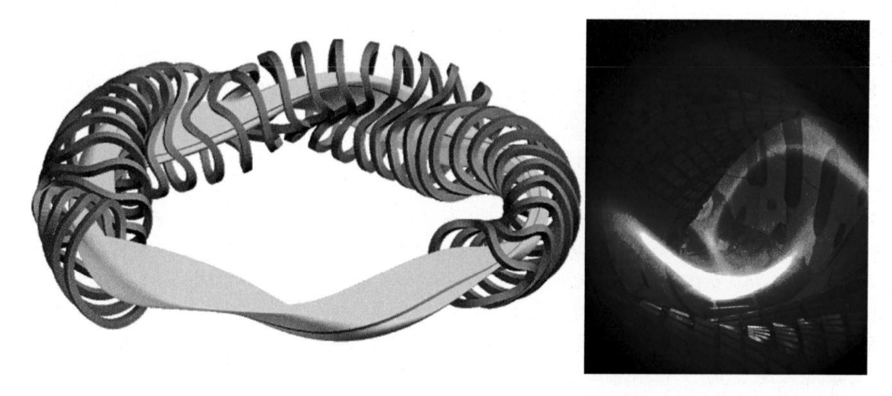

Fig. 8.13 The *left diagram* is how the Wendelstein 7-X stellarator plasma confinement device is designed. The *blue rings* are the supermagnets, while the *yellow* represents the plasma. The right photograph is from February, 2016 when the first hygrogen plasma was generated in the stellarator. The temperature is 80 million degrees Celsius. The stellarator shape is intended to maximize plasma stability and length of fusion generation. (*Image credit* Max-Planck-Institut fur Plasmaphysik)

A device called a stellarator was actually invented before the tokamak, way back in the pre-*Star Trek* 1950s. They were gradually replaced by tokamaks in the 1970s based on better confinement performance. The main difference between a stellarator (named so because it might allow mankind to harness the power of stellar objects) and a tokamak is the shape of the confinement field and therefore the shapes of the confinement magnetic coils. The new W7-X stellarator in Wendelstein, Germany has coils that confine the plasma in a twisted torus, something that resembles a cross between a Mobius strip and a conventional torus (Fig. 8.13). The new magnet designs and the improved stellarator shape have led to improved confinement instruments in the US, Japan, and Europe; fusion reactions can be held for over 30 min in stellarators. In addition to energy generation via these devices, understanding the behavior of plasmas via tokamaks and stellarators better will also increase fusion reactor efficiency. Unfortunately, there will be no tokamak or stellarator present to mediate nuclear fusion at a transporter destination site. Tremendous energies are involved in even the smallest fusion processes, so for now it is mere speculation to talk of harnessing these mechanisms to build anything other than the smallest atoms on demand.

8.6.4 Building Matter from Energy

Transporting the vast amount of energy that is equivalent to a person or large object will require an awfully large pipe, and once at the destination, the conversion back to matter will have to be near instantaneous. Changing energy back to matter follows $E = mc^2$ just as does converting matter to energy. In the basic sense, massless particles (photons) are converted to massed particles (fermions), so the reacting photon must have more energy than the resting mass energy of the particles to be produced. The process is called "pair production" and can take the form of energy conversion to an electron and a positron, or if you start with more energy packed into the photons, almost any particle-antiparticle pair you wish. Since energy is related to matter, even massed particles with enough energy can be used for pair production. Since electrons are less massive than quarks or whole protons, the energy need to produce an electron-positron pair is about 1800 fold less than to produce a proton-antiproton pair.

Due to reasons of conservation of momentum, it is impossible for a photon to just be converted to a fermion pair, there would be nothing there to accept the momentum of the lost photon; therefore, pair production would occur near a nucleus, or include the collision of a nucleus. In proton pair production for example, two protons, or even two electrons (Achard and L3 Collaboration 2003), with sufficient kinetic energy are collided to form a proton-antiproton pair, yet still maintain the two original particles at lower energy levels; it is their associated energy that is converted to the two new massed particles. Prior to 2014, the way to go about pair production with high-energy photons was to induce a cloud of atoms hit them with a very high-density beam of photons in a particle collider like the LHC and hope for the best. That's the only way it was *thought* that it could be done, until one afternoon when four scientists were having coffee together. In short order, they worked out a method using existing technologies wherein a couple of high-energy photons, produced by running a electron into a gold target, could be fired at one another and give an electron-positron pair (Pike et al. 2014). This method uses a vacuum hohlraum tube (cavity whose walls are in radiant equilibrium with the light confined in it, so none is absorbed) to focus and guide the photons and a magnetic field to separate the electron and positron after production so they won't annihilate one another. To date, no one has built such a photon-photon collider, but the ability to do so is entirely plausible.

Science has even produced entire atoms from energy, the only problem being that said atom is antimatter, anti-hydrogen specifically (Fig. 8.14). The atom is made up of antiproton and a positron, produced by one of several methods deduced by different teams at different collider facilities in the last decade and a half. Production of cold anti-hydrogen was first achieved at CERN (Amoretti et al. 2002), followed the first trapping of the antimatter atom in 2010 (ALPHA Collaboration 2011). In 2013, CERN scientists improved the method for production and collection of more anti-hydrogens through improvement of the trapping mechanism (Kuroda et al. 2014). This advance allows for finer in-flight spectroscopic comparisons of hydrogen and anti-hydrogen in a search for possible violation of symmetry that would give clues as to why our universe is overwhelmingly matter rather than antimatter. In a 2015 study, researchers in Australia and the UK used a positronium (a bound positron and electron) hit with an antiproton. By manipulating the energy of the positronium, the researchers found that they could greatly increase the efficiency of anti-hydrogen production (Kadyrov et al. 2015).

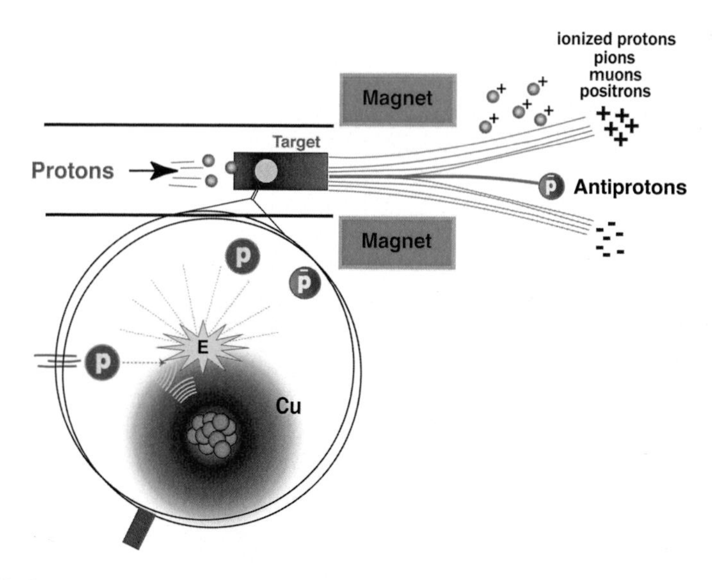

Fig. 8.14 One way to stimulate proton pair production is to run high speed protons into a copper target. The *inset* shows the pair production of a proton and antiproton from a single high speed proton. This is a rare event. The products can be separated by magnets placed downstream of the target. The products may be the initial protons if nothing happens, ionized protons, breakdown products from the collisions (pions, muons, positrons), and rarely—an antiproton. (*Image credit* Rolf Landua/CERN)

Such developments can be seen as first steps in the production of larger atoms from controlled collider events. Just as with exotic particle production that might serve as a basis for future work to build atoms, lower energy methods to produce anti-hydrogen atoms, the possibility of photon-photon colliders, and mechanisms to control and modulate different forms of pair production offer a glimpse of hope that conversion of energy to matter might someday be harnessed. Once this is done, the next issue concerning production of larger atoms from smaller ones in controlled fusion events will be addressed. Only at that point will everything come back to how one might best, fastest, and at the lowest energy cost, assembly the atoms back into the person or object that was teleported. Perhaps chemistry, perhaps molecular assembly (Sect. 3.5.2) or nanobots (Sect. 9.6.3) will be mature by that time. One thing is nearly certain, the problems of how to put something together at a destination devoid of tools for assembly and using a pattern that has no way to be read at the site will seem trivial when science has learned to harness $E = mc^2$ for everyday use and colliders that release megatons of energy are small enough to fit in a transporter pad.

8.7 Conclusion

When one considers the process of teleportation within the *Star Trek* universe, it seems so compact, so well explained, so sensible. For nearly every question that arises, the franchise has an answer, developed over time through the work of producers, writers, and scientific consultants. A few internal inconsistencies appear, such as whether energy or matter is transported, although these seem small when you see the process work, week in and week out. Unfortunately, when one digs into the mechanisms described for the transporter, some very real problems do arise. Interestingly, these problems don't involve the concept of quantum teleportation, science has already shown that to be not only possible, but practical for communication and security. No rules of quantum or classical physics are broken when teleporting quantum state information from one place to another, the technical issues are all that stand between today's research problems and tomorrow's commercial quantum computing and teleportation. It is only when considering the goal of transporting macroscopic objects and humans do the laws of physics get stressed to the point to the point of head shaking and comments about real science being mixed with science fiction. No doubt, it will be a long, hard road to transport a human from one place to another without ever existing in between the two positions, if it turns out to be possible at all.

Take your pick of which issue will be the hardest to overcome—Heisenberg, receiver-less transmission, transmitting matter, detecting and saving a quantum pattern, turning matter to energy or energy to matter—they all appear insurmountable. Or are they? The uncertainty principle might have weak spots that can be exploited, and fusion reactors are real-life links to both building and tearing down matter. Considering that it has been less than a decade since quantum computing has been realized on any experimental level, who is to say what the limits on entanglement or coherence times might be. In different research areas, the problem is that science doesn't know what it doesn't know; in this case, the rules are known, it is the exceptions and loopholes that have to be uncovered. The worst case scenario for "scattering McCoy's atoms all over the galaxy" is that the rules of physics can't be bent and there is no way to send matter or all the energy and information that equates to a living being from one place to another. The computer issues look more and more like it may not be the primary problem. Targeting is possible now and will only get better with time, although a line of sight problem could affect the system is entanglement switches (with no direct path needed) are seen to be limited. The energy requirements are tremendous, so the mastering of large-scale fusion reactors will be a must. Taking a realistic view, the work on quantum computing, matter-energy conversions, targeting, and entanglement/coherence efficiency will all move forward and at some point a researcher will try to link them to transport larger and larger packages with more information and degrees of freedom. Whether they ever get to the point of sending a person via quantum mechanisms is more plausible now than it was just ten years ago, yet is still daunting to say the least.

At the point when macroscopic teleportation becomes something more than conjecture or reasoned speculation, then the final issue will need to be addressed; is the person who arrives at the destination the same person who left the origin? Within the franchise, this question is brought up and debated from time to time (*ENT: Daedalus*), with the majority of people assenting to transport due to a philosophical agreement with the concept or simply because of the practicality of the technology. In many cases, getting from a ship to the planet in a timely fashion is just untenable using any other technology. This won't be the case on Earth; as science gets closer to transporter technology, the metaphysical discussions about the preservation of the life and the soul will become paramount.

References

R Aaij, and LHCb Collaboration. Observation of the two new Ξ_b^- baryon resonances. *Physical Review Letters* 114(6); 06204, 2015. doi: 10.1103/PhysRevLett.114.062004. http://journals.aps.org/prl/abstract/10.1103/PhysRevLett.114.062004

M Ablikim, and BESIII Collaboration. Observation of a charge charmoniumlike structure in $e^+e^- \rightarrow \pi^+\pi^- J/\psi$ at $\sqrt{s} = 4.26$ GeV. *Physical Review Letters* 110(25); 252001, 2013. doi:10.1103/PhyRevLett.110.252001. http://journals.aps.org/prl/abstract/10.1103/PhysRevLett.110.252001

P Achard, and L3 Collaboration. Proton-antiproton pair production in two photon collisions at LEP. *Physics Leters B* 572(1-2); 11-20, 2003. doi:10.1016/j.physlettb.2003.05.005. http://www.sciencedirect.com/science/article/pii/S0370269303009286

L Adamczyk and STAR Collaboration. Beam-Energy Dependence of Directed Flow of Protons, Antiprotons and Pions in Au Au Collisions. *Physical Review Letters*, 112(16); 162301, 2014. doi:10.1103/PhysRevLett.112.162301. http://journals.aps.org/prl/abstract/10.1103/PhysRevLett.112.162301

A Adare, and the PHENIX Collaboration. Cross section and parity-violating spin asymmetries of W ± Boson Production in polarized p + p collisions at $\sqrt{s} = 500$ GeV. *Physical Review Letters* 106(6); 062001, 2011. doi:10.1103/PhyRevLett.106.06201. http://journals.aps.org/prl/abstract/10.1103/PhysRevLett.106.062001

AF Ali, M Faizal, MM Khalil. Absence of black holes at LHC due to gravity's rainbow. *Physics Letters B* 743; 295-300, 2015. doi:10.1016/j.physletb.2015.02.065. http://www.sciencedirect.com/science/article/pii/S0370269315001562

ALPHA Collaboration. Confinement of antihydrogen for 1,000 seconds. *Nature Physics* 7; 558-564, 2011. doi:10.1038/nphys2025. http://www.nature.com/nphys/journal/v7/n7/full/nphys2025.html

M Amoretti, C Amsler, G Bonomi, A Bouchta, P Boew, et al. Production and detection of cold antihydrogen atoms. *Nature* 419; 456-459, 2002. doi:10.1038/nature01096. http://www.nature.com/nature/journal/v419/n6906/full/nature01096.html

F Belgiorno, SL Cacciatori, M Clerici, V Gorini, G Ortenzi, L Rizzi, E Rubino, VG Sala, and D Faccio. Hawking Radiation from ultrashort laser pulse filaments. *Physical Review Letters* 105 (20); 203901, 2010. doi:10.1103/PysRevLett.105.203901. http://journals.aps.org/prl/abstract/10.1103/PhysRevLett.105.203901

G Benford. A scientist's notebook: living in an eleven-dimensional world. *The Magazine of Fantasy and Science Fiction* 103(4-5); 187, 2002.

D Borghino. Quantum computers inch closer to reality thanks to entangled qubits in silicon. Gizmag. November 16, 2015. Accessed February 14, 2016. http://www.gizmag.com/advance-programmable-silicon-quantum-computers/40420/

D Boschi, S Branca, F De Martini, L Hardy, S Popescu. Experimental realization of teleporting an unknown pure quantum state via dual classical and Einstein-Podolsky-Rosen Channels. *Physical Review Letters* 80(6); 1121, 1998. doi:10.1103/PhysRevLett.80.1121. http://journals.aps.org/prl/abstract/10.1103/PhysRevLett.80.1121

MW Choptuik, and F Pretorius. Ultrarelativistic particle collisions. *Physical Review Letters* 104 (11); 11101, 2010. doi:10.1103/PhysRevLett.104.11101. http://journals.aps.org/prl/abstract/10.1103/PhysRevLett.104.111101

SM Clegg, R Wiensa, AK Misrab, SK Sharma, J Lambert, S Bender, R Newell, K Nowak-Lovato, S Smrekar, MD Dyar, and S Maurice. Planetary geochemical investigations using raman and laser-induced breakdown spectroscopy. *Applied Spectroscopy* 68(9); 925-936, 2014. doi:10.1366/13-07386. http://asp.sagepub.com/content/68/9/925.short

JP Dehollain, S Simmons, JT Muhonen, R Kaira, A Laucht, F Hudson, KM Itoh, DN Jmaieson, JC McCallum, AS Dzurak, and A Morello. Bell's inequality violation with spins in silicon. *Nature Nanotechnology* 11; 242-246, 2016. doi:10.1038/nnano.2015.262. http://www.nature.com/nnano/journal/v11/n3/full/nnano.2015.262.html

DO Collaboration. Observation of a new $B^0_S\pi^{\pm}$ state. Submitted February 24, 2016. http://www.arxiv.org/abs/1602.07588

WE East, and F Pretorius. Ultrarelativistic blck hole formation. *Physical Review Letters* 110(10); 101101, 2013. doi:10.1103/PhysRevLett.110.10101. http://journals.aps.org/prl/abstract/10.1103/PhysRevLett.110.101101

J Erhart, S Sponar, G Sulyok, G Badurek, M Ozawa, and Y Hasegawa. Experimental demonstration of a universally valid error–disturbance uncertainty relation in spin measurements. *Nature Physics*, 8; 185-189, 2012. doi: 10.1038/nphys2194. http://www.nature.com/nphys/journal/v8/n3/full/nphys2194.html

R Genzel. Inward bound: high-resolution astronomy and the quest for black holes and extrasolar planets. In: *Visions of Discovery: New Light on Physics, Cosmology, and Consciousness*. Cambridge: Cambridge University Press, 2010, pg. 310.

D Golze, M Icker, and S Berger. Implementation of two-qubit and three-qubit quantum computers using liquid-state nuclear magnetic resonance. *Concepts in Magnetic Resonance, Part A* 40a (1); 25-37, 2012. doi:10.1002/cmr.a.21222. http://onlinelibrary.wiley.com/doi/10.1002/cmr.a.21222/abstract

DW Hahn, and N Omenetto. Laser-induced breakdown spectroscopy (LIBS), Part I: review of basic diagnostics and plasma-particle interactions: still-chellenging issues within the analytical plasma community. *Applied Spectroscopy* 64(12); 335a-366a, 2010. doi:10.1366/000370210793561691. https://www.osapublishing.org/as/abstract.cfm?uri=as-64-12-335A

R Harrington. RHIC Run 14; A "flawless run of firsts." Brookhaven National Laboratory features. Brookhaven National Laboratory Features, August 4, 2014. Accessed March 12, 2016. https://www.bnl.gov/newsroom/news.php?a=25026

B Hensen, H Bernien, AE Dreau, A Reiserer, N Kalb, MS Blok, J Ruitenberg, RFL Vermeulen, RN Schouten, C Abellan, W Amaya, V Pruneri, MW Mitchell, M Markham, DJ Twitchen, D Elkhouss, S Wehner, TH Taminiau, and R Hanson. Loophole-free Bell inequality violation using electron spins separated by 1.3 kilometres. *Nature* 526;682-686, 2015. doi:10.1038/nature15759. http://www.nature.com/nature/journal/v526/n7575/full/nature15759.html

CA Hutchison, R-Y Chuang, VN Noskov, N Assad-Garcia, TJ Deerinck, MH Ellisman, J Gill, K Kannan, BJ Karas, L Ma, JF Pelletier, Z-Q Qi, RA Richter, EA Strychalski, L Sun, Y Suzuki, B Tsvetanova, KS Wise, HO Smith, JI Glass, C Merryman, DG Gibson, and JC Venter. Design and synthesis of a minimal bacterial genome. *Science* 351(6280); aad6253, 2016. doi:10.1126/science.aad6253. http://science.sciencemag.org/content/351/6280/aad6253

TH Jiang, A Rudenko, M Kurka, KU Kuhnel, Th Ergler, L Foucar, M Schoffler, S Schossler, T Havermeier, M Smolarski, K Cole, R Dorner, S Dusterer, R Treusch, M Gensch, CD Schroter, R Moshamer, and J Ullrich. Few-photon multiple ionization of N2 by extreme ultraviolet free-electron laser radiation. *Physical Revie Letters* 102(12); 123002, 2009.doi:10.1103/PhysRevLett.102.123002. http://journals.aps.org/prl/abstract/10.1103/PhysRevLett.102.123002

AS Kadyrov, CM Rawlins, AT Stelbovics, I Bray, and M Charlton. Antihydrogen formation via antiproton scattering with energized positronium. *Physical Review Letters* 114(8); 183201, 2015. doi:10.1103/PhysRevLett.114.183201. http://journals.aps.org/prl/abstract/10.1103/PhysRevLett.114.183201

FO Kirchner, A Gliserin, F Krausz, and P Baum. Laser streaking of free electrons and 25 keV. *Nature Photonics* 8:52-57, 2014. doi:10.1038/nphoton.2013.315. http://www.nature.com/nphoton/journal/v8/n1/full/nphoton.2013.315.html?WT.ec_id=NPHOTON-201401

DJ Knobloch, E Lobkovsky, and PJ Chirik. Dinitrogen cleavage and functionalization by carbon monoxide promoted by a hafnium complex. *Nature Chemistry* 2; 30-35, 2009. doi:10.1038/nchem.477. http://www.nature.com/nchem/journal/v2/n1/full/nchem.477.html

S Koike, H Takahashi, H Yonezawa, N Takei, SL Braunstein, T Aoki, A Furusawa. Demonstration of quantum telecloning of optical coherent states. *Physical Review Letters* 96 (6); 060504, 2006. doi:10.1103/PhysRevLett.96.060504. http://journals.aps.org/prl/abstract/10.1103/PhysRevLett.96.060504

V Kooper. *George Clayton Johnson, Fictioneer: from Ocean's Eleven, through the Twilight zone, to Logan's run, and beyond.* Page 150. Albany, GA: BearManor Media.

LM Krauss. *The Physics of Star Trek.* New York: Basic Books, 1995.

Y Kubo, C Grezes, A Dewes, T Umeda, J Isoya, H Sumiya, N Morishita, H Abe, S Onoda, T Ohshima, V Jacques, A Dreau, J-F Roch, I Diniz, A Auffeves, D Vion, D Esteve, and P Bertet. Hybrid quantum circuit with a superconducting qubit coupled to a spin ensemble. *Physical Review Letters* 107(22); 220501, 2011. doi:10.1103/PhysRevLett.107.220501. http://journals.aps.org/prl/abstract/10.1103/PhysRevLett.107.220501

G Kurizki, P Bertet, Y Kubo, K Molmer, D Petrosyan, P Rabi, and J Schmiedmayer. Quantum technologies with hybrid systems. *Proceedings of the National Academy of Sciences USA* 112 (13); 3866-3873, 2015. doi:10.1073/pnas.1419326112. http://www.pnas.org/content/112/13/3866.full

N Kuroda, S Ulmer, DJ Murtagh, S Van Gorp, Y Nagata, et al. A source of antihydrogen for in-flight hyperfine spectroscopy. *Nature Communications* 53089, 2014. doi:10.1038/ncomms4089. http://www.nature.com/ncomms/2014/140121/ncomms4089/full/ncomms4089.html

KC Lee, MR Sprague, BJ Sussman, J Nunn, NK Langford, X-M Jin, T Champion, P Michelberger, KF Reim, D England, D Jaksch, and IA Walmsley. Entangling macroscopic diamonds at room temperature. *Science* 334(6060); 1253-1256, 2011. doi:10.1126/science.1211914. http://science.sciencemag.org/content/334/6060/1253.abstract

LHCb Collaboration. Observation of J/ψp resonances consistent with pentaquark states in $\Lambda 0b \rightarrow J/\psi K - p$ decays. *Physical Review Letters* 115(7); 072001 2015. doi:10.1103/PhysRevLett.115.072001. http://journals.aps.org/prl/abstract/10.1103/PhysRevLett.115.072001

T Li, and Z-Q Yin. Quantum superposition, entanglement, and state teleportation of a microorganism on an electromechanical oscillator. *Science Bulletin* 61(2); 163-171, 2016. doi:10.1107/s11434-015-0990-x. http://link.springer.com/article/10.1007/s11434-015-0990-x

ZQ Liu, and Belle Collaboration. Study of $e^+e^- \rightarrow \pi^+\pi^-$ J/ψ and observation of a charged charmoniumlike state at Belle. *Physical Review Letters* 110(25); 25202, 2013. doi:10.1103/PhysRevLett. 110.252002. http://journals.aps.org/prl/abstract/10.1103/PhysRevLett.110.252002

VL Lyuboshitz, and VV Lyuboshitz. Spin correlations in the ΛΛand ΛΛsystems generated in relativistic heavy-ion collisions. *Physics of Atomic Nuclei* 73(5); 805-814, 2010. doi:10.1134/S106377881005008X. http://link.springer.com/article/10.1134/S106377881005008X

X-S Ma, T Herbst, T Scheidl, D Wang, S Kropatschek, W Naylor, B Wittmann, A Mech, J Kofler, E Anisimova, V Makarov, T Jennewein, R Ursin, A Zeilinger. Quantum teleportation over 143 kilometres using active feed-forward. *Nature* 489; 269-273, 2012. doi:10.1038/nature11472. http://www.nature.com/nature/journal/v489/n7415/full/nature11472.html

ST Muzzin. For one tiny instant, physicists may have broken a law of nature. *Yale News* March, 19, 2010. Accessed January 14, 2016. http://news.yale.edu/2010/03/19/one-tiny-instant-physicists-may-have-broken-law-nature

M Okuda, D Okuda, and D Mirek. *Star Trek Encyclopedia, Third Edition*. New York: Pocket Books, 1999. page 20.

C Ott, A Kaldun, L Argenti, P Raith, K Meyer, M Laux, Y Zhang, A Blattermann, S Hagstotz, T Ding, R Heck, J Madronero, F Martin and T Pfeifer. Reconstruction and control of a time-dependent two-electron wave packet. *Nature* 516; 374-378, 2014. doi:10.1038/nature14026. http://www.nature.com/nature/journal/v516/n7531/full/nature14026.html

M Ozawa. Universally valid reformulation of the Heisenberg uncertainty principle on noise and disturbance in measurement. *Physical Review A* 67(4); 042105, 2003. doi:10.1103/PhysRevA.67.042105. http://journals.aps.org/pra/abstract/10.1103/PhysRevA.67.042105

RB Patel, J Ho, F Ferreyrol, TC Ralph, and GJ Prude. A quantum Fredkin gate. *Science Advances* 2(3); e1501531, 2016. doi:10.1126/sciadv.1501531. http://advances.sciencemag.org/content/2/3/e1501531

TG Philbin, C Kuklewicz, S Robertson, S Hill, F Konig, and U Leonhardt. Fiber-Optical Analog of the Event Horizon. *Science* 319; 1367-1370, 2008. doi:10.1126/science.1153625., http://science.sciencemag.org/content/319/5868/1367

O Pike, F Mackenroth, EG Hill, and SJ Rose. A photon-photon collider in a vacuum hohlraum. *Nature Photonics* 8; 434-436, 2014. doi:10.1038/nphoton.2014.95. http://www.nature.com/nphoton/journal/v8/n6/full/nphoton.2014.95.html

L.A. Rozema, A. Darabi, D.H. Mahler, A. Hayat, Y. Soudagar, and A.M. Steinberg. Violation of Heisenberg's measurement-disturbance relationship by weak measurements. *Physical Review Letters* 109 (10); 100404, 2012. doi:10.1103/PhysRevLett.109.100404. http://journals.aps.org/prl/abstract/10.1103/PhysRevLett.109.100404

K Saeedi, S Simmons, JZ Salvail, P Dluhy, H Riemann, NV Abrosimov, P Becker, H-J Pohl, JJL Morton, and MLW Thewalt. Room temperature quantum bit storage exceeding 39 minutes using ionized donors in silicon-28. *Science* 342(6160); 83-833, 2013. doi:10.1126/science.1239584. http://science.sciencemag.org/content/342/6160/830

CP Shen, and Belle Collaboration. Evidence for a new resonance and search for the *Y*(4140) in the $\gamma\gamma \rightarrow \phi J/\psi$ process. *Physical Review Letters* 104 (11): 112004, 2010. doi:10.1103/PhysRevLett. 104.112004. http://journals.aps.org/prl/abstract/10.1103/PhysRevLett.104.112004

M Shiddiq, D Komijani, Y Duan, A Gaita-Arino, E Coronado, and S Hill. Enhancing coherence in molecular s;in qubits via atomic clock transitions. *Nature* 531(7594); 348, 2016. doi:10.1038/nature16984. http://www.nature.com/nature/journal/v531/n7594/abs/nature16984.html

II Smolyaninov, E Hwang, E Narimanov. Hyperbolic metamaterial interfaces: Hawking radiation from Rindler horizons and spacetime signature transitions. *Physical Reviews B* 85(23); 23512, 2012. doi:10.1103/PhysRevB.85.235122. http://journals.aps.org/prb/abstract/10.1103/PhysRevB.85.235122

H Song, SA Bass, U Heinz, T Hirano, and C Shen. 200 A GeV Au + Au collisions serve a nearly perfect quark-gluon liquid. *Physical Review Letters* 106(19); 192301, 2011. doi:10.1103/PhyRevLett.106.192301. http://journals.aps.org/prl/abstract/10.1103/PhysRevLett.106.192301

A Streltsov, U Singh, HS Dhar, M N Bera, and G Adesso. Measuring quantum coherence with entanglement. *Physical Review Letters* 115(2); 020403, 2015. 10.1103/PhysRevLett.115.020403. http://journals.aps.org/prl/abstract/10.1103/PhysRevLett.115.020403

AI Taylor, VB Pinheiro, MJSmola, AS Morgunov, S Peak-Chew, C Cozens, KM Weeks, P Herdewjin, and P Holliger. Catalysts from synthetic polymers. *Nature* 518; 427-430, 2015. doi:10.1038/nature13982. http://www.nature.com/nature/journal/v518/n7539/full/nature13982.html

P van Loock, and SL Braunstein. Telecloning of continuous quantum variables. *Physical Review Letters* 87(24); 247901, 2001. doi:10.1103/PhysRevLett.87.247901. http://journals.aps.org/prl/abstract/10.1103/PhysRevLett.87.247901

S van Velzen, GE Anderson, NC Stone, M Fraser, T Wevers, BF Metzger, PG Jonker, AJ van der Horst, TD Staely, AJ Mendez, JCA Miller-Jones, ST Hodgkin, HC Campbelll, and RP Fender. A radio jet from the optical and x-ray brght stellar tidal disruption flare ASASSN-14li. *Science* 351(6268); 62-65, 2016. doi:10.1126/science.aad1182. http://science.sciencemag.org/content/351/6268/62

M Veldhorst, JCC Hwang, CH Yang, AW Lenstra, B de Ronde, JP Dehollain, JT Muhonen, FE Hudson, KM Itoh, A Morello, and AS Dzurak. An addressable quantum dot qubit with fault-tolerant control-fidelity. *Nature Naontechnology* 9; 981-985, 2014. doi:10.1038/nnano.2014.216. http://www.nature.com/nnano/journal/v9/n12/full/nnano.2014.216.html

X-L Wang, X-D Cai, Z-E Su, M-C Chen, D Wu, L Li, N-L Liu, C-Y Lu, and J-W Pan. Quantum teleportation of multiple degrees of freedom of a single photon. *Nature* 518; 516–519, 2015. doi:10.1038/nature14246. http://www.nature.com/nature/journal/v518/n7540/full/nature14246.html

SE Whitfield and G Roddenberry. *The Making of Star Trek*. New York: Ballantine Books, 1968. pages 43-44.

JM Zadrozny, J Niklas, OG Poluektov, and DE Freedman. Millisecond coherence time in a tunable molecular electronic spin qubit. *ACS Central Science* 1(9); 488-492, 2015. doi:10.1021/acscentralsci.5b00338. http://pubs.acs.org/doi/full/10.1021/acscentsci.5b00338

K Zhao, Q Zhang, M Chini, Y Wu, X Wang, and Z Chang. Tailoring a 67 attosecond pulse through advantageous phase-mismatch. *Optics Letters* 37(18); 3891-3893, 2012. doi:10.1364/OL.37.003891. https://www.osapublishing.org/ol/abstract.cfm?uri=ol-37-18-3891

Z Zhao, A-N Zhang, X-Q Zhou, Y-A Chen, C-Y Lu, A Karlsson, and J-W Pan. Experimental realization of optimal asymmetric cloning and telecloning via partial teleportation. *Physical Review Letters* 95(3); 030502, 2005. doi:10.1103/PhysRevLett.95.030502. http://journals.aps.org/prl/abstract/10.1103/PhysRevLett.95.030502

Chapter 9
The Tricorder: A High Tech Multitool

Do you believe in an afterlife?
I accept there are things in the universe that can't be scanned with a tricorder.

—First officer Chakotay
VOY: Barge of the Dead

9.1 Introduction

The technologies of *Star Trek* are analogous to the franchise itself. The five year mission is to gain knowledge and information about the universe and the intelligent life therein. Likewise, many of the technologies are information gatherers as well. The universal translator takes in communication signals from aliens and the navigational shields work with long-range sensors to figure out what things might be in the way. Even Geordi's visor is a translator of sorts, it takes the available EM energy data and presents it to his brain for interpretation. This mountain of this sensor data serves two purposes within the series; one, to be interpreted for the explicit purpose for which it was gathered, and two, to store it into some large repository for future use, as in how the universal translator stores all previous encounters with a species so that their next meeting it will be able to draw upon more visual, cultural and lingual information.

Even with all these different sources of sensor data, there is one piece of technology that contributes more data to the Federation's computer systems than any other: the tricorder. This unassuming device has a basic but difficult job, gather whatever information is requested. What is that black obelisk made of? Is it going to rain? What caused this ulcer on my lip? How do I get back to the ship from here? A handheld device that can answer any question you have about the things going on around you really captures the audience's attention.

Since the tricorder makes sense to us and can open our world up in practical ways, it isn't surprising that engineers and scientists have started mimicking the tricorder. Smartphones and their peripherals are obvious examples, a single smartphone has dozens of sensors that provide data points either continuously or at

M.E. Lasbury, *The Realization of Star Trek Technologies*,
DOI 10.1007/978-3-319-40914-6_9

short intervals. It is just a matter of whether those data points are being recorded and stored, and who or what machines have access to that data. However, the most apt demonstration of the sensing and interpreting power of the tricorder is the Internet of Things (IoT) and machine to machine communication (M2M). Devices of all types are being connected to the internet, and by extension, to one another. These meshing networks can transfer sensor data from one networked machine to another in a direct fashion, but most sensor data will be uploaded to large databases that can be accessed in real time or at anytime in the future. The key is to make the most of the data that is being captured.

Factories sense the workings of their machines and report on temperatures, wear, speeds, demand for products, inventory stores, etc. Google Maps and Facebook locator match images to locations and events. Smart buildings sense traffic patterns, water use, video data, etc. All this primary data is out there to be gathered and used as secondary data. One can combine something that one device senses with some piece of environmental, geologic, cultural, chemical, or biological data that one of your devices detects. With the proper algorithms to put those data together, answers can be had for just about any question—providing not only knowledge, but also practical solutions to problems and interpretations of where we are and what is going on around us.

The marketplace has already started to pick up on the IoT's ability to amass primary sensor data. Companies like OpenSensors.io, Basho, and Datafloq allow you to manage the sensor data that your devices collect and to access and use public real-time sensor data for fun and profit. Crowdsourcing funds is very similar to a new trend of crowdsensing—people or companies allow the sensors from their devices to dump data into a repository for access in exchange for some benefit to themselves down the road, or for altruistic reasons. This isn't to say all crowd-sensing efforts are entered into *knowingly* by the sensor owner. Many have been negotiated for you—that app you bought for ninety-nine cents may very well give the company access to your smartphone sensor data and camera, and who knows what they might do with the data they collect. Read your agreements carefully.

All this data is useless without algorithms to turn it into usable information. Number crunching, averages, converting numbers to patterns and predicting future trends, scanning large databases and comparing outcomes—all these applications of big sensor data require (at least for the time being) humans to assign meaning to them. Several sources may be combined to provide more accurate data—and this is where the tricorder shines. It is a primary data-gathering device, with the added ability to combine sensor information from different sensors and repositories of data to put together complex interpretations. Need a cool example? How about ten million dollar prize for a team building a medical tricorder that can diagnose diseases by running lab tests, monitoring vital signs, taking histories and symptoms, and the comparing all that data with millions of medical records? Not bad for a technology that appears in less than a third of the episodes of all the series.

9.2 Star Trek Tricorders

Dr. McCoy stands over a young man in a red uniform, waving a short metal tube across his prostrate body. Bones looks up at Kirk, "He's dead Jim." The doctor utters some version of that phrase no fewer than twenty times in the original series —twice over Kirk's own body. Interestingly, the good doctor never says it in any of the movies. But the point is made; the tricorder is indispensible for medical personnel. And this has led to some confusion amongst casual *Star Trek* viewers. There are actually several different types of tricorders, although there is some overlap among them as well. No matter if it is a standard, heavy-duty, medical or psychotricorder, what they do well is sense the environment and report their findings.

9.2.1 Tricorder Functions

The standard Starfleet tricorder is a handheld device, with or without a separate wand sensor. The first question to be answered is why they are called tricorders. It doesn't have just three functions or capabilities; the original series places no limits on its use. Even though there might be occasions when the tricorder might not register a signal, no character ever says that the tricorder is useless in some situation. The typical use for the tricorder is medical, yet they can do so much more. It had uses in metallurgy, materials science, geology, geography, physics, botany, biology, meteorology, and countless other-ologies. The tri-portion of the name is said to refer to the fact that it detects, records, and analyzes primary sensory data (Palestine 1977, p. 143).

This is a legitimate explanation, yet the *Next Generation* TR-560 model clearly shows three buttons on the console face marked GEO, MET, and BIO for geology, meteorology, and biology. Many of the sensory functions can be loosely lumped into one of these categories and the implication is that banks of sensors or analysis software must be activated; the tricorder doesn't sense all things all the time.

The standard Starfleet tricorder is usually issued to officers, although they can also be found in equipment lockers in several of the series. It has the capabilities that are most often needed on board and away missions, and it can be modified for specialized tasks using peripheral attachments and database chips. The *Star Trek: The Next Generation Technical Manual* states that the standard tricorder has 235 sensors for mechanical, electromagnetic, and other phenomena, with 115 sensors pointing forward and 120 that are omnidirectional (p. 119). The stored handheld sensor has only 17 different sensors, but can scan a much narrower field.

Stepping outside the *Star Trek* universe, the tricorder is an important storytelling device. Whatever information is needed to move the story along is available from the tricorder, so its functions vary over time. It analyzes substances, such as when Spock adjusted his tricorder to pick up silicon specifically (*TOS: The Devil in The*

Dark). It can map areas (*TNG: Chain of Command, part 1*) or detect life forms, even if they have already left the area (*Voy: Phage*). Environmental conditions (*TOS: The Apple*) can be monitored and assessed, and information is recorded (*TNG: Booby Trap*)—whatever information is required at that very moment in the episode is supplied the tricorder. Modern culture in most countries have a tricorder-like device to provide communication, advice, data, or analysis at any second. It's called the smartphone, although I doubt Mr. Scott ever wasted an hour playing "Angry Birds of Prey 2" on his tricorder when he should have been working.

With new applications and peripherals available every day, the smartphone is quickly becoming a formidable sensing device. The "corder" part of the tricorder name implies that the device can record the information that it senses, be it audio (recorder), video (camera), or other data it can collect—again the smartphone comes to mind. In essence, the tricorder and the smartphone can add to the information contained in computer databanks every moment they are turned on, and this data can be accessed and combined with other sensory data—*Star Trek* basically invented crowdsensing and smartphones are helping it play out in the real world. *Star Trek* even predicted the addiction that many people have to their smartphone. Seven of Nine once shot the tricorder out of the bad guy's hand and said that he could then be easily be apprehended (*VOY: Relativity*). How many people would be just as lost without their smartphone constantly in their hand?

9.3 Smartphones as Tricorders

The average smartphone has a dozen or more sensors, from the predictable radio waves and microwave sensors for communications to accelerometers, GPS, and fingerprint identification sensors. Much of these data are collected and may be used for crowdsensing projects, some with great importance. For instance, the US Geological Survey is investigating using the position and accelerometer sensors in huge numbers of smartphones as early feedback information for earthquake detection (Minson et al. 2015). Not as crucial yet still impressive, insurers can track driving performance using sensors in smartphones and merge the data with location and traffic via GPS to see if you are driving safely (Agero, Medford, MA). The analyzed data can be used to reward good drivers, penalize bad drivers, and pinpoint locations for emergency assistance. Thankfully, not all smartphone apps are so serious; one app even lets your smartphone look and sound like a tricorder, with lots of blinking LED's, beeping, and whirring (Moonblink, Sunnyvale, CA). This app doesn't *do* anything with the sensor data, it just shows you the information that your phone is tracking. But that's O.K., there are lots of apps that have genuine tricorder-like functions; some of them are even useful.

9.3.1 Peripheral Sensors for GEO Tricorder Functions

The tricorder, as used in *Star Trek*, is a location aware device. It provides a comprehensive picture of the local environment, from the physical layout to the materials found there. Of all these functions, detailed mapping of the area is probably most important because it gives a framework on which to hang much of the other data sensed from the environment. Mapping is of significant help to away teams, locating signals and finding their way around. Captain Picard tells Dr. Crusher to, "Make sure your tricorder is keeping a precise map of the route. We could easily get lost in here," (*TNG:Chain of Command, part 1*), while Miles O'Brien uses a tricorder to find his way through a cave (*DS9: Whispers*). This is no different than the myriad smartphone applications that let one track a bicycle ride or hike on a map using GPS.

This "sense of place" provided by the *Star Trek* tricorder was instrumental in Google CTO Michael Jones' idea of what Google Earth should aspire to be. Jones started his talk at the Where 2.0 Conference in 2007 (San Jose, CA) with a sound that radiated through the audience. Asked if anyone in the arena could identify that sound many people shouted out, "It's a tricorder!" Jones saw the tricorder as a way to achieve that "sense of place;" one of his talks the previous year had been entitled exactly that. Just how mapping was accomplished in a tricorder is a bit foggy, as location-sensitive mapping for smartphones and other devices uses external sources of information to triangulate a precise position. The app uses stored information via Google Earth and myriad other crowdsensed information points on the internet to put that location in context with the surrounding environment. All we know about the tricorder version is that it ended up with the same output.

The tricorder not only tells you where you are, it senses what that place is made of, how it relates to the places around it and how its history has led to the geography at this time. Several new apps help to put location in perspective as well. The FLIR One (FLIR Systems, Wilsonville, OR) is a thermal imaging camera peripheral that combines heat imaging with standard camera images of the surroundings (Fig. 9.1), while Spike (ike^{GPS}, Wellington, NZ), is one of several mapping packages that excel at distancing objects from the phone. With Spike, a peripheral device for your smartphone contains a laser and a compass. Using your smartphone's camera, Spike can laser distance objects to ± 20 cm and can integrate the measurements with GPS.

All the information to describe a local environment in total is what Jones wanted for Google Earth, just as Spock used stored data along with current visual mapping to follow technological progress over time in a specific area (*TOS: Errand of Mercy*). One smartphone app has just caught up to *Star Trek* in this respect. The Pivot (Pivot the World, Cambridge, MA) app uses GPS, a smartphone camera, a database of historical photographs, and augmented reality software to project a historical image of whatever place you capture on screen, as long as a relevant photograph is available in the crowdsourced database. Pivot is beta testing as of early 2016, having already surpassed its Kickstarter crowdfunding goal.

Fig. 9.1 The FLIR One contains two cameras; one takes a traditional image while the other takes an image in infrared. There is also a device where the smartphone camera takes the traditional image and the dongle takes only the thermal image. The software extracts the pertinent line and edge information from the traditional image and superimposes it on the thermal image to make the thermal image easier to read. The image show is traditional on the bottom, while the extracted information and the thermal image are shown above. (*Image credit* FLIR Systems, Wilsonville, OR)

9.3.2 Environmental Smartphone Sensors

Smartphone sensing platforms have become very popular for atmospheric conditions. Companies large and small have produced dongles and wireless peripherals that will sample air, water, soil, gases, etc. and report what they detect. The single largest commercial platform comes from Variable Corp. in Chattanooga, TN. Its Node sensing platform is based on an oversized lipstick-shaped, handheld probe that looks suspiciously like the original tricorder probe, except that this one has interchangeable sensors on one end. The unit connects wirelessly to any mobile device for analysis, reporting, and recording of the data. Its "oxa" sensor can track air quality (carbon monoxide, nitric oxide, nitrogen dioxide, chlorine, sulfur dioxide, carbon dioxide, and hydrogen sulfide), and other dongles are used for color matching and temperature sensing.

The "clima" sensor attachment simultaneously and continuously monitors barometric pressure, room temperature, ambient light levels and humidity. There is also a professional level infrared temperature sensor. All this climate data seems to be ripe for a crowdsensing application, yet Variable VP for Marketing John Kowalski says that they have no present plans to amass Node sensed values on the cloud. He told me that it would not be difficult to write an app that throws all the clima data up to an IoT site, and that if someone did (he suggested I do it), that they would consider adding an opt-in button to allow the user to choose whether the data and location information would be posted to the crowdsensing database.

On the water quality end of environmental sensing, a group of students from the University of Illinois introduced the Mobosens in 2013. Led by Dr. Logan Liu, the team was awarded a $120,000 at the first Nokia Sensing X-Challenge and has used the money to expand their senor capabilities. The first units detected water nitrate levels only, but Dr. Liu told me that the team is extending the sensor functions to nitrites, phosphates, bacteria, arsenic, and heavy metal ions. The dongles are designed to be cheap and easy to use so that underprivileged communities with water safety issues will be able to closely monitor their water supplies. The data is uploaded to the cloud along with location information so that water quality maps for an area can be crowdsensed over time and distance.

9.3.3 Smartphones as Health Monitors

Perhaps the greatest number of sensor peripherals and applications for smartphones involve health monitoring or other biological data. Literally thousands of sensors can be worn or attached to bicycles, running shoes or your body to collect data on heart rate, speed, distance, etc. Because medical applications of the tricorder and its surrogate, the smartphone, are the focus of the remaining portion of the chapter, here we will quickly describe a few biologic peripherals and applications that may not be strictly related to medicine and laboratory testing. For example, the Lapka PEM peripherals (Airbnb, San Francisco, CA) are a set of four sensor dongles, one of which measures the nitrite levels in food. This is a direct marker of nitrogen containing fertilizer use and can therefore be used to test the "organicity" of food—whether the food was grown without pesticides or herbicides.

Other peripherals have similar diet-based health functions. Soon to be on the market is the Nima (6SensorLabs, San Francisco, CA), a wireless peripheral that tests prepared food for gluten. The wireless unit is shaped like a triangle and has a small testing chamber where prepared foods are tested for the presence of gluten. If gluten is detected, the unit shows a frowny face and transmits the data to the smartphone app. The app transmits the information, along with the entered dish title, restaurant and location into a freely accessible database gathered from all the other Nima tests. After the gluten test debuts in mid-2016, the company intends to add functionalities for peanut and milk testing in 2017.

The functions of biological testing peripherals for use with smartphones goes well beyond food testing for allergens. The MobiUS SP1 (Mobisante, Redmond, WA) is an ultrasound device that emits very high frequency sound waves and detects the waves that bounce off of structures in their path. The handheld transceiver looks very much like that of a conventional ultrasound machine, but is plugged into a smartphone, nothing more is needed. Medical imaging with ultrasound is one use for the technology; however, not every use of ultrasound needs to be so serious. Ultrasonics emitted from a wireless floating peripheral (FishHunter, Canada) can transmit returning signal patterns to a phone many feet away—it's a fish finder hooked to your line as a bobber. Not to be outdone in the area of weird sensors, there are peripherals to monitor bad breath (Breathometer Mint, Burlingame, CA) or breath alcohol level (Lapka BAM or Breeze from Breathometer). We shall see below that breathalyzers like these could become very important in the next few years in terms of non-invasive, non-contact diagnostic tools.

9.4 Achieving a Medical Tricorder

When you ask the casual *Star Trek* fan what a tricorder is used for, you will mostly likely get an answer related to medicine. We most often envision it in the hands of the Bones or Beverly Crusher, telling us what's wrong with a crewmember or alien. While only the medical staff is issued medical tricorders, even the standard tricorders can discern a great deal about a life form. As a result, the audience sees medical uses of tricorders even outside the sickbay and being used by other crewmembers. For example, Spock is able to determine that Mister Flint is human using his standard tricorder and he notes some peculiarities in Flint's body systems, including the fact that he seems to be six thousand years old (*TOS: Methuselah*). Spock is no doctor, yet he can gain significant medical information just using the tricorder.

We are all used to having blood drawn or supplying a urine specimen, having the test done in a laboratory and waiting for the doctor to call with the results. A non-invasive device with immediate diagnosis sounds very appealing. This is probably what intrigues the audience most; the medical tricorder can sense and interpret signals from every body system of most any sort of body, yet how it does this is a bit of a mystery. The doctor or yeoman passes the tricorder over a body part and tells the crewmember what's wrong. Making this real is a current goal of the XPrize Foundation (Culver City, CA)—a medical device useable by consumers that can monitor health and diagnose disease.

9.4.1 The Qualcomm Tricorder XPrize

In 2013, Qualcomm (San Diego, CA) sponsored an XPrize of ten million dollars for the teams that could best develop a medical tricorder. The name of the competition was specifically chosen to reflect the capabilities of the device from the *Star Trek* franchise, and reflects the idea that most people believe the tricorder to be primarily a medical device. Their reasons for sponsoring the competition were many, but a desire for consumers to have better control over monitoring their own health was a primary concern (Chandler 2014), where that monitoring can include diagnosis, treatment, and follow up. Starfleet tricorders led the way in terms of suggesting treatment and follow up, as in Deep Space Nine when Lisa, a stranded Starfleet officer tells Dr. Bashir, "And, yes, I've been giving myself fifteen cc's of triox every four hours … just like it says in my medical tricorder," (*DS9: The Sound Of Her Voice*), so the XPrize committee decreed that winning devices will also diagnose disease and render medical advice to the user. The winning team will receive between four and nine million dollars, depending on how the scores come out, with category winners or overall runners-up garnering one to two million dollars each. If the overall winner also wins the vital signs monitoring category award, they could walk away with the entire ten million dollars.

The tricorders of the XPrize competition are required to assess five vital signs and diagnose the presence or absence of thirteen conditions or diseases. To keep them small, the devices must weigh less than five pounds (2.27 kg) total including peripherals. However, all the functions need not be housed in a single unit. Importantly, the tests must be selected, samples collected, and processes run by the consumers themselves. No medical personnel may be involved in the protocol, not even as an advisor by phone or on social media. The data obtained from a medical history and symptom information supplied by the user, as well as the data from the laboratory tests, must be uploaded to the internet for analysis and storage in real time.

Thirty-four teams took part in the preliminary rounds in 2013 and 2014. In 2015 seven finalists were announced. The teams still in the running for the prizes include the Dynamical Biomarkers group from Taiwan, CloudDx from Canada, and five teams from around the United States. The teams submitted thirty identical devices for clinical testing June 30, 2015, with each machine given to patients with one of the required or optional conditions or a healthy volunteer. Using healthy volunteers is important because the tricorder must be able to confirm the absence of a disease as well as its presence. Multiple tests were run on each machine and the interface and algorithms for diagnosis and advice were put to the test. In the end, the judges determined that all the teams could use extra time to refine their testing and data algorithms; none of the devices met the rigorous standards to merit the award after that first testing period. Therefore, Qualcomm awarded each team an additional seven months time to perfect their designs. Final judging will now take place in August of 2016, with consumer testing of the machines in September. The winners will be announced in early 2017.

9.4.2 Team Aezon, Undergraduate Tricorder XPrize Finalists

Perhaps the most inspiring of all the finalist teams is Aezon Health, nineteen undergraduate students from Johns Hopkins University. They have taken on the challenge of producing an XPrize tricorder not for class credit nor for the chance of an internship; they do all this work in addition to their normal classes and must find time and funding on their own simply for the joy of doing science and making healthcare more accessible for all. Neil Rens, a fourth year biomedical engineering student, told me that in order to work most efficiently the team is broken up into several small groups, each focusing on a part of the project. His specialty is the user experience, making the tricorder easy for the medically untrained patients and volunteers to use and understand the information that the algorithms return.

Despite the fact that this is the *Tricorder* XPrize and team Aezon's leader, Tatiana Rypinski, has posed for publicity shots in a *Next Generation* uniform holding an original series tricorder, Neil said that most of the team members are not *Star Trek* fans. He thinks that the influence of the shows is not as great for Aezon as for other researchers because of their young age. He says that few of his teammates mentioned watching the show growing up and that because he was too young to be influenced by most of the series (*Voyager* went off the air when most of them were less than seven years old) it had little to do with his decision to go into science and engineering. This trend is supported by discussions with other researchers; the older they are, the more they talk about the influence that *Star Trek* has had on their work and career path. It may also highlight the importance of the first two series with their more thorough introduction to most of the gadgets and their plausible mechanisms of action. What is encouraging is that those older scientists are raising their children on *Star Trek* streaming video, so a new generation of scientists with *Star Trek* aspirations may be just coming into their teenage years now.

Mr. Rens said that the competition provided its own incentive, beyond the fact that it conjured up images of *Star Trek* devices. He and his teammates were intrigued by the idea of empowering consumers in the management of their own health and the possibilities of using technology and mass communication devices to help in this challenge. Being children of the information age, the interfaces and user experience issues flowed from them like water; they have been steeped in these person-machine relationships for their entire lives and this makes them uniquely qualified to set up a user friendly, immersive experience that draws on all the capabilities of the internet and connectivity.

On the other hand, the young age and inexperience of the team members does make parts of the project more difficult. Raising money was focused on a crowd-funding web campaign and a couple of Johns Hopkins translational science research grants and generous funding from the Abell Foundation (Baltimore, MD), and though cash is always an issue, Neil is pleased that when they have needed cash at different points, they have managed to find the funds somehow, some way. Being constantly cash strapped makes them extremely frugal in their choices, including using some of the Hopkins machining laboratories to 3D print stands and other parts for their testing apparatus. Adding together their money woes, issues of relative inexperience in the field, and demands on their time for schoolwork, it is an amazing achievement that they earned a place in the finals.

9.4.3 Vitals Functions Monitoring

The first challenge for each team was to find a way to continuously monitor heart rate, temperature, respiration rate, blood pressure and blood oxygenation, and to move that data to a cloud-based platform in real time. Normally, these vital signs are monitored once during a visit, using different instruments for each—a thermometer, a stethoscope, a blood pressure cuff (sphygmomanometer), a pulse oximeter, and a set of human eyes to count breaths over a period of time. Tricorders that incorporated all these would already weigh more than five pounds. Luckily, medical entrepreneurs have shown how these various instruments can be combined with reduced mass.

Box 9.1 The Apple Watch and the Heart Rate Monitor

The Apple watch (Apple, Cupertino, CA) was introduced in 2015 with many included many applications and security features that rely on detecting the heartbeat.

Detecting a heartbeat:

- informs the watch that it is being worn
- turns on health monitoring applications
- is required for stress monitoring applications
- enables Apple Pay
- allows notifications to show up on screen

Heartbeat detection is achieved by:

- flashing a green light from the bottom of the watchcase through the skin of the wrist hundreds of time per second
- green light is absorbed by the blood in the vessels, it reflects red light and is why we see blood as red
- a heartbeat causes the vessels to expand to hold more blood
- more green light is absorbed during a heartbeat
- less green light is bounced back to photodiodes on the undersides of the watchcase during a heartbeat
- cyclic changes in returned green light represent heart rhythm.

While at rest, a second system is used to monitor heart rate:

- infrared (IR) light is flashed through the skin by same photodiodes
- IR light is reflected by blood
- More IR light is reflected during a heartbeat because the vessels expand and contain more blood
- Cyclic changes in IR light detection correspond to the heart rhythm
- IR is used at rest because it is less energy consuming than green light and is sufficiently accurate under rest conditions.

Apple confirms wrist tattoos may interfere with heart rate monitor:

- Traditional black colored tattoos absorb green light and IR light
- Solid colored tattoos of most colors absorb green light
- Tattoos with red ink absorb the most green light
- Tattoos that cover the wrist prevent green and perhaps IR light from reaching vessels and being reflected
- No heart rhythm is detected
- Several applications on Apple watch are not enabled

Apple does not anticipate a fix for the problem, but has provided work-arounds. The heart beat monitor can be turned off, although this reduces the security of the device, and one could wirelessly connect the watch to a chest band heart monitor, if one is so inclined.

No fewer than half a dozen devices to monitor multiple vital signs are on the market, with most extensive systems as of 2016 being the ViSi from Sotera Wireless (San Diego, CA) and the Scanadu Scout (Scanadu, Moffett Field, CA). Approved for use in the United States by the FDA in 2013, the ViSi mobile system monitors both diastolic and systolic blood pressure, skin temperature, heart rate, blood oxygenation, and respiration rate, with a built in electrocardiogram device to monitor heart rhythm. The user wears one sensor on their thumb and one around their chest, and the data are sent wirelessly to a collection device, probably a smartphone or tablet. The Scout is a hockey puck-shaped apparatus with no attached sensors. It assesses the same parameters as the ViSi, but does so by pressing the entire unit against the left temple; this completes a circuit by pressing a thumb on the bottom of the Scout. The index finger is placed on top and covers the pulse oximeter sensor to assess blood oxygenation. All the data are wirelessly moved to a smartphone or other mobile device, analyzed on the web, and returned via the interface application for assessment. It is important to note that the competition requires *continuous* vitals monitoring, a task the ViSi is designed to do, while the single measurement Scout would require some adaptation. This is indicative of the issues that even the teams that had technology in hand as the contest began had to overcome.

The Aezon team originally planned to develop a single vitals monitoring device for all five parameters. Housed in a neck piece worn as a yoke and open on the front, the Arc (designed by Aegle, Baltimore, MD), as they called it, would use a thermometer to detect surface temperature, optical sensors to read oxygen levels as a clampless pulse oximeter, an accelerometer to filter out chest movements not related to respiration, and multiple electrodes around the neck for an electrocardiograph (EKG). An EKG machine detects the electrical signals that control the heartbeat and rhythm by sensing minute changes in skin electrical charge that correspond to the different parts of the heartbeat. The EKG can easily be used to monitor heart rate, it shows each beat on the Y axis and the X axis is time; the interesting part is that EKG can be used to monitor blood pressure (BP) without an inflatable cuff. The EKG registers the pulse at the heart and the pulse oximeter registers the pulse at the fingertip. The time between these two pulses for the same heartbeat is an indirect measure of arterial stiffness and can be used as a continuous estimate of both systolic and diastolic BP, using the heart rate to better account for feedback loops in BP regulation (Wang et al. 2014).

Despite these advances, monitoring vital signs isn't always as easy as it sounds. Julia Harris of the finalist Basil Leaf Technologies states that vitals monitoring is one of the most challenging aspects of the competition because even though there are several products on the market already, many monitors are not sufficiently accurate or reliable to be useful with diagnostic algorithms. Activity monitoring products such as heart rate monitors for athletics have begun to be questioned as to their usefulness in health monitoring programs because of the inherent errors and the variability of their outputs; medical diagnostic software that incorporates continuous vital function monitoring requires a higher level of accuracy. Therefore, many of the teams have devised their own monitoring devices.

9.4.4 Tricorders for Diagnosis of Disease

Disease is defined as any change from normal in a living thing that harms the function of the organism in some way. To diagnose a disease or condition, the change must first be seen or measured. The key is to find a biologic change that is specific to the condition or disease for which you are testing, called a biomarker or analyte, and then devise a way to detect the biomarker. The Tricorder XPrize committee placed quite a challenge in front of the teams, as the conditions they are required to detect are varied, including infectious diseases (pneumonia and ear infections), lung disease (COPD), physiologic dysfunctions (anemia and sleep apnea), and chemical imbalances (diabetes)—each of which would have a different type of marker or markers.

Finding and tracking a single biomarker for each disease is difficult enough, but the tricorder has to detect them all as well as be correct when it says a biomarker is not present. To detect the analytes, biosensors are used in conjunction with patient history information and self-reported symptoms. The biosensor is the testing device; one component of the device is biologic, a protein or some other organic material that acts as a probe. It's called a probe because it searches out the specific biomarker—it probes the sample for the marker. The second component of the biosensor is a physicochemical or electromechanical detector, a transducer. Together, the parts of the biosensor bind the analyte that is going to be detected. A biomedical engineer wants to design biosensors that manifest simple, small, sensitive, and reliable tests that are specific for a single disease biomarker or that can distinguish between several biomarkers. On the other hand, some tests measure the marker directly, by imaging or detecting the affect the biomarker has on light passing through the sample (one form of spectroscopy).

The sample/specimen is some tissue or fluid in which analyte can be found if it is present (blood, urine, cerebrospinal fluid, sputum, etc.) or altered in concentration. The volume of sample is important; it must be sufficient to measure the marker, yet small enough to make it easy to obtain and not harm the patient. Many testing systems in use today require as little as 1–5 μl (1–5 thousandths of a milliliter) of blood. The small volume helps keep the testing systems small and light, although the situation may be complicated by the need to process the sample, like removing red cells from blood specimens.

Box 9.2 Methods to Detect Markers of Disease

Tomography—an assortment of various techniques that 3D image sections via penetrating waves. *X- ray imaging* is not a form of tomography since it gives a 2D projection of a 3D structure and the image is parallel to the axis of the film.

- *X-ray computed tomography* (CT scan) uses X-rays to produce section images that are perpendicular to the axis of the object.
- *Magnetic resonance imaging* (MRI), uses radio waves and a strong magnetic field to align hydrogen atoms with the field to show soft tissues well.
- *Positron emission tomography* (PET scan) uses positron-electron annihilations via specific radioactive labels to track functional activity in tissues.

These are powerful techniques but would be difficult to incorporate in a tricorder because they use transmission of waves; the target must be placed between a source and a detector.

Spectroscopy—various methods using EM radiation to stimulate, pass through, or reflect off of molecules or atoms. Different portions of the EM spectrum are tuned to returning different types of information.

- *X-ray spectroscopy* is good for determining bond lengths and angles while ultraviolet rays quantify nucleic acids in solution.
- *Absorption spectroscopic* techniques, specific substances will absorb energy at specific wavelengths, and these wavelengths will be missing from the energy that passes through the sample and is detected on the other side.
- *Emission spectroscopic* techniques capture the light that is scattered from a sample. The EM radiation that is absorbed and re-emitted will have a characteristic wavelength when it returns to the detector.

Chromatography—separating different constituents based on some characteristic.

- *Analytical chromatography* can be used to identify a marker based on the timing or conditions of the separation. The separation can be on a column or a planar substrate like paper. There is an immobile phase that provides retention for the compounds based on some characteristic (size, charge,

etc.) and a mobile phase (gas or liquid) that moves the sample through the immobile phase.
- *Liquid chromatography-mass spectroscopy* (LC-MS) uses chromatography for separation and spectroscopy for detection.

Ligand Binding—to identify a particular target in a sample using a ligand-binding molecule (antibody, protein, receptor) that interacts specifically with the target.
- *Label* is a fluorescent, radioactive or enzymatic molecule which is trackable when combined with the ligand. The label can be *direct* (attached to the ligand binding molecule) or *indirect* (attached to a third molecule that will bind the target or the ligand-binding molecule).
- *Enzyme linked immunosorbent assays* (ELISAs) and *competitive receptor binding assays* are forms of ligand binding assays.
- *Immunoaffinity chromatography* uses a chromatography column to separate many compounds and a ligand binding assay to identify the target.

Once the specimen is in a form that can be used for the specific test, it is combined with the biosensor device. There are many different sorts of biosensor probes; proteins that bind specific markers, metals, vitamins or cofactors, antibodies that bind specific antigens, cell receptors that binds specific ligands, or nucleic acids that can bind other nucleic acids. Each test is designed to use a biosensor that specifically binds one marker—nonspecific binding gives false positive tests, while a biosensor that doesn't bind the marker strongly or reliably provides false negative results. With a small sample, there is little room for error in binding efficiency, so if the sample is small, then so must be the biosensor.

After the marker is bound to the biosensor and any unbound sample is washed away, the presence of the analyte has to be detected and/or measured. Binding of the marker to the sensor brings a change to the sensor; it is the change that is detected by the transducer element. The change can be of many types; position, production of light or a colored chemical reaction product, or the closing of a circuit to induce electrical current flow. The key is that very small changes can be detected and measured. The smaller the measured change, the more sensitive the test. With small samples and small biosensors, the detection devices can also be small. This wave of miniaturization will help produce a small but powerful tricorder. Currently, the problem with building a tricorder according to the above schema isn't that we are lacking methods detect biomarkers, it's that there are too many (Box 9.1), each requiring different equipment and reagents. A handheld tricorder will benefit from development of a single testing platform that will work for all the different biomarkers.

9.5 Point of Care Testing and the Medical Tricorder

One of the guiding principles of the Tricorder XPrize competition is that the tools produced by the teams be easily used by consumers without medical training. The printed guidelines cite sources that highlight issues such as the 21 day average to obtain an appointment in a doctor's office and that 75 % of people have difficulty receiving after hours advice or care without making a trip to an emergency room (Qualcomm Tricorder XPrize Competition Guidelines, 2015. p. 3). Therefore, testing, vitals monitoring, and using the algorithms for advice and diagnosis must be made clear, simple and with as few steps and procedures as possible. This is what is referred to as the user experience.

One step along this continuum from doctor- to consumer-centered medicine is the advent of point of care (POC) testing. When tests can be run at the bedside, in the clinic, or even at home, it is better for the patient in terms of time, money, and sense of involvement in their own care. There are issues to be addressed with POC testing, the knowledge of the person running the test and the experience of the person analyzing the results will be reduced. A bedside test may be run by a nurse, who will have less training than a physician, and less training in running laboratory tests than a medical technologist. A test run at home will require a caregiver or the patient themselves to both run the test and interpret the result. Tests designed for POC medicine must be small, inexpensive, and easy to use and interpret. Neil Rens of Team Aezon says that this was both the team's greatest challenge and biggest accomplishment. Many of the XPrize teams have taken advantage of innovative POC testing platforms in designing their entries.

9.5.1 Microfluidics

Microfluidics is not a testing method, nor is it a device or instrument. Microfluidics is a distribution system for small volumes of liquid that can be used for POC testing. Microfluidics started with Siemens' inkjet printer heads of the 1980s, but it has found its largest market within medicine and medical research. Small samples for high-pressure liquid chromatography (HPLC) and gas chromatography (GC) (Box 9.1) brought microfluidics into the biological research and medicine realm. Though HPLC and GC instruments are large, they use small samples and test chambers, so microfluidics became more important as these technologies grew in the 1980s and 90s. Miniature detectors, computer chips, and circuitry combined with microfluidics in the 2000s to give the "lab-on-a-chip" (Fig. 9.2).

Microfluidics involves the movement of biologic samples through small channels and into microscopic chambers of various substrates like paper, fleece, nitrocellulose, or plastic or glass slides in order to run individual chemical tests. Basic medical research laboratories, by in large, do not rely on miniaturized, user-friendly testing platforms; they have space and expertise. On the other hand, hospital wards,

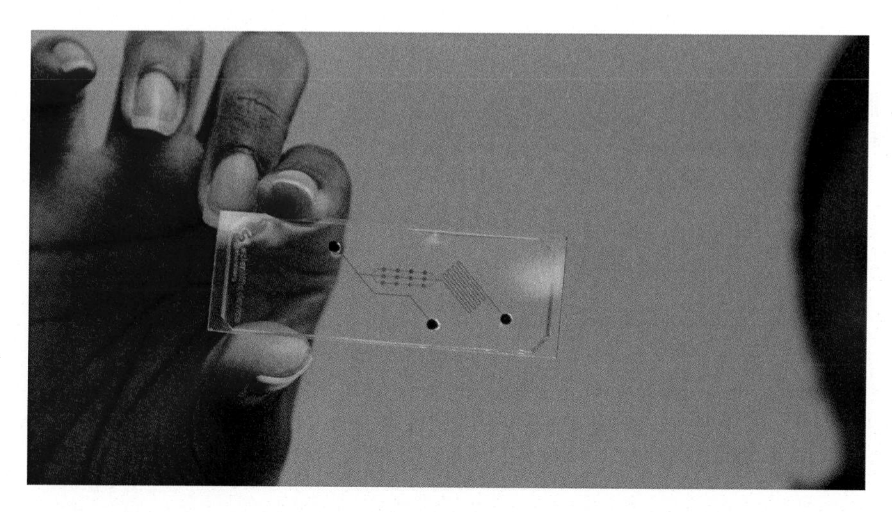

Fig. 9.2 There is no typical microfluidics chip to show in an image. They are all designed for a specific function or test. This particular microfluidics chip is utilized in pathological testing requiring only a small sample. Dye has been introduced into the sample chambers, fluid passages and mixing wells in order to make them visible. (*Image credit* Scientific Device Laboratory, Des Plaines, IL)

medical offices and homes require mobile and cost-effective tests that can be administered by nonprofessionals; these uses are greatly affected by microfluidics and are a perfect model for a designing a tricorder. Unfortunately, POC testing has also created a problem for tricorder design; research by many groups has led to many possible testing platforms, materials and detecting systems. Running one test from each of several systems is costly and consumes a lot of space. Test developers have pursued enzymatic reactions, antibody binding, receptor ligand interactions, nucleic acid (like DNA or RNA) hybridizations, and protein-protein interactions. Transduction of the binding into measureable signals makes use of microelectrical systems (piezoresistance, conductance, capacitance), optical systems for visible or fluorescent light, color producing reactions, and even systems for detection of positional changes. Without reinventing the wheel, how can an XPrize team make use of chip testing of such varied designs?

Team Aezon first set out to have many of their tests run in parallel in a multiwell microfluidics chip, all using a single detection system. There would be one sample for many tests, and the computer would pick out the data needed based on the history, symptoms and test results, making the user experience easier by reducing the choices that the consumer would have to make. This strategy is a sure thing in the future; perhaps hundreds of tests will be run in parallel, all based on one small sample, one testing process, and one detection system. However, the conditions and timing of the XPrize wouldn't allow Team Aezon to pursue a single sample to be processed in a single way for all the tests—a blood sample for anemia and cell counts is processed differently than for hemoglobin or glucose testing. As their entry evolved, Aezon came to realize that for financial reasons and ease of use by the consumer, all their tests would need to run independently, with little crossover. This may be the same for other groups, but Neil pointed out that his team has little to no information on the methodologies and strategies being used by the other finalists.

9.5.2 Lateral Flow Testing

The home pregnancy test is a type of low-end microfluidics test called a lateral flow dipstick test. Lateral flow tests are a good because they are light, can be adapted to many different markers, and because they are easily read by either the patient or a healthcare worker using their eyes or a low cost detector. Urine dipstick tests use lateral flow, even if is only to move the sample throughout the individual test squares by capillary action. Glucose strips and meters for diabetics are a good example of the plethora of choices that developers must make in designing POC tests. The strips can be made of different materials, some use chemical reactions with colored products while others use gold as an electrical conductor. In some cases the chemical reactions are different, while others have the same reactions but measure different products.

In the majority of strips and glucometers, an electrical current is the measure of glucose concentration. The glucose in the blood interacts with an enzyme called glucose oxidase held on the strip. The enzymatic reaction converts glucose to gluconic acid, with the production of hydrogen peroxide (H_2O_2) as a byproduct. A second reaction between the H_2O_2 and ferricyanide also bound to the strip transfers two electrons to the iron compound, reducing it to ferrocyanide, which can then give up an electron to the gold electrode and create a closed circuit for electrical current. The more glucose there is the sample, the more H_2O_2 is produced and more electrons will be available to the electrode. This means that electrical current is directly proportional to the glucose concentration. The meter's computer chip plots the current reading against a standard for glucose concentration and reports the data as mg/dl of blood.

Box 9.3 Not Always Flowing Like Water

The first large-scale use of lab-on-chip technology was with DNA analyses. Using hundreds of spotted DNAs, the presence or activity of specific genes could be rapidly determined. These microarrays don't really use microfluidics; the entire slide (chip) is overlaid with sample and chemical reactants. However, microfluidics and lab chips emerged from microarray testing when ways to move smaller samples through channels to individual reaction were devised.

Fluids moving on small scales behave differently from fluids on a macro scale. The rules of physics do not change, it's just that different forces become more important when the channels have nanometer (1×10^{-9} m) to micrometer (1×10^{-6} m) widths.

Viscosity is the shear stress as a liquid flows past itself - is important for mixing two fluids. As diameter of channel decreases, viscosity increases relatively and flow becomes laminar with no turbulence. No turbulence means no mixing. Therefore, microfluidics chips often include porous chambers filled with pillars to force mixing of sample fluids and reactants.

Non-Newtonian fluids are viscoelastic; their flow is not dependent solely on temperature and viscosity. For Newtonian fluids an increase in pressure for flow also increases the shear stress. To make a squirt gun squirt faster you must squeeze the trigger much harder. Non-Newtonian fluids do not show this direct relationship between flow rate and viscosity

- *Shear-thinning non-Newtonian fluids*—If you struggle in quicksand, you increase the shear stress and the quicksand becomes less viscous causing you to sink further.
- *Shear-thickening non-Newtonian* fluids—water and cornstarch (oobleck) turns solid as stress is applied. So does the Polish Moratex Institute for Security Technologies' liquid body armor. The shear stress from penetration of a bullet increases the viscosity of the fluid to the point of being as strong as Kevlar (Reuters New Service, April 2, 2015). This isn't the first attempt at such armor; efforts in Britain, Iran, and the United States have been ongoing since the 2000's.

Body Fluid specimens—many biologic fluid samples are non-Newtonian fluids in which viscoelasticity is time-dependent. The timing of the force for flow will give a viscosity change and some elastic characteristics wherein the fluid will deform and bounce back to its original position. The elastic number generally increases as the microfluidic channel or chamber size decreases, making forward flow of blood unsteady and backward flow a larger component (Sousa 2010). Therefore, relaxation times and flow rectifying designs must be taken into account when designing medical microfluidic chips.

Other glucose tests detect color instead of electrical current. The chemical reaction of glucose and glucose oxidase still produces H_2O_2, but instead of reacting this with an iron compound, tetramethylbenzidine is oxidized by the peroxide to produce a blue chemical. The visible spectrophotometer within the glucometer detects the blue color at 650 nm. A third method for glucose monitoring uses an NAD^+ cofactor glucose dehydrogenase enzyme to produce gluconic acid and NADH. The accumulation of NADH can be detected using UV spectroscopy rather than visible light. Glucose monitoring is just one example where several testing systems exist for one lateral flow POC testing platform, each with a different detection system. One can imagine that a tricorder tasked with testing for dozens of different markers might have to incorporate several different platforms, unless a team develops a single system for a large number of samples and markers. This might take years of work.

9.5.3 *Quantitative Polymerase Chain Reaction in POC Testing*

Polymerase chain reaction (PCR) is a laboratory technique that is beginning to show up in POC testing systems. Its mechanism of action is to mimic cellular DNA replication using artificial means to produce billions of copies of very specific patient DNA target sequences. The sample from the patient might be blood, tissues cells or a fluid that is suspected of containing disease causing pathogens, with the specific sequence amplified enough times to allow for detection *only* if that specific sequence is present in the patient sample. If the PCR test amplifies a certain mutated DNA sequence, then the patient is positive for that mutation and the consequences it might bring. If PCR for a mutated sequence does not result in billions of copies of the mutated DNA, then the patient does not carry the mutation.

The issue is a little more complicated for infectious organisms or viruses. Humans harbor many bacteria and viruses and yet do not show symptoms of disease. The mere presence of a specific bacterial DNA sequence does not mean that the individual is infected and has the disease caused by that pathogen. An additional step is needed to see if the number of bacteria or virus is increased to the point that infection is present. To determine this, the *number* of DNA sequence copies produced by PCR is evaluated. Early in the test reaction, a single target sequence will be replicated and two will result. The number of copies will double with every cycle of replication because the copies get copied themselves in the following rounds of replication, so after a dozen or more cycles there may be millions or billions of copies. The key is to assess the reaction early enough that the number of copies is still proportional to the number of target sequences in the original sample. This is quantitative PCR or qPCR.

Diagnostic tests using qPCR are good for determining the number of target sequences within a patient sample in addition to the presence or absence of a specific gene sequence. Unfortunately, the size of the sample can muddle the results. A larger sample size may have more target sequences simply because there are more cells, so the qPCR result would give a falsely high result. Therefore qPCR tests include at least one additional PCR target sequence, a control sequence that is known to occur a specific number of times in a cell's genome. The test copy number is compared to the control copy number; the higher the ratio, the more copies were present in the original sample regardless of sample size. Properly performed, qPCR is very good at diagnosing genetic predispositions, genetic diseases, or infections, and many targets can be amplified and differentially detected in a single test (based on a different color probe for each target). However, the equipment is expensive, somewhat cumbersome, requires a lot of electricity, and the samples require significant processing.

Despite these limitations, some POC qPCR tests have been developed, including for Ebola and for HIV and certain lung infections. Enigma Diagnostics (Salisbury, UK) has a PCR machine with automated sample processing to isolate DNA and test kits for detection of several organisms. Their kit to detect influenza and respiratory syncytial virus were found to be more than 99 % sensitive and specific in a recent study (Goldenberg and Edgeworth 2015). However, the device itself is 22 lb (10 kg), so incorporating it into a Tricorder XPrize entry is not possible. A newer version of portable qPCR machine from Ubiqutiome (Otago, NZ) weighs in at only 5.3 lb (2.4 kg) and can run four different samples at the same time, so it will not be long before qPCR methodology at the bedside or in the field is common.

Team Aezon's initial diagnostic device was based on qPCR tests for many of the thirteen disease targets called for in the XPrize guidelines, but the team discovered that the heating/cooling elements would be difficult to incorporate into a small footprint and the user experience was difficult, both in getting the patients to choose the correct test and collecting/processing the sample properly. Many of the methodologies used by Team Aezon in their final entry are off the shelf POC testing methods, with another portions use novel testing designs or are based on symptoms, vital signs, and history alone. This makes their computer algorithm a key factor in translating the information into a diagnosis and medical advice.

9.5.4 Differential Diagnoses from Computers

The software portion of each Tricorder XPrize entry is crucial in the effort to make tools that consumers can use, relegating Bones, Crusher, and Bashir to the sidelines until treatment is needed. The user experience begins the moment the consumer picks up the interface device and starts inputting information and making decisions based on the feedback they receive. It continues through the time they describe their symptoms, have tests recommended by the computer and then submit samples and to be tested. The results of the tests, along with the history and symptom information are then coalesced into a *differential diagnosis*, a hierarchy of possible conditions that match the information that the tricorder has detected or has been given.

To keep the user experience simple, Team Aezon split their tests into different packages and gave them identifying names. When their tricorder suggests a test to a patient, it does so by the identifier, so that mistakes in testing are reduced. Each suggested test is housed in a different small package with a different name, with one patient needing to perform one or more tests in succession. On the other hand, a finalist team from the DNA Medicine Institute (DMI) in Cambridge, MA has a device that can run hundreds of tests (projected) from a single drop of blood. Their device, called rHEALTH, was the 2014 winner of the Nokia Sensing X Challenge, and is the heart of their Tricorder XPrize entry as well. Due to the compactness and ease of use of the rHEALTH, DMI has partnered with NASA to test the use of microfluidics in the microgravity of space (Phipps et al. 2014). The aim is to develop testing mechanisms for monitoring the health of astronauts while in space, instead of taking samples and then returning them to Earth for testing and interpretation.

The user experience ends only when they have a possible or probable result from the interface and know what they should do next. To achieve a differential diagnosis, a computer algorithm puts all the collected data together and compares it to a database. Just how the teams collate this data is a major part of each entry. Team Aezon worked with with SymCat, a database system of health information and patient records that can be used to compare history, symptom and testing information in order to render a possible diagnosis. The interpretation piece of the algorithm is complicated; symptoms can suggest more than one disease and the reasons for a certain test result can differ by age, gender or health state. Additionally, self-reported symptom and history information is notoriously inaccurate without proper follow up questions and requests for additional information. All these items must be taken into account and then comparison made via large data sets and probabilistic Bayesian algorithms made and interpreted. Later in the design phase, Aezon developed a proprietary algorithm for symptom analysis that integrated the continuous vital signs monitoring and focused only on the diseases set forth in the competition guidelines. This alteration was necessary because the SymCat software could not incorporate the vitals signs data.

One finalist team, Basil Leaf technologies (Paoli, PA) states that both their sensor system and the artificial intelligence algorithms of their diagnostic piece (called DxtER) are all of their own design. The strategy behind DxtER is to reduce time and input for a needed diagnosis, as this system is based on the needs of emergency room physicians. As Julia Harris of the Basil Leaf team says, the intention is to, "surpass the *Star Trek* tricorder—to incorporate Dr. McCoy into the machine itself." To achieve this goal, tests must become more sensitive, easier to use, and algorithms more comprehensive and easy to interpret. Progress is being made in all areas, including some very inventive ways to expand testing on a single platform with increased sensitivity and reduced cost.

9.6 Emerging Platforms for POC Testing

Increasing the sensitivity of a laboratory test can mean several things. Perhaps it means that less sample is needed, so the test is less invasive. It might also mean that the threshold of detection is lower, so rare biomarkers can still be identified and a reliable result achieved. Finally, sensitivity might mean that a weaker output signal can be detected and measured, allowing for a smaller testing device and more tests in a given footprint. Several new systems are being developed that make progress in each of these areas.

9.6.1 Microcantilevers

When you stand on the end of a diving board, it bends (deflects) downward and oscillates. If a detector was either attached to or pointed at the diving board, one could know when someone was standing on it. If the detector was sensitive enough, one could tell how much that person weighed or whether there was more than one person on the board. Now imagine shrinking the diving board to the length of a cell or smaller, have it be only 1 μm (4×10^{-5} in.) thick, and then placing many of them in a 1–2 μl sample (a drop of water is 50 μl). This is how microcantilevers can be used in the rapidly expanding field of microfluidics–based tests. The microcantilever is a device that can act as a physical, chemical, or biological sensor. When a marker is bound to the ligand on the cantilever, there is a bending of the cantilever or a change in the vibrational oscillation frequency of the tiny diving board. What creates the bend or change in vibration? The binding of even a single marker molecule to biosensors adsorbed to the cantilever surface is enough to change the mass of the system and cause a change in the bending of the cantilever (Fig. 9.3).

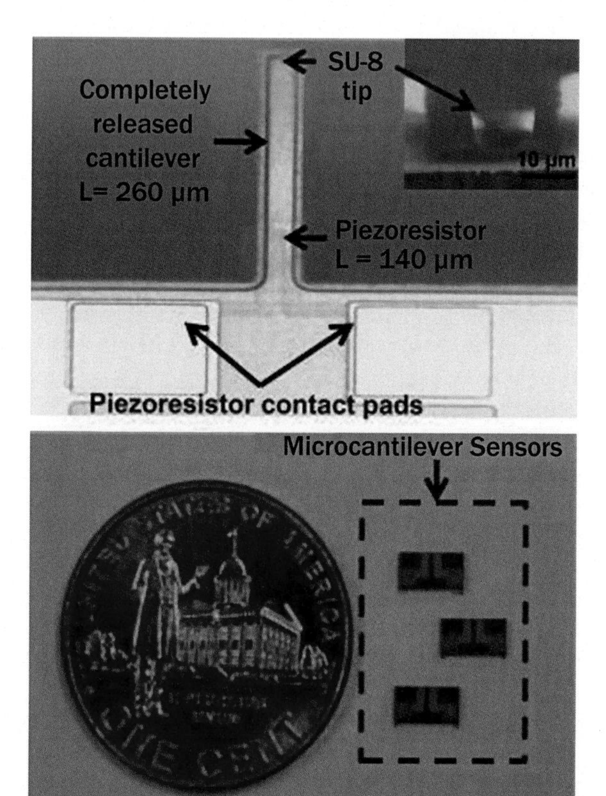

Fig. 9.3 These micrographs give the reader an idea about the size of some microcantilever sensors being used in POC testing. The cantilever show in the *top image* is 260 μm long (0.01 in.) and only 10 μm wide (0.0004 in.). The SU-8 tip is sensitive to breast cancer tissue while the piezoresistor responds to the bend in the cantilever by altering a current in the contact pads. The entirely assembly, as shown below, is a fraction of the size of a penny, yet each one contains 50 of the microcantilever units. (Reprinted from Pandya et al. Accurate characterization of benign and cancerous breast tissues: aspecific patient studies using piezoresistive microcantilevers. *Biosensors and Bioelectronics* 63; 414–424, 2015. Copyright (2015) with permission from Elsevier)

The change induced will be proportional to the number of marker molecules that bind to the surface, so microcantilevers can be quantitative sensing systems. The deflection might be downward, corresponding to the added weight on the cantilever, but it doesn't have to be. If just the top face of the cantilever is adsorbed with the ligand binding molecules, and the adsorption can create stress on the cantilever, it is possible that the binding of the marker will relieve the tension and cause the cantilever to deflect upward. Detecting just the movement makes them label-free detectors (Box 9.1), which reduces the cost and complexity of the chip tests.

Small samples and cantilevers are coupled with detection methods that can sense small displacements, making them accurate and sensitive. As with any biosensor system, the methods of detecting the cantilever deflection are several. Laser optical detection uses a laser to read the position of the cantilever before and after addition of the sample. The difference in position is reflective of the amount of the marker present in the sample. In a separate technique, a piezoresistor can be embedded in the biosensor-containing surface of the cantilever. When the marker binds to the receptors, the deflection of the cantilever creates strain in the piezoresistor and this translates to a change in the resistance to an electrical current running through the cantilever.

Choices must always be made; the readout system included on the chip makes the piezoresistor a smaller system, but the optical laser system is 10× more sensitive to changes in deflection. However, if the microcantilever is one plate of a plane capacitor, then absorption of a marker to the ligand will change the capacitance capability of the plates and the electrical current will reflect the change with a sensitivity equal to the optical laser system. Additional output systems include magnetic, electrostatic, thermoelectrical, and even electrochemical; however, no matter what methodology used, they all work to transduce the binding of the marker into a signal that can be detected and recorded.

The cantilevers are so small that you can have a chip with an array of many different microcantilever chambers. Each chamber can detect a different marker, making one multichamber chip a sensitive test for many conditions. All would use the same methodology so detections could be streamlined and kept small. Being small, reliable, sensitive, with a built in detection system would make this ideal for a tricorder system. While no microcantilever microfluidics tests have been approved by the FDA to date, sensitive tests have ben developed to detect prostate serum antigen and the protein breast cancer marker HER2 (Kierny et al. 2012). The technology can also be used to monitor drug levels in patients, accurately, simply, and quickly (Huang et al. 2014, 2015).

9.6.2 Quantum Dots

Along with the search for a testing platform that can detect many biomarkers using one methodology, piece of equipment and labeling system, there is also a desire for a simple testing system that can be used for many biomarker types. PCR can be used for many different DNA targets at the same time, but is no good for protein or cells. If a traceable particle could be labeled with different ligands (protein, lipid, DNA, RNA, cell, crystal, metal, hormone, etc.), then a single reading system could be used for many different kinds of biomarkers. If the particles could be specifically labeled for different output signals, then the particles could be used for different tests in the same system at the same time. Quantum dots (QDs), nanoparticle crystals made of semiconductive materials that can give off light, offer all these possibilities and more. QDs are just 2–10 nm wide and made up of only 10–50 atoms, yet can produce enough visible light to be used as biomarker labels. A QD is basically an electron trapped in a small cage of atoms.

The small size and specific 3D shape of QDs is very important, as below a certain threshold charge carriers experience something called quantum confinement. The energy transmitted through them via electrons is constrained in all three directions so that unique photo-properties result. Quantum dots are essentially artificial atoms; they are larger than atoms yet display single atom behaviors. When light hits a valence electron in a molecule, a specific wavelength of energy will be absorbed and the electron will move to a specific higher energy level. When the electron falls back to its ground state, energy of a specific wavelength is emitted; if that energy is in the visible range, it is called fluorescence. Quantum dots are unique fluorophores, they can absorb light at *any wavelength* and emit fluorescent light in a very narrow range of wavelengths, determined by the energy difference between the *average* radius of the electron at its ground state (valence band) and the *average* radius in its higher energy state (the conductance band).

If the energy difference or "band gap" is larger than the radius of the quantum dot as a whole, quantum confinement of the electrons comes into play and greatly narrows the possible wavelengths of emitted light. Instead of a range of emitted wavelengths, the fluorescent light will be of a precise wavelength, equivalent to one pure color (Fig. 9.4). This makes QDs easily detected; their emitted energy will be in very precise ranges. What is more, there is a predictable inverse relationship between QD size and the wavelength (color) of the emitted light; smaller quantum dots emit violet light while large ones emit red light because the band gap decreases as quantum dot's size increases. This tuneability means that different colored light can be stimulated from quantum dots made from the same material, based solely on size. This is beyond even the properties of individual atoms. The photic characteristics of QDs make them great for building efficient solar panels (Chuang et al. 2014), LED-driven, LCD televisions, and as qubits for quantum computers (Sects. 8.3.4 and 8.3.5). These amazing potentials aside, QDs will probably be used most in medicine and laboratory medical testing.

Biocompatible QDs of different sizes can be conjugated to many different ligands and used as probes at the same time because they will emit different colored light (Massey et al. 2015; Wang et al. 2015). Simply introduce the QDs into a specimen, allow them to bind their specific marker, wash away the unbound QDs and then shine a light on them. The wavelengths of emitted light will tell you which markers were present in the sample. This has recently been used as the basis of microfluidics tests (Kwon et al. 2015) and DNA detection platforms without the

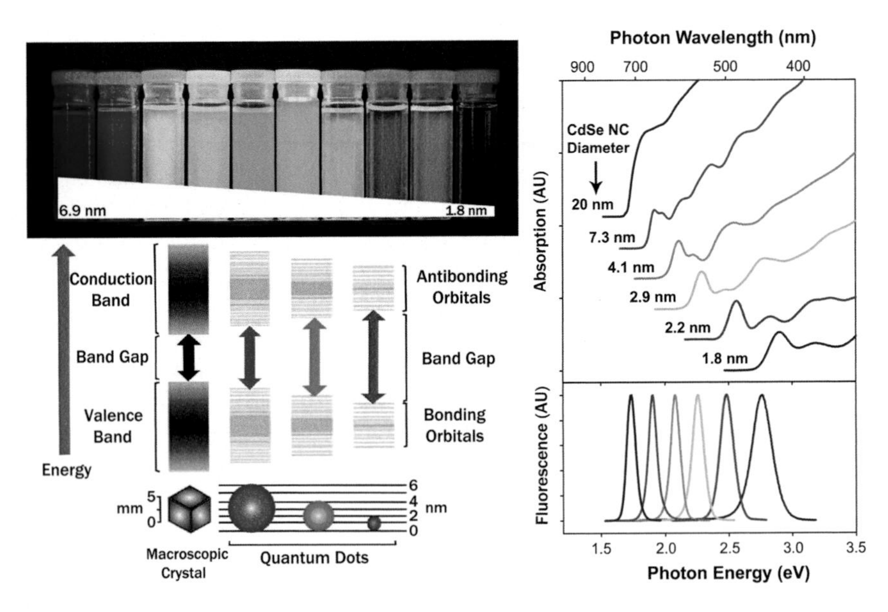

Fig. 9.4 The *top left* image shows the color of fluorescent quantum dots based on their size. The *colors* are vibrant and very pure, so they are being used in new plasma television screens. The *bottom left* image shows that the band gap, the energy difference between the normal valence band and the excited conduction band, increases as the size of the quantum dot decreases. Wider gaps lead to higher energy and frequency emission, so the color moves from red to violet. The *right graph* shows that the different size quantum dots emit light of very specific and narrow energies, which is why the colors are so pure. (Image credit (Chou and Dennis 2015) by Creative Commons License CC BY 4.0, http://creativecommons.org/licenses/by/4.0/legalcode, and reprinted from Smith et al. 2010. Copyright 2016 ACS)

need for amplification (Loo et al. 2016), although researchers have hardly scratched the surface of their possibilities. In the body, one can introduce the QDs tagged to bind rectal cancer cells and then use a camera and photomultiplier to identify the site and size of a tumor or even individual malignant cells (Park et al. 2014). In essence, any marker in a sample that can be bound to a ligand will be detectable using this one testing platform, one visible light spectrometer and one population of specifically sized QDs conjugated to the correct ligand. It is conceivable that one of the Tricorder XPrize finalist teams has developed a massively parallel testing system based on QDs, and if they haven't, someone will soon.

9.6.3 Magnetic Nanoparticles and Nanorobots

Nanoparticles of a different sort are at the heart of a concerted tricorder effort outside the XPrize realm. Google X Life Sciences recently spun off a medical diagnostic platform project as a subsidiary of Alphabet, Inc. (Mountain View, CA), to develop a portable continuous monitoring system for thousands of biomarkers. Google X's scientific director Andrew Conrad continues to lead the effort, and has given many interviews touting the power of the system. Google's tricorder will use magnetic iron oxide core nanoparticles as substrates on which biomarker ligands such as proteins or antibodies are bound, making them specific for many different markers. After introducing the nanoparticles into the body, a magnet placed on the wrist will gather the magnetic nanoparticle (MNPs) near the surface and the fluorescence can be read for detection and quantitation.

The use of different fluorescent labels on each type of MNP will allow for simultaneous monitoring of many different markers, while the small size (4–10 nm diameter) of the particles gives them access to the entire body. They can leave the blood vessels, infiltrate the tissues and find bind the specific markers for which they were designed. Over time, they could make their way back into the circulation, unless they are bound to their particular marker and held in the tissue. For example, nanoparticles could be designed to give off green fluorescent light and bind lung cancer cells of a specific type. If they encounter the cancer cells on their transits through the body, they will stay bound and not return to the blood stream. In this case, the decrease in a green fluorescent signal when the magnetic nanoparticles are gathered at the wrist would be an indicator that there is lung cancer.

Then there is the possibility of nanorobots, also called nanobots or nanoids, for wireless retrieval of internal data, much like the swallowed cameras for gastrointestinal imaging that exist today. True nanorobotics would use micromotors, pumps, and other devices to direct nanoids to specific tissues and carry out chemical and mechanical functions there, while reporting back to monitors wirelessly. However, the state of this technology is immature at best (Sect. 3.5.1), with designs and proposals for use far outnumbering actual devices. Nanoparticles, nanorobots, and QDs will be great diagnostic tools in the future, but they are still somewhat invasive as they are introduced into the body. Luckily, new non-invasive techniques are being developed as well, very similar to Bones passing a tricorder sensor over a body and getting all the information he needs.

9.6.4 Smelling Disease

Perhaps science is looking in the wrong direction for advances in sensing for medicine. Instead of looking to the nano or quantum realms, some researchers are looking to nature—in particular, man's best friend. Apparently Fido knows what cancer smells like. Dogs can smell about 1 part in 2 trillion, more than 50,000× as sensitive as the human olfactory sense. The common analogy is that dogs can smell a teaspoon of sugar in the equivalent of two million Olympic sized pools of water. Obviously, dogs are smelling many things that man is not, and this includes disease. There are many studies that show trained dogs can specifically and sensitively detect cancers in humans by sensing volatile organic compounds (VOC's), including prostate cancer in urine samples (Taverna et al. 2015), and lung cancer in exhaled air (Amundsen et al. 2014).

The prostate is intimately involved in the urogenital system, so having prostate cancer VOC markers in the urine is logical, as is having cancer VOCs for lung cancer in one's breath. Less straightforward is how dogs can smell breast cancer on the breath of women with early and late stage disease (McCulloch et al. 2006) and ovarian cancer in blood samples (Horvath et al. 2013). Tumors in these locations don't have a direct communication with breath, so the reason that dogs can smell them in these samples is less clear. The answer is blood. The unique organic compounds produced by cancerous tissues enter the blood as it passes through the tumor tissue, and this same blood makes a pass through the lungs during every trip through the body. Since the VOCs can evaporate in air, they move from the blood compartment to the air compartment of the lungs when they come into contact with the air sacs and are then exhaled.

Science is now trying to develop devices that can match or surpass the dog's nose in detecting VOCs, since trained dogs can distinguish between the breath of people with early cancer even before the tumor would show up on the best current X-ray or MRI machines. Laboratory instruments like gas chromatography units can detect and identify VOCs with slightly different chemical formulae, including identifying early stage rectal cancers (de Boer et al. 2014), but these devices are costly and huge. What is needed is a small, portable, accurate device to diagnose disease and health like a *Star Trek* tricorder or in this case, an electronic nose (e-nose).

NASA is very interested in detecting gas leaks on spacecraft, detecting the contamination of foodstuffs, as well as monitoring the health of the astronauts using electronic noses with microelectromechanical sensors (MEMS). One NASA spinoff technology for detecting VOCs is called the Cyranose, developed for commercial use by Cyrano Sciences (Pasadena, CA). The product is designed to track the quality of chemical industry products and foodstuffs based on how they smell. The handheld Cyranose 320 device has onboard pattern recognition software to identify and warn of abnormal conditions based on an array of sensors in a polymer that expands when bound by particular VOCs. The expansion changes the polymer's electrical properties and a change in electrical current through the circuit is detected.

Other laboratories are pursuing chemical detection via volatile chemicals, but the leading research is being performed in the Israeli Technion University laboratory of Dr. Hossam Haick. He and his team have developed a system using microelectromechanical sensing (MEMS) to discriminate between many different types of cancer, just using a breathalyzer (Fig. 9.5). A polyurethance strip inside the breathalyzer device contains printed nanoelectrodes with organic compound-coated gold nanoparticles spanning the electrode gaps. Depending on what organics are coated on to the nanoparticles, they bind different VOCs in the breath of the patient. When they bind specifically to the appropriate nanoparticles, the strength of the electrical current flowing between the electrodes changes (chemiresistor sensors). The more compound that is bound, the bigger the change in the resistance to the current.

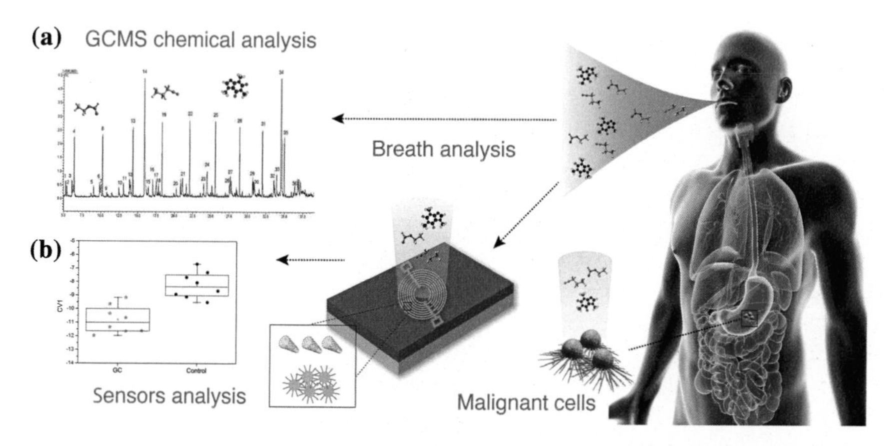

Fig. 9.5 Volatile organic compounds (VOCs) produced by stomach cancer cells can be detected in the breath of individuals with the disease. The VOCs can be detected using two methods. Traditional breath analysis uses gas chromatography and mass spectroscopy (**a**), which require time, large pieces of equipment, and significant expertise. Or, the VOCs can be captured on sensors made of VOC receptors bound to gold particles and carbon nanotube substrates that induce detectable electrical changes when specifically bound (**b**). This method is fast, requires little expertise, yet can still differentiate between cancer induced VOCs and normal VOCs (sensor analysis). (*Image credit* H. Haick and H. Amal)

By recording which electrical currents are affected, Haick's devices can identify which volatile organics are in the breath sample and distinguish gastric cancer from ulcers (Xu et al. 2013) or late cancer from precancerous lesions (Amal et al. 2016). Some of the newest chemiresistor sensors to come out of Haick's lab are flexible and self-healing (Fig. 9.6). If they break in two or get scratched, they can heal themselves in less than 48 h and start working again (Huynh et al. 2016). Dr. Haick has at least two commercial products in development that use the MEMS arrays, the NaNose (for lung and gastric cancers) and the Sniffphone. The Sniffphone is a breathalyzer that you plug into your smartphone and simply breathe across, it is noncontact and therefore more sanitary (Fig. 9.7). In a recent test of a similar Haick device using the flexible gold nanoparticle sensor arrays, the instrument predicted which 17 of 43 women volunteers actually had ovarian cancer with 82 % accuracy (Kahn et al. 2015).

Beyond cancer, infrared and e-nose breathalyzers are being developed for diagnosing pre-diabetes and diabetes (Novak et al. 2007) based on the fruity smell of ketones produced by burning fat. Diseases like asthma (de Vries et al. 2015), infections in cystic fibrosis (Joensen et al. 2014) and liver disease (Wlodzimirow et al. 2014) each have a distinctive breathprint which can be identified by both dogs and by e-noses. But the technology isn't just for sick people, the total number different volatile organic compounds given off by an individual adds up to their unique "smell print." It may be possible that changes from your normal odor could pinpoint a specific health problem or change before any symptoms develop or just

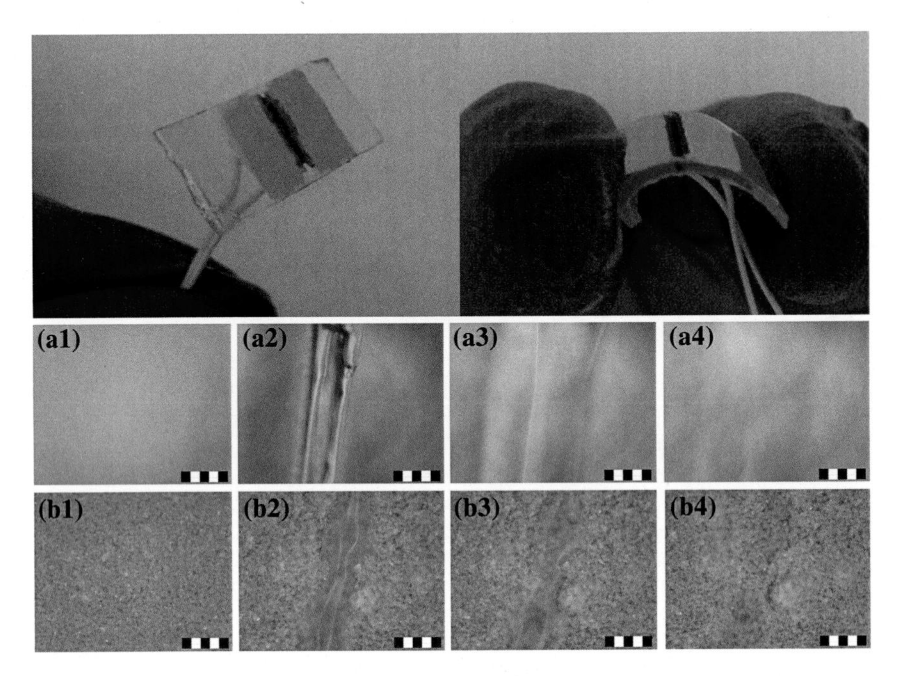

Fig. 9.6 The *upper left image* shows the substrate and volatile organic sensing material for the latest enose chip of Dr. Hossam Haick. The *top right image* shows how this chip is flexible, which opens up new opportunities for its use. The *bottoms images* demonstrate that the polymer substrate (a images) and gold-containing electrodes (b images) of the sensor are self-healing. The *left most images* before cutting (a1, b1) them, after cutting them (a2, b2) and after 30 min and 16 h (a3 and a4) or 10 min and 30 min (b3 and b4). (*Image credit*: Huynh and Haick 2016)

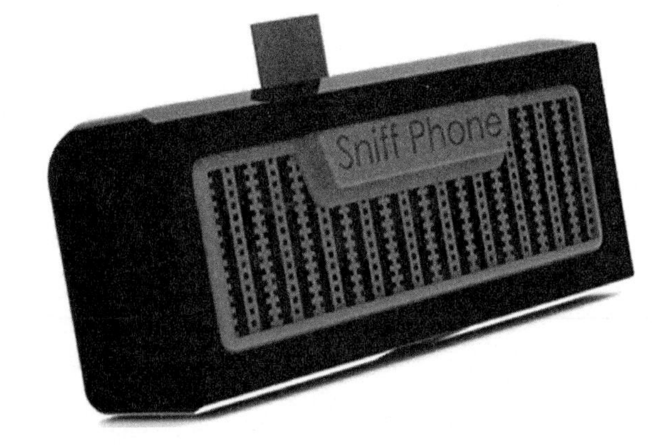

Fig. 9.7 The sniffphone is a dongle that attaches to smartphones and will be able to detect volatile organic compounds (VOCs) associated with a variety of diseases and is anticipated to be a method of monitoring health status non-invasively. (*Image credit* Sniffphone/H. Haick)

for identification; future tricorders might well smell a person, know who they are and deduce their health status. It would be non-contact, non-invasive, and comprehensive—a tricorder of which Bones McCoy could be proud.

9.7 Conclusion

Star Trek couldn't have predicted the huge amount of primary sensor information that is being dumped onto the internet everyday, although the communication of the universal translator with the computer or the computer with the tricorder does suggest that Roddenberry and his writers had an inkling of the coming Internet of Things. Data is stored on the ship's computer, yet can be accessed by the universal translator, transporter or tricorder—is this a parallel to a server or web farm as a repository of information?

Smartphones and other mobile devices interact with the central computer in two directions, either supplying it with primary sensory or processed data or by drawing on data stored on the computer—again similar to the processes on *Star Trek*. Working together, the entire system becomes more powerful and is able to provide immediate information to users about their environment, medical issues, or anything else they need to know. Current technologies are just now beginning to take advantage of the primary sensory data available from the millions of connected devices, and this will only increase logarithmically in the years to come.

The tricorder supplies more of this primary sensory data than any other device in *Star Trek*, and this is one of the reasons it has captured the imagination of the viewing public. It is accessible technology, and with a few advancements, will be obtainable as well. Many would argue that smartphone peripherals and applications are a sign that engineering has already produced a tricorder. However, if that were so, would Qualcomm and the XPrize committee need to offer ten million dollars in prizes for producing a medical tricorder? Driven by point of care testing and the need for increased consumer autonomy in health care, the medical tricorder is beginning to be realized, but the difficulties of the finalist teams and the additional time provided to them to meet the goals of the committee show that it remains a considerable challenge.

Advances in laboratory testing materials, understanding of biologic processes, and detection systems are bringing and will continue to stimulate increased sensitivity, reduced equipment size, and lower cost of the testing platforms. These are precisely the features needed to put all the appropriate technology in a single tricorder device and still have it be user friendly. Present day testing platforms are amazing; no one working molecular biology fifteen years ago would have predicted a qPCR device being used in the field and reporting real time results to an iPad or a PET scanner that a person can wear as a hat as they have their brain functions displayed on their smartphone. Yet, these are nothing compared to the coming technologies for medical diagnosis.

The most interesting example may be the e-noses that will smell our disease on us like…well, like a dog can smell fear. And as always, it is nature that has given humans the best models to follow. Just as the Earth's magnetosphere and ionosphere demonstrates to us how astronauts might be protected from cosmic radiation or how cephalopod skin has led us to active camouflage systems, the olfactory talents of dogs led science to pursue the idea of smellprints in disease conditions.

What are the things we need to keep this trend going? People willing to observe and study nature to uncover its secrets, people willing to convert nature's talents into technologies, and people who can see the possibilities in nature and technology and show us how they might be used to improve our lives. Dedicated scientists and engineers have the first two items covered, and no one has shown us the possibilities better than *Star Trek*.

References

H Amal, M Leja, K Funka, R Skapars, A Sivins, G Ancans, I Liepniece-Karele, I Kikuste, I Lasina, and H Haick. Detection of precancerous gastric lesions and gastric cancer through exhaled breath. *Gut* 65; 400-407, 2016. doi:10.1136/gutjnl-2014-308536. http://gut.bmj.com/content/65/3/400.full

T Amundsen, S Sundstrøm, T Buvik, OA Gederaas, and R Haaverstad. Can dogs smell lung cancer? First study using exhaled breath and urine screening in unselected patients with suspected lung cancer. *Acta Oncol.* 53(3); 307-315, 2014. doi:10.3109/0284186X.2013.819996. http://www.tandfonline.com/doi/full/10.3109/0284186X.2013.819996

DL Chandler. A doctor in the palm of your hand: how the Qualcomm Tricorder X-Prize could help to revolutionize medical diagnosis. Pulse, IEEE. 5(2); 50-4, 2014. doi:10.1109/MPUL.2013.2296803. http://ieeexplore.ieee.org/xpl/login.jsp?tp=&arnumber=6763301&url=http%3A%2F%2Fieeexplore.ieee.org%2Fxpls%2Fabs_all.jsp%3Farnumber%3D6763301

KF Chou, and AM Dennis. Foerster resonance energy transfer between quantum dot donors and quantum dot acceptors. *Sensors* 15(6); 13288-13325, 2015. doi:10.3390/s150613288. http://www.mdpi.com/1424-8220/15/6/13288/htm

CH Chuang, PR Brown, V Bulovic, and MG Brown. Improved performance and stability in quantum dot solar cells through band alignment engineering. *Nature Materials* 13(8); 796-801, 2014. doi:10.1038/nmat3984. http://www.nature.com/nmat/journal/v13/n8/full/nmat3984.html

NK de Boer, TG de Meij, FA Oort, I Ben Larbi, CJ Mulder, AA van Bodegraven, and MP van der Schee. The scent of colorectal cancer: detection by volatile organic compound analysis. *Clinical Gastroenterology & Hepatology*. 12(7); 1085-9, 2014. doi:10.1016/j.cgh.2014.05.005. http://www.cghjournal.org/article/S1542-3565(14)00715-0/abstract

R de Vries, P Brinkman, MP van der Schee, N Fens, E Dijkers, SK Bootsma, FH de Jongh, and PJ Sterk. Integration of electronic nose technology with spirometry: validation of a new approach for exhaled breath analysis. *J Breath Research* 9(4); 046001, 2015. doi:10.1088/1752-7155/9/4/046001. http://iopscience.iop.org/article/10.1088/1752-7155/9/4/046001/meta

SD Goldenberg, and JD Edgeworth. The Enigma ML FlusAB-RSV assay: a fully automated molecular test for the rapid detection of influenza A, B and respiratory syncytial virsues in respiratory specimens. *Expert Rev Mol Diagn* 15(1); 23-32, 2015. doi:10.1586/14737159.2015.983477. http://www.tandfonline.com/doi/full/10.1586/14737159.2015.983477

G Horvath, H Andersson, and S Nemes. Cancer odor in the blood of ovarian cancer patients: a retrospective study of detection by dogs during treatment, 3 and 6 months afterward. *BMC Cancer*. 13; 396, 2013. doi:10.1186/1471-2407-13-396. http://bmccancer.biomedcentral.com/articles/10.1186/1471-2407-13-396

LS Huang, C Gunawan, YK Yen, and KF Chang. Direct determination of a small-molecule drug, valproic Acid, by an electrically-detectedmicrocantilever biosensor for personalized diagnostics. *Biosensors (Basel).* 5(1); 37-50, 2015. doi:10.3390/bios5010037. http://www.mdpi.com/2079-6374/5/1/37

LS Huang, Y Pheanpanitporn, YK Yen, KF Chang, LY Lin, and DM Lai. Detection of the antiepileptic drug phenytoin using a single free-standing piezoresistive microcantilever for therapeutic drug monitoring. *Biosens Bioelectron.* 59; 233-238, 2014. doi:10.1016/j.bios.2014.03.047. http://www.sciencedirect.com/science/article/pii/S0956566314002280

T-P Huynh, and H Haick. Self-healing, fully functional, and multiparametric flexible sensing platform. *Adv Mater* 28(1); 138-143, 2016. doi:10/1002/adma.201504104. http://onlinelibrary.wiley.com/doi/10.1002/adma.201504104/abstract

O Joensen, T Paff, EG Haarman, IM Skovgaard, PO Jensen, T Bjarnsholt, and KG Nielsen. Exhaled breath analysis using electronic nose in cystic fibrosis and primary ciliary dyskinesis patients with chronic pulmonary infections. *PLoS One* 9(12); e115584, 2014. doi:10.1371/journal.pone.0115584. http://journals.plos.org/plosone/article?id=10.1371/journal.pone.0115584

N Kahn, O Lavie, M Paz, Y Segev, H Haick. Dynamic Nanoparticle-Based Flexible Sensors: Diagnosis of Ovarian Carcinoma from Exhaled Breath. *Nano Letters* 15(10); 7023-7028, 2015. doi: 10.1021/acs.nanolett.5b03052. http://pubs.acs.org/doi/10.1021/acs.nanolett.5b03052

MR Kierny, TD Cunningham, and BK Kay. Detection of biomarkers using recombinant antibodies coupled to nanostructured platforms. *Nano Rev.* 3, 2012. doi:10.3402/nano.v3i0.17240. http://www.nano-reviews.net/index.php/nano/index

S Kwon, CH Cho, ES Lee, and JK Park. Automated measurement of multiple cancer biomarkers using quantum-dot-based microfluidic immunohistochemistry. *Anal Chem* 87(8); 4177-4183, 2015. doi:10.1021/aces.analchem.5b00199. http://pubs.acs.org/doi/abs/10.1021/acs.analchem.5b00199

AH Loo, Z Sofer, D Bousa, P Ulbrich, A Bonanni, and M Pumera. Carboxylic carbon quantum dots as a fluorescent sensing platform for DNA detection. *ACS Appl Mater Interfaces* 8(3); 1951-1957, 2016. doi:10.1021/acsami.5b10160. http://pubs.acs.org/doi/abs/10.1021/acsami.5b10160

M Massey, M Wu, EM Conroy, and WR Algar. Mind your P's and Q's: the coming of age of semiconducting polymer dots and semiconductor quantum dots in biological applications. *Curr Opin Biotechnol.* 34; 30-40, 2015. doi:10.1016/j.copbio.2014.11.006. http://www.sciencedirect.com/science/article/pii/S0958166914001931

M McCulloch, T Jezierski, M Broffman, A Hubbard, K Turner, and T Janecki. Diagnostic accuracy of canine scent detection in early- and late-stage lung and breast cancers. *Integrative Cancer Therapies.* 5(1); 30-39, 2006. doi:10.1177/1534735405285096. http://ict.sagepub.com/content/5/1/30.long

SE Minson, BA Brooks, CL Glennie, JR Murray, JO Langbein, SE Owen, TH Heaton, RA Iannucci, and DL Hauser. Crowdsourced earthquake early warning. *Science Advances* 1(3); e1500036, 2015. doi:10.1126/sciadv.1500036. http://advances.sciencemag.org/content/1/3/e1500036.full

BJ Novak, DR Blake, S Meinardi, FS Rowland, A Pontello, DM Cooper, and PR Galassetti. Exhaled methyl nitrate as a noninvasive marker of hyperglycemia in type 1 diabetes. *Proc Natl Acad Sci USA* 104 (40); 15613-15618, 2007. doi:10.1073/pnas.0706533104. http://www.pnas.org/content/104/40/15613.full

E Palestine. *Star Fleet Medical Reference Manual.* New York: Star Fleet Production, distributed by Ballantine Books. pg. 143, 1977.

HJ Pandya, R Roy, W Chen, MA Chekmareva, DJ Foran, and JP Desai. Accurate characterization of benign and cancerous breast tissues: aspecific patient studies using piezoresistive microcantilevers. *Biosensors and Bioelectronics* 63; 414-424, 2015. doi:10.1016/j.bios.2014.08.002. http://www.sciencedirect.com/science/article/pii/S0956566314005867

Y Park, Y-M Ryu, T Wang, T Baek, Y Yoon, SM Bae, J Park, S Hwang, J Kim, E-J Do, S-Y Kim, E Chung, KH Kim, S Kim, and S-J Myung. Spraying quantum dot conjugates in the colon of live animals enabled rapid and multiplex cancer diagnosis using endoscopy. *ACS Nano* 8(9); 8896-8910, 2014. doi:10.1021/nn5009269. http://pubs.acs.org/doi/abs/10.1021/nn5009269

WS Phipps, Z Yin, C Bae, JZ Sharpe, AM Bishara, ES Nelson, AS Weaver, D Brown, TL McKay, D Griffin, and EY Chan. Reduced-gravity environment hardware demonstrations of a prototype miniaturized flow cytometer and companion microfluidic mixing technology. *J Vis Exp* 93: e51743, 2014. doi:10.3791/51743. http://www.jove.com/video/51743/reduced-gravity-environment-hardware-demonstrations-prototype

AM Smith, and S Nie. Semiconductor nanocrystals: structure, properties, and band gap engineering. *Acc Chem Res* 43(2); 190-200, 2010. doi:10.1021/ar9001069. http://pubs.acs.org/doi/abs/10.1021/ar9001069

PC Sousa, FT Pinho, MSN Oliverio, and MA Alves. Efficient microfluidic rectifiers for viscoelastic fluid flow. *Journal of Non-Newtonian Fluid Mechanics* 165: 652-671, 2010. doi: 10.1016/j.nnfm.2010.03.005. http://paginas.fe.up.pt/ceft/pdfs/JNNFM2010_rectificadores.pdf

G Taverna, L Tidu, F Grizzi, V Torri, A Mandressi, P Sardella, G La Torre, G Cocciolone, M Seveso, G Giusti, R Hurle, A Santoro, and P Graziotti. Olfactory system of highly trained dogs detects prostate cancer in urine samples. *J Urol.* 193(4); 1382-1387, 2015. doi:10.1016/j.juro.2014.09.099. http://www.jurology.com/article/S0022-5347(14)04573-X/abstract

C Wang, F Hou, and Y Ma. Simultaneous quantitative detection of multiple tumor markers with a rapid and sensitive multicolor quantum dots based immunochromatographic test strip. *Biosens Bioelectron.* 68; 156-62, 2015. doi:10.1016/j.bios.2014.12.051. http://www.mdpi.com/1424-8220/10/7/6623

R Wang, W Jia, Z-H Mao, RJ Sclabassi, and M Sun. Cuff-free blood pressure estimation using pulse transit time and heart rate. *Proceedings of the 12th International Conference on Signal Processing (ICSP)* 115-118, 2014. doi:10.1109/ICOSP.2014.7014980. http://www.ncbi.nlm.nih.gov/pmc/articles/PMC4512231/pdf/nihms706152.pdf

KA Wlodzimirow, A Abu-Hanna, MJ Schultz, MA Maas, LD Bos, PJ Sterk, HH Knobel, RJ Soers, and RA Chamuleau. Exhaled breath analysis with electronic nose technology for detection of acute liver failure in rats. *Biosens Bioelectron* 53; 129-134, 2014. doi:10.1016/j.bios.2013.09.047. http://www.sciencedirect.com/science/article/pii/S0956566313006659

Z-q Xu, YY Broza, U Tisch, L Ding, H Liu, Q Song, Y-y Pan, F-x Xiong, K-s Gu, G-p Sun, Z-d Chen, M Leja, and H Haick. A nanomaterial-based breath test for distinguishing gastric cancer from benign gastric conditions. *British Journal of Cancer* 108; 941-950, 2013. doi:10.1038/bjc.2013.44. http://www.nature.com/bjc/journal/v108/n4/full/bjc201344a.html

Bibliography

Books

S Aaronson. *Quantum Computing Since Democritus*. Cambridge: Cambridge Press, 2013.
A Asherman. *The Star Trek Interview Book*. New York: Pocket Books, 1988.
JA Barad and E Robertson. *The Ethics of Star Trek*. New York: Harper Collins, 2000.
JF Callan and FM Raymo. *Quantum Dot Sensors: Technology and Commercial Applications*. Singapore: Pan Stanford, 2013.
FF Chen. *Introduction to Plasma Physics and Controlled Fusion*. New York: Plenum Press, 1984.
M Clark. *Star Trek FAQ: Everything Left to Know About the First Voyages of the Starship Enterprise*. Montclair, NJ: Applause Theatre & Cinema Books, 2012.
J Ecklar. *Star Trek: The Kobayashi Maru #47*. New York: Pocket Books, 1989.
N Gisin, and S Lyle. *Quantum Chance: Nonlocality, Teleporation, and Other Quantum Marvels*. Cham: Springer, 2014.
DJ Griffiths. *Introduction to Quantum Mechanics*. Upper Saddle River, NJ: Pearson Prentice Hall, 2005.
S Ings. *A Natural History of Seeing: The Art and Science of Vision*. New York: WW Norton, 2008.
S Johnson. *Star Trek: Mr. Scott's Guide to the Enterprise*. New York: Pocket Books, 1987.
F Joseph. *Star Trek – Starfleet Technical Manual: Training Command, Starfleet Academy*. New York: Ballantine Books, 2006.
M Kaku. *Physics of the Impossible*. New York: Doubleday, 2008.
V Kooper. *George Clayton Johnson, fictioneer: from Ocean's Eleven, through the Twilight zone, to Logan's run, and beyond*. Albany, GA: BearManor Media.
LM Krauss. *The Physics of Star Trek*. New York: Basic Books, 1995.
PCH Li. *Fundamentals of microfluidics and lab on a chip for biological analysis and discovery*. Boca Raton: CRC Press/Taylor & Francis, 2010.
D Mack. *The Starfleet Survival Guide*. New York: Pocket Books, 2002.
CJ Nitta, MK Farrens, and V Akella. *On-Chip Photonic Interconnects: A Computer Architect's Perspective*. San Rafael, CA: Morgan & Claypool, 2014.
N-T Nguyen and ST Wereley. *Fundamentals and Applications of Microfluidics*. Boston: Artech House, 2006.
M Okuda, D Okuda, and D Mirek. *Star Trek Encyclopedia, Third Edition*. New York: Pocket Books, 1999.
E Palestine. *Star Fleet Medical Reference Manual*. New York: Star Fleet Production, distributed by Ballantine Books, 1977.
S Perkowitz. *Slow Light: Invisibility, Teleportation, and Other Mysteries of Light*. London: Imperial College Press, 2011.
B Robinson, and M Riley. *USS Enterprise: Owner's Workshop Manual*. New York: Gallery Books, 2010.

M.E. Lasbury, *The Realization of Star Trek Technologies*,
DOI 10.1007/978-3-319-40914-6

M Schroeder. *Fractals, Chaos, Power Laws: Minutes from an Infinite Paradise*. New York: W.H. Freeman, 1991.

W Shatner and C Walter. *Star Trek: I'm Working on That*. New York: Pocket Books, 2004.

R Sternbach and M Okuda. *Star Trek The Next Generation: Technical Manual*. New York: Pocket Books, 1991.

R Sternbach, and M Okuda. *Star Trek Voyager Technical Manual, V1.0. Internal Paramount Picture document*. Published by Paramount pictures, 1994. Accessed December 28, 2015.

DK Yeomans. *Near-Earth Objects: Finding Them Before They Find Us*. Princeton, NJ: Princeton University Press, 2013.

SE Whitfield and G Roddenberry. *The Making of Star Trek*. New York: Ballantine Books, 1968.

H Zimerman, R Sternbach, and D Drexler. *Star Trek, Deep Space Nine: Technical Manual*. New York: Pocket Books, 1998.

Journal Articles

M Alcubierre. The warp Drive: hyper-fast travel within general relativity. *Classical and Quantum Gravity* 11(5); L73-L77, 1994. http://iopscience.iop.org/article/10.1088/0264-9381/11/5/001/pdf

G Benford. A scientist's notebook: living in an eleven-dimensional world. *The Magazine of Fantasy and Science Fiction* 103(4-5); 187, 2002.

J Breuer, and P Hommelhoff. Laser-induced acceleration of nonrelativistic electrons at a dielectric structure. *Physical Review Letters* 111 (13); 134803, 2013. doi: 10.1103/PhysRevLett.111.134803. http://journals.aps.org/prl/abstract/10.1103/PhysRevLett.111.134803

JS Choi, and JC Howell. Paraxial Ray Optics Cloaking. *Optics Express* 22(24); 29465-29478, 2014. doi: 10.1364/OE.22.0293465. https://www.osapublishing.org/oe/fulltext.cfm?uri=oe-22-24-29465&id=304785

KF Chou, and AM Dennis. Foerster resonance energy transfer between quantum dot donors and quantum dot acceptors. *Sensors* 15(6); 13288-13325, 2015. doi: 10.3390/s150613288. http://www.mdpi.com/1424-8220/15/6/13288/htm

G Du, Q Fang, JM den Toonder. Microfluidics for cell-based high throughput screening platforms – a review. *Anal Chim Acta* 903; 36-50, 2016. doi: 10.1016/j.aca.2015.11.023. http://www.sciencedirect.com/science/article/pii/S0003267015013823

N G, A Tan, Y Farhatnia, J Rajadas, MR Hamblin, PT Khaw, and AM Seifalian. Channelrhodopsins: visual regeneration and neural activation by a light switch. *New Biotechnology* 30(5); 461-474, 2013. doi: 10.1016/j.nbt.2013.04.007. http://www.sciencedirect.com/science/article/pii/S1871678413000599

D Issadore, YI Park, H Shao, C Min, K Lee, M Liong, R Weissleder, and H Lee. Magnetic sensing technology for molecular analyses. *Lab Chip* 14(14): 2385-2397, 2014. doi: 10.1039/C4LC00314D. http://pubs.rsc.org/en/Content/ArticleLanding/2014/LC/c4lc00314d#!divAbstract

TZ Lauritzen, J Harris, S Mohand-Said, JA Sahel, JD Dorn, K McClure, and RJ Greenberg. Reading visual braille with a retinal prosthesis. *Front Neurosci* 6; 168, 2012. doi: 10.3389/fnins.2012.00168. http://journal.frontiersin.org/article/10.3389/fnins.2012.00168/abstract

H Li, B Ma, and KA Lee. Spoken language recognition: from fundamentals to practice. *Proceedings of the IEEE* 101(5); 1136-1159, 2013. doi: 10.1109/JPROC.2012.2237151. http://ieeexplore.ieee.org/stamp/stamp.jsp?arnumber=6451097

HJ Pandya, R Roy, W Chen, MA Chekmareva, DJ Foran, and JP Desai. Accurate characterization of benign and cancerous breast tissues: aspecific patient studies using piezoresistive microcantilevers. *Biosensors and Bioelectronics* 63; 414-424, 2015. doi: 10.1016/j.bios.2014.08.002. http://www.sciencedirect.com/science/article/pii/S0956566314005867

H Saberkari, HB Ghavifekr, and M Shamsi. Comprehensive performance study of magneto cantilevers as a candidate model for biological sensors used in lab-on-a-chip applications. *J Med Signals Sens* 5(2); 77-87, 2015. http://www.ncbi.nlm.nih.gov/pmc/articles/PMC4460669/

AM Smith, and S Nie. Semiconductor nanocrystals: structure, properties, and band gap engineering. *Acc Chem Res* 43(2); 190-200, 2010. doi: 10.1021/ar9001069. http://pubs.acs.org/doi/abs/10.1021/ar9001069

M Stanisavljevic, S Krizkova, M Vaculovicova, R Kizek, and V Adam. Quantum dots-fluorescence resonance energy transfer-baser nanosensors and the application. *Biosens Bioelectron* 74; 562-574, 2015. doi: 10.1016/j.bios.2015.06.076. http://www.sciencedirect.com/science/article/pii/S0956566315302451

J Sun, E shah Hosseini, A Yaacobi, DB Cole, G Leake, D Coolbaugh, and MR Watts. Two dimensional apodized silicon photonic phased arrays. *Optics Letters* 39(2) 367-370, 2014. doi:10.1364/OL/39/000367. https://www.osapublishing.org/ol/abstract.cfm?uri=ol-39-2-367

P Zhang, S Wang, Y Liu, C Lu, Z Chen, and X Zhang. Plasmonic Airy beams with dynamically controlled trajectories. *Optics Letters* 36(16); 3191-3193, 2011b. doi: 10.1364/OL.36.003191. https://www.osapublishing.org/ol/abstract.cfm?uri=ol-36-16-3191#Abstract

Online Articles

http://ieeexplore.ieee.org/stamp/stamp.jsp?arnumber=6451097

http://www.theverge.com/2013/9/17/4596374/machine-language-how-siri-found-its-voice, talks about recording for Siri and others.

http://www.scientificamerican.com/article/how-are-elements-broken-d/, short piece on energy needed to break down atoms by S Reucroft and JD Swain of Northeastern University for Scientific American.

M Nielsen. Quantum computing for everyone. http://michaelnielsen.org/blog/quantum-computing-for-everyone/

http://www.theverge.com/2015/5/29/8672371/learn-esperanto-language-duolingo-app-origin-history, 2015 article on Esperanto and internet, includes Duolingo course and language hack

Star Trek: Original Series Writer's Guide, third Edition, 1967. http://leethomson.myzen.co.uk/Star_Trek/1_Original_Series/Star_Trek_TOS_Writer%27s_Guide.pdf

LMK Vandersypen, IL Chuang, and D Suter. Liquid-state NMR quantum computing. Accessed March 31. 2016. https://e3.physik.uni-dortmund.de/~suter/eprints/153_Liquid_NMR-QIP.pdf

Websites

Air Force Research Laboratory. http://www.af.mil/AboutUs/FactSheets/Display/tabid/224/Article/104463/air-force-research-laboratory.aspx

Brookhaven National Laboratory, Relativistic Heavy Ion Collider. https://www.bnl.gov/rhic/

Chrissie's Transcript Site. http://www.chakoteya.net/

Defense Advanced Research Projects Agency. http://www.darpa.mil/

European Radiation Superconducting Shield (SR2S). http://www.sr2s.eu/

European Space Agency. http://www.esa.int/ESA

Fluent in three months. http://www.fluentin3months.com/

Google Translate. https://translate.google.com/

International Astronomical Union. http://www.iau.org/

Memory Alpha. http://memory-alpha.wikia.com/wiki/Portal:Main

National Aeronautic and Space Administration. https://www.nasa.gov/

Office of Naval Research. http://www.onr.navy.mil/

Official US Government information about the Global Positioning Satellite System (GPS) and related topics. http://www.gps.gov/

Qualcomm Tricorder XPrize. http://tricorder.xprize.org/
Skype Translator. http://www.skype.com/en/translator-preview/
Star Trek Script Search. http://scriptsearch.dxdy.name/
United States Army Research Laboratory. http://www.arl.army.mil/www/default.cfm

Index

M.E. Lasbury, *The Realization of Star Trek Technologies*,
DOI 10.1007/978-3-319-40914-6